PHYLOGENETIC ANALYSIS OF DNA SEQUENCES

PHYLOGENETIC ANALYSIS OF DNA SEQUENCES

Edited by

Michael M. Miyamoto
University of Florida
Gainesville, Florida

Joel Cracraft
University of Illinois
Chicago, Illinois

New York *Oxford*
OXFORD UNIVERSITY PRESS
1991

Oxford University Press
Oxford New York Toronto
Delhi Bombay Calcutta Madras Karachi
Petaling Jaya Singapore Hong Kong Tokyo
Nairobi Dar es Salaam Cape Town
Melbourne Auckland

and associated companies in
Berlin Ibadan

Published by Oxford University Press, Inc.,
200 Madison Avenue, New York, New York 10016

Library of Congress Cataloging-in-Publication Data
Phylogenetic analysis of DNA sequences / edited by
Michael M. Miyamoto, Joel Cracraft.
p. cm. Includes bibliographical references and index.
ISBN 0-19-506698-7
1. DNA—Evolution. 2. Nucleotide sequence. 3. Molecular
evolution. I. Miyamoto, Michael M. II. Cracraft, Joel.
QH371.P47 1992 574.87'3282—dc20 90-28771

9 8 7 6 5 4 3 2

Printed in the United States of America
on acid-free paper

Contents

Preface

The comparative analysis of DNA sequences is becoming increasingly important in systematic and evolutionary biology and will continue to do so as faster and more efficient methods for collecting these data are developed. Large amounts of comparative sequence data will be required to answer most molecular systematic questions, but this labor-intensive effort will only be the first of several problems faced by the systematist. Although the use of DNA sequences in systematics is still in its infancy, already a healthy mixture of opinion exists about the most appropriate methods for reconstructing phylogenetic history from nucleotide data. Moreover, some question whether DNA sequences will prove to be more informative in all cases when compared to more traditional data-bases. Thus, in using DNA sequences comparatively, the biologist is confronted by staggering complexities that are often not appreciated even by the expert systematist or molecular evolutionist.

This volume has assembled an internationally recognized group of investigators representing different theoretical viewpoints and disciplines to address critically a diversity of questions about DNA systematics. The book begins with an introduction by Miyamoto and Cracraft, followed by 14 additional chapters emphasizing data acquisition, sequence analysis, and the broader systematic importance of nucleotide information. Contributors on data acquisition have focused on improved techniques for obtaining comparative sequence information by manual (Slightom et al.) and automated (Ferl et al.) approaches. With regard to data analysis, authors have concentrated on methodological problems dealing with sequence alignment (Waterman et al. and Mindell) and different tree-building algorithms (Nei, Sidow and Wilson, Fitch and Ye, and Penny et al.). Finally, contributors have focused on more general issues having broad implications within systematics. Specifically, their chapters have concentrated on the evaluation of phylogenetic reliability and information content of different sequences and data sets (Cracraft and Helm-Bychowski, Li and Gouy, and Hillis), on the relationship between molecular evolutionary bias and phylogeny reconstruction (Larson), and on the application of consensus and congruence approaches in systematics (Swofford and Wheeler).

This book has its roots in the symposium "Recent Advances in Phylogenetic Studies of DNA Sequences," which was part of the special centennial celebration of the American Society of Zoologists, held in conjunction with the Society of Systematic Zoology, on December 26–30, 1989 in Boston, Massachusetts. Researchers from different disciplines and approaches presented papers at the symposium, but rather than just describe their methods or dwell on particular groups, each participant concentrated on the strengths, limitations, and assumptions of their approaches relative to others. The diversity of topics and viewpoints represented at the symposium constituted its greatest strength, and in turn, has now become the most important quality of this book.

The following people and organizations are profoundly acknowledged for their assistance. The American Society of Zoologists and Society of Systematic Zoology contributed administrative and financial assistance for the symposium. J.S. Farris, J. Felsenstein, A.G. Kluge, T.D. Kocher, J.A. Lake, and A. Meyer presented papers at the symposium, but chose not to contribute chapters. External reviews of the individual chapters were provided by M.W. Allard, J.M. Carpenter, J. Felsenstein, D.H.A. Fitch, W.M. Fitch, D.M. Hillis, R. Holmquist, R.L. Honeycutt, B.F. Koop, W.-H. Li, D.R. Maddison, D.P. Mindell, C.J. Nairn, J.L. Patton, D.R. Siemieniak, J.L. Slightom, D.L. Swofford, B.S. Weir, W.C. Wheeler, R. Wilson, C.-I. Wu, and E.A. Zimmer. W.F. Curtis of Oxford University Press is recognized for his encouragement and for his help in seeing this volume through the initial stages of production. The book's index was compiled by M.R. Tennant, and K. Lee, and G. Kiltie, and A. McClaughry helped with all aspects of the secretarial work. Both editors were supported by National Science Foundation awards during the organization of the symposium and completion of the book. Financial assistance was also provided by our respective Departments and Universities. All of these individuals and institutions are gratefully thanked for their help and support.

Finally, Michele R. Tennant and Terry Root are especially thanked for their continuous love and support.

M.M.M.
J.C.

Contributors

Ernesto Almira
Interdisciplinary Center for Biotechnology Research
University of Florida
Gainesville, Florida 32611

Joel Cracraft
Departments of Anatomy and Cell Biology and Biological Sciences
University of Illinois
Chicago, Illinois 60680

Mark Eggert
Departments of Mathematics and Molecular Biology
University of Southern California
Los Angeles, California 90089

Robert J. Ferl
Interdisciplinary Center for Biotechnology Research and Department of Vegetable Crops
University of Florida
Gainesville, Florida 32611

Walter M. Fitch
Department of Ecology and Evolutionary Biology
University of California
Irvine, California 92717

Manolo Gouy
Laboratoire de Biométrie
Université Lyon I
69622 Villeurbanne
Cedex, France

Kathleen Helm-Bychowski
Department of Anatomy and Cell Biology
University of Illinois
Chicago, Illinois 60680

Michael D. Hendy
Department of Mathematics and Statistics
Massey University
Palmerston North, New Zealand

David M. Hillis
Department of Zoology
The University of Texas
Austin, Texas 78712

Jana Joyce
Departments of Mathematics and Molecular Biology
University of Southern California
Los Angeles, California 90089

Allan Larson
Department of Biology
Washington University
St. Louis, Missouri 63130

Wen-Hsiung Li
Center for Demographic and Population Genetics
University of Texas
Houston, Texas 77225

David P. Mindell
Department of Biological Sciences
University of Cincinnati
Cincinnati, Ohio 45221

Michael M. Miyamoto
Department of Zoology
University of Florida
Gainesville, Florida 32611

C. Joseph Nairn
Departments of Vegetable Crops and Botany
University of Florida
Gainesville, Florida 32611

Masatoshi Nei
Department of Biology and Institute of Molecular Evolutionary Genetics
Pennsylvania State University
University Park, Pennsylvania 16802

David Penny
Department of Botany and Zoology
Massey University
Palmerston North, New Zealand

Jin-Xiong She
Division of Basic Sciences
Department of Pathology and Laboratory Medicine
University of Florida
Gainesville, Florida 32611

Arend Sidow
Department of Molecular and Cellular Biology
University of California
Berkeley, California 94720

David R. Siemieniak
Molecular Biology—Unit 7242
The Upjohn Company
Kalamazoo, Michigan 49007

Leang C. Sieu
Molecular Biology—Unit 7242
The Upjohn Company
Kalamazoo, Michigan 49007

Jerry L. Slightom
Molecular Biology—Unit 7242
The Upjohn Company
Kalamazoo, Michigan 49007

Michael A. Steel
Fakultät für Mathematik
Universität Bielefeld
Postfach B640
D-4800 Bielefeld 1, Germany

David L. Swofford
Illinois Natural History Survey, Urbana
607 East Peabody Drive
Champaign, Illinois 61820

Edward Wakeland
Division of Basic Sciences
Department of Pathology and Laboratory Medicine
University of Florida
Gainesville, Florida 32611

Michael S. Waterman
Departments of Mathematics and Molecular Biology
University of Southern California
Los Angeles, California 90089

Ward C. Wheeler
Department of Invertebrates
American Museum of Natural History
Central Park West at 79th St.
New York, New York 10024

Allan C. Wilson
Department of Molecular and Cellular Biology
University of California
Berkeley, California 94720

Jia Ye
Department of Biology
University of Southern California
Los Angeles, California 90089

PHYLOGENETIC ANALYSIS OF DNA SEQUENCES

1

Phylogenetic Inference, DNA Sequence Analysis, and the Future of Molecular Systematics

MICHAEL M. MIYAMOTO AND JOEL CRACRAFT

The Emergence of Phylogenetic Systematics

Comparative biology has had a long and noble history (Rieppel, 1988). Its central goal has been to understand the bewildering diversity of form observed across the world's organisms. Concepts such as taxa (in particular, species) and homology have been the core intellectual instruments for facilitating this understanding. Thus, comparative biologists have labored to sort the world's organisms into species (however construed), largely based on their characteristics of form (but viewed broadly, including any intrinsic attribute), and to describe their similarities and differences. For several hundred years now, this effort to compare has led to the realization that similarities and differences among species are best ordered in terms of a hierarchy of relationships, which we now take to represent the primary pattern of life's history as it has diversified over the last 4 billion years.

Systematics is the science of comparative biology and the primary goal of systematists is to describe taxic diversity and to reconstruct the hierarchy, or phylogenetic relationships, of those taxa (Hennig, 1966; Eldredge and Cracraft, 1980; Nelson and Platnick, 1981; Wiley, 1981). The overall importance of this goal is that corroborated hypotheses of relationship are the prerequisites for all inferences regarding what needs to be explained (i.e., various patterns) within the context of an historical perspective. This might involve repeated patterns of geographic distribution from one group to another (Nelson and Platnick, 1981), the distribution across lineages of certain behavioral, ecological, or morphological attributes, or the sharing among taxa, or perhaps among spatially segregated populations, of variation at the level of DNA. Just as importantly, it is only by having an hypothesis for the hierarchical pattern of shared attributes—with its implications for understanding character polarity—that one can begin to formulate explanations for derived novelties that have arisen and become fixed in populations.

Procedures for constructing phylogenetic hypotheses have been most fully developed by the discipline of phylogenetic systematics, or cladistics (Hennig, 1966), which presently dominates the field of systematics (Hull, 1989). The emergence of cladistics can be related to its direct connection to genealogy in that only shared derived characters (synapomorphies) are used as evidence to support hypotheses about phylogenetic relationships. With respect to those hypotheses, similarities due to the retention of primitive features (symplesiomorphies) are ignored because they are uninformative regarding relationships. The cladistic approach is thus conceptually coherent in that only synapomorphies are considered to provide evidence for monophyly, and only monophyletic groups have objective reality as historical entities.

The increasing acceptance of cladistics can also be attributed to its placement of systematic methodology within a broader scientific context. Thus, one of the more important philosophical advances has been to require that phylogenetic hypotheses conform to observations as closely as possible (Farris, 1983). Stated differently, those hypotheses which make the fewest ad hoc assumptions about shared patterns of character-state distributions are to be preferred (Eldredge and Cracraft, 1980; Wiley, 1981), yet the hypotheses themselves will always remain vulnerable to testing by additional observations. This philosophical and methodological approach, known as parsimony, has become an essential component of modern systematic methodology (Farris, 1983; Sober, 1983, 1988), and as we shall discuss below, also provides the rationale for another important component of phylogenetic inference: congruence analysis.

The Growth of Comparative Sequence Analysis

Prior to the 1960s, most systematic studies utilized morphological characters as evidence for relationships. However, over the last 20 to 30 years, the contributions of molecular approaches to phylogenetic research have increased steadily (Hillis, 1987; Patterson, 1987; Doolittle, 1990; Hillis and Moritz, 1990). At this time, DNA sequences are becoming the preferred database for molecular systematics. Nuclear genomes, coupled with their extranuclear counterparts, are exceedingly large and complex, thereby offering the systematist an almost endless array of characters with different structural/functional properties, mutational/selectional biases, and evolutionary rates (Li et al., 1985; Nei, 1987; Li and Graur, 1991). Therefore, the potential exists for using the evolutionary characteristics of particular genomic regions to match those sequences to specific systematic questions. This permits the use of sequences to address a broad range of systematic questions. Thus, studies of population variation can be investigated with mitochondrial DNA (mtDNA) sequences from the noncoding control region, a part of the molecule which often shows significant variation, even among individual organisms (Greenberg et al., 1983). At the other extreme, the phylogenetic relationships within and among kingdoms can be evaluated with highly conserved stretches of coding DNA, for example, those for ribosomal RNAs (rRNAs) (Field et al., 1988; Lake, 1988; Sogin et al., 1989; Mindell and Honeycutt, 1990). The tremendous flexibility in their resolving power ensures the future importance and widespread acceptance of DNA sequences in comparative research.

Several recent advances have now made it possible for systematists, even those with little formal training in molecular biology, to obtain DNA sequence data on a routine basis. Although DNA sequencing has become a simple laboratory procedure (Sanger et al., 1977; Sanger, 1981), the polymerase chain reaction (PCR), without question, is most responsible for the increased interest in using DNA sequences in comparative work (Saiki et al., 1988; Erlich, 1989; Kocher et al., 1989). Pairs of oligonucleotide primers are chosen to flank the region of DNA that is to be amplified by PCR. Following cycles of DNA denaturation by heat, primer annealing by cooling, and strand extension with a thermostabile enzyme such as *Taq* polymerase, microgram quantities of double-stranded DNA can be synthesized from nanogram amounts of template. By modifying the ratios of the two primers, additional amplification cycles can produce single-stranded DNA (ssDNA), which can then be directly sequenced to obtain nucleotide data (Gyllensten and Erlich, 1988). As very little template is required (theoretically, only one molecule), the original samples for PCR need not be fresh or frozen tissue (Pääbo et al., 1988; Kocher et al., 1989). Thus, DNA can be amplified from alcoholic and formalin specimens, as well as from skins, feathers, bones, single hairs, and mummified material (Higuchi et al., 1984; Pääbo et al., 1988, 1989; Pääbo, 1989). PCR has therefore greatly enhanced the value of samples which were previously not accessible to the molecular systematist. The ability to use museum specimens is of special importance, particularly since natural biological diversity continues to be lost at an alarming rate. Furthermore, PCR studies of museum specimens collected at different times promise to add a temporal component to phylogenetic and evolutionary analysis (Thomas et al., 1990).

PHYLOGENETIC INFERENCE USING DNA SEQUENCES: METHODOLOGICAL PROBLEMS

Despite the spectacular technical advances in obtaining comparative sequence data, many difficulties still exist for the investigator, not only during the process of acquiring the sequence information itself, but also in subsequent analyses undertaken to generate phylogenetic hypotheses for the taxa of interest. Some of these difficulties and limitations are discussed below.

Acquisition of Comparative Sequence Data

Most comparative sequence data that have been used in molecular systematics were obtained by conventional cloning procedures (Sambrook et al., 1990), although many investigators have now adopted asymmetrical PCR as the method of choice (Gyllensten and Erlich, 1988; Allard et al., 1991). While PCR is relatively simple in theory and practice, many laboratories attempting to obtain comparative sequence data using asymmetrical amplification have often experienced variable success in producing high quality ssDNA, especially when trying to amplify fragments over 800 to 1000 base pairs (bp) in length. Yet, other factors may be as important as fragment length in influencing the quality of single-stranded product from asymmetrical PCR, and reaction conditions must be carefully optimized (Gyllensten, 1989). These difficulties may be only transitory, however, since new procedures for obtaining single-stranded template from double-stranded PCR reactions have the potential for greatly facilitating large-scale comparative studies (Higuchi and Ochman, 1989; Mitchell and Merril, 1989).

Only after it becomes possible to obtain large amounts of comparative sequence information rapidly, accurately, and at reasonable cost, will DNA sequences fulfill their potential for systematics. In order to answer systematic questions, the most important objective is to obtain character-state data, which can often be accomplished most effectively by aligning, in tandem, those segments of sequence that are easiest to collect; for example, by using several so-called universal PCR primer-pairs (Kocher et al., 1989). Thus, it will sometimes be the case that systematist may not bother to sequence both complementary strands of DNA, much less, complete genes. In instances such as this, the goal of the systematist will frequently conflict with that of the molecular evolutionist who may be interested in different questions involving the evolutionary patterns seen across entire genes. Such conflicts are regrettable, given the general importance of DNA sequences to fields other than molecular systematics (Fickett and Burks, 1989).

Once DNA sequences have been determined, they are deposited in permanent data banks that can be accessed by others for many different purposes. Given their general utility, the accuracy of these sequences should therefore be verified by checking both complements against one another. For a similar reason, nucleotide sequencing using RNA templates should be avoided in favor of direct DNA analysis (Hillis et al., 1990).

A major concern of molecular evolution is the relationship between the higher-level structure of genes and their products (including, for example, proteins and structural RNAs), and their primary sequences (Gerbi, 1985; Gautheret et al., 1990; Johnson et al., 1990). Such information is also relevant to molecular systematics. For example, knowledge about the secondary structure of rRNA may provide an important basis for assigning weight to different regions of the molecule in a phylogenetic analysis (Wheeler and Honeycutt, 1988; Smith, 1989). Consequently, obtaining complete gene sequences across a spectrum of taxa becomes important, because data of this kind will permit the detailed analyses of molecular structure, function, and evolution necessary to justify the weighting (Irwin et al., 1991).

Acquiring sequence data is labor intensive; therefore, molecular systematists generally use only single individuals to represent each taxon in their study. As a consequence, the assumption is made (usually implicitly) that polymorphism does not affect the phylogenetic analysis because within-group variation is negligible. However, this assumption may not be well-founded, especially when the branch points under consideration are of a relatively recent age, as would be true in the case of a recent radiation of species within a clade (Nei, 1986). Under these circumstances, the sequences may not reflect the true species relationships of the group, even if a single branching pattern is strongly supported by the data. Instead, the strong support might reflect a gene phylogeny that may not be congruent with the cladistic history of the taxa themselves. Such misleading situations arise because of random sorting of ancestral polymorphic alleles by drift after cladogenesis. The only way to test whether a gene phylogeny is congruent with the species phylogeny is to compare the current results against data from other individuals and for unlinked genes (Nei, 1986; Felsenstein, 1988).

Sequence Alignment

The first step in analyzing nucleotide data is to align the sequences against one another (Waterman, 1984, 1989; see also papers in Doolittle, 1990). This initial step is crucial, as all subsequent analyses are dependent on the final alignment. In many instances (e.g., involving protein-coding sequences of mtDNA), sequence alignment is easy and can be done by eye, without the aid of an alignment algorithm. On the other hand, many kinds of sequences vary so much across taxa that computer-assisted alignment is essential to minimize the differences among them. Most computer procedures use some measure of similarity (or dissimilarity) to search for the best alignment for a given pair of sequences. Different pairwise comparisons are then combined to produce the final overall result. Because of their indirect approach to the problem, such strategies may not reveal the optimal alignment for multiple sequences. Algorithms for the simultaneous comparison of multiple sequences related by a given tree topology have been developed, but are limited by their dependence on the tree topology itself and by the need for large amounts of computer time (Sankoff, 1975; Sankoff and Cedergren, 1983; see also Waterman et al., and Mindell, this volume).

In molecular systematics, sequence alignment is essentially a problem of homology[1] of character-state data that are the individual nucleotides at each base position. Morphologists have traditionally recognized homolo-

[1] Here, sequence homology is equated with orthology (common ancestry by speciation) and paralogy (ancestry by gene duplication) is not included in the definition (Fitch, 1970; Patterson, 1988). In phylogenetic analysis, paralogous comparisons are avoided, because they provide evidence on the order of gene duplications, but not of speciation. Thus, only orthologous sequences are of importance in reconstructing species relationships.

gies by the criteria of similarity and congruence (Patterson, 1988). Homologies themselves are hypotheses of synapomorphy that are first postulated on the basis of observed similarity (Eldredge and Cracraft, 1980). After phylogenetic reconstruction, these assessments are then evaluated according to their agreement, or disagreement, with the most-parsimonious solution (i.e., with respect to their congruence with other characters). Those similarities which conflict with the most-parsimonious distribution of the entire suite of characters are reinterpreted as homoplasies (convergences, parallelisms, and reversals).

Homology has been largely equated with similarity by comparative molecular biologists (see Patterson, 1988). Unfortunately, this emphasis on similarity has obscured the conceptual and methodological relationship between homology and character support for a genealogical hypothesis. Even though methods of sequence alignment may maximize overall similarity prior to a phylogenetic analysis, hypotheses of base-position homology nevertheless remain falsifiable by character congruence on the tree that most parsimoniously explains the distribution of all the data across the taxa. The same tests of similarity and congruence that apply to morphological characters, therefore, apply to sequence data as well.

The importance of sequence alignment to comparative analysis is highlighted by the use of large-subunit rRNA sequences to resolve relationships among prokaryotes and eukaryotes. The original study of these groups by Lake (1988) emphasized small-subunit rRNA sequences and concluded that eocytes and eukaryotes are each other's closest living relatives. A major implication of this arrangement was the possibility that eukaryotes originated from a sulfur-metabolizing, thermophilic, anucleate ancestor. However, both Lake (1990) and Gouy and Li (1990) have noted that the results for large-subunit rRNA sequences are dependent on the particular alignment used as well as the choice of species. The instability of the large-subunit rRNA results reemphasizes the need for more rigorous and efficient methods of aligning multiple sequences in which the tasks of alignment and phylogenetic analysis are more fully integrated (Hein, 1989, 1990; see also Waterman et al., and Mindell, this volume).

Reconstructing Phylogenetic History

That nucleotide sequences provide a rich source of data for resolving phylogenetic relationships is no longer disputed. What is being debated, rather, is not so much how phylogenetic hypotheses can be constructed from those data (many procedures are capable of generating such hypotheses), but which method should be preferred and how we might objectively assess the veracity, or reliability, of the result. These questions define a host of critical problems within systematics in general, many of which bear on the nature of comparative evidence and its application to evaluating alternative hypotheses of relationship. Entangled with these problems are a series of debates over methods of phylogenetic inference that take as their focus what investigators either assume or infer about the evolutionary dynamics of nucleotide sequences. None of these controversies, however, appear to be unique to molecular data. For years, morphologists have debated whether it is desirable or possible to use assumptions or inferences about the evolutionary characteristics of morphological features and whether one can apply these to evaluate the reliability of those characters as evidence for phylogenetic inference.

A good deal is known about the evolutionary dynamics of DNA (e.g., see the summaries of Li et al., 1985; Nei, 1987; Li and Graur, 1991), and assumptions about the nature of its evolution have been incorporated into all methods of phylogenetic inference in the form of weighting schemes, correction factors for multiple mutations, and so forth. Thus, it is not the use of these assumptions *per se* that would lead one to prefer one method over another inasmuch as it is generally possible to incorporate any given assumption into most tree-building algorithms (except in those special cases when certain assumptions, such as constant rates, are a logical corollary of the method). Justification for a particular method will therefore have to come from elsewhere, and various possibilities have been proposed. Felsenstein (1988), for example, identifies two approaches to justification: "hypothetico-deductive," which he equates with the application of parsimony; and statistical. Other investigators have sought to justify the choice of method by simulation analysis in which DNA sequences evolve under various assumptions and according to "known" phylogenies. The method of phylogenetic inference that recovers the "true" phylogeny, given the particular *a priori* assumptions of the simulation, is preferred over those which do not (Sourdis and Krimbas, 1987; Sourdis and Nei, 1988; Saitou and Imanishi, 1989).

There are at least two general sets of methods of phylogenetic inference that can be applied to sequence data (we appreciate the fact that others may categorize these differently) (Felsenstein, 1981, 1988; Swofford and Olsen, 1990). The first set utilizes discrete character data and includes two well-known approaches; Hennigian cladistics (often called the parsimony method, but this term can be applied more broadly to other procedures as well; see below), and maximum likelihood. A second set of procedures clusters intertaxon similarity/dissimilarity distance measures derived from paired comparisons of the sequences. Categorizing methods in this way mirrors some of the most intense debates seen in the recent systematic literature, and also emphasizes certain theoretical and methodological problems that should concern all workers undertaking phylogenetic inference.

It is probably the case that the majority of nucleotide sequence data sets have been analyzed using distance methods. Surprisingly, however, few investigators who apply distance algorithms have addressed the extensive literature criticizing and defending the use of distances in phylogenetic inference (Farris, 1981, 1983, 1985, 1986a, b; Felsenstein, 1984, 1986). Distance analysis receives its justification from the argument that phylogenetic inference must be treated as a case of statistical inference (Felsenstein, 1984, 1986), yet as the debates have highlighted, this is not entirely self evident (Farris, 1983). Indeed, the conflict between using methodological parsimony[2] and statistical inference as the basis for justifying any tree-building method will undoubtedly remain a controversial issue within systematics for some years to come.

If one chooses to apply a statistical approach, then the data must possess certain properties in order for the statistical procedures to have validity. This is the basis for much of the disagreement: it is well known that systematic data, including DNA sequences, often do not satisfy the underlying assumptions of statistical models (Sanderson, 1989; Swofford and Olsen, 1990). This appears to be a problem for the statistical viewpoint but not for methodological parsimony. In principle, at least, methodological parsimony may be applied no matter what the underlying characteristics of the data. It is certainly the case that the statistical structure of data is very much a concern for all methods of analysis—even in cladistics one assumes independence of characters—but the exact nature of that structure is itself transparent to methodological parsimony. This forms the basis for the suggestion that methodological parsimony is a more general approach to hypothesis evaluation than is statistical inference.

Methodological parsimony is a general criterion in science for ajudicating the effectiveness of alternative hypotheses in accounting for data (Farris, 1983; Sober, 1983). It applies to all methods of phylogenetic inference which rely on an optimality criterion (Swofford and Olsen, 1990). Some quantity is being minimized or maximized in all of these methods, and the decision to choose the minimum or maximum solution forms the basis for the application of parsimony. It must be understood, however, that this approach does not address the truth of the resulting hypotheses. The application of statistical methods has been offered as one way to evaluate the efficacy, or reliability, of hypotheses relative to others. If one chooses not to apply a statistical approach, however, the most-parsimonious solution remains the single best hypothesis for explaining the data, given the criterion forming the basis for the parsimony decision. To claim otherwise is to argue that our choice of hypothesis does not have to conform to evidence, a claim that will lead scientific discovery nowhere.

As noted, it has become popular to use simulations to compare the properties of various tree-building algorithms. The reliability of a method is then judged on whether it accurately identifies the "true" tree used in the simulation. No one would suggest that a given approach will always find the true phylogenetic relationships for a set of real taxa, and few would suggest that simulations are not informative regarding when a particular

[2] We distinguish here between parsimony methods and methodological parsimony. The former is generally equated with cladistic algorithms in which the parsimony criterion is the minimization of homoplasies. Methodological parsimony is a more general, philosophical criterion, and keeping the distinction in mind may help clarify the debate over the role of statistics versus parsimony in phylogenetic inference.

procedure might fail to find the correct tree. But there are some fundamental philosophical and empirical differences between simulations of fictitious taxa and their DNA sequences, on the one hand; and real-world taxa and their sequence characteristics, on the other. These differences place important limitations on the general significance of simulated results.

The results of simulation are logically dependent upon the model of nucleotide evolution used to generate the sequences, along with the parameters of the model such as generation time, time to cladogenesis, and sequence length, as well as the algorithm used in the specific analysis. The evolutionary models used in many simulation studies are exceedingly simple, and even though they will surely become more sophisticated (e.g., more "realistic") in the future, such studies will still face a credibility gap. There is no single evolutionary model, no matter how complicated, that can mimic all of the historical complexities of sequence data. Even as models increase in sophistication, no one who studies DNA would expect that a particular model of sequence change will necessarily hold across different clades of taxa. If the assumptions of these models lack veracity with respect to real-world situations, their relevance to empirical cases will continue to be questioned.

In short, because of the immense complexities of the real world, simulation studies can only be expected to identify some of the limitations of a given method of phylogenetic inference. It is unrealistic to expect any computer model to capture all of the nuances of the molecular evolutionary process. Thus, the successful reconstruction of a simulated phylogeny by a particular method does not necessarily guarantee similar success when the method is applied to actual data. Some important parameter of the actual data may have been overlooked in designing the simulation, thereby limiting the general significance of its conclusions. There is a need, therefore, for an approach to evaluate methods and phylogenetic results in the absence of truth about the real world. In the next section, we suggest that congruence analysis, an extension of methodological parsimony, offers one such approach for comparing the reliability of competing methods and/or empirical results.

CONGRUENCE ANALYSIS AS ARBITER

Congruence analysis is a time-honored tradition of science, in general, and of systematics, in particular. Unfortunately, its importance has been overlooked within molecular systematics, especially by those searching for a means to evaluate the reliability of phylogenetic methods. As a principle, the notion of congruence applies broadly, from instances of pattern recognition to expectations of theory. Thus, it is congruence of observations that signifies the presence of a pattern and implies the need for a common explanation; it is congruence of evidence that allows us to prefer one hypothesis over another. In both instances, the use of concordance can be viewed as an extension of the application of methodological parsimony.

In principle, the true phylogenetic relationships for a given set of taxa will never be known with certainty. In the absence of such knowledge, systematics has long relied on studies of congruence among data sets to assess the reliability of its phylogenetic hypotheses. Well-corroborated patterns of relationship are accepted as the best estimates of the true phylogeny because it is more parsimonious to accept a common explanation—common descent, for example—than to conclude that two (or more) identical results were obtained independently. For such reasons, studies of congruence have been viewed by systematists as a powerful way to resolve difficult phylogenetic questions (Penny et al., 1982; Kluge, 1989).

Congruence analysis provides an important mechanism with which to evaluate the reliability of different tree-building methods (see also Mickevich, 1978). Given a common data set, those approaches which lead to congruent results should be preferred over those that do not. Furthermore, such comparisons can then serve as a basis for investigating why a particular method has failed to converge on a congruent pattern that is supported by different sets of data and/or other methods of phylogenetic inference. Studies of congruence, therefore, can provide insights into the limitations and assumptions of different tree-making algorithms.

We suggest that empirical investigations of congruence among actual data sets using different tree-building procedures, along with varying the assumptions of those methods, will likely provide more insight into the efficacy of these approaches than will simulation analysis, which is generally far removed from the real world. One approach might rely on phylogenetic hypotheses of major groups of organisms that are highly corroborated by a variety of molecular and nonmolecular data, such as higher-level relationships within the primates (Miyamoto and Goodman, 1990). Those methodological approaches which lead to congruence with well-corroborated hypotheses should be preferred over those that rarely do. In this way, the reliability of different tree-building methods can be evaluated against our best estimates of phylogenetic relationship without interference from oversimplifications of the real world and from individual bias during the selection of parameters in simulation analyses.

Congruence analysis also permits one to evaluate the reliability of different weighting schemes of character transformations for use in a phylogenetic study. In their phylogenetic investigation of cervid and other artiodactyl mtDNA sequences, Kraus and Miyamoto (1991) found that cladistic analysis of all mutations (transitions, transversions, and gap events) resulted in a phylogeny supporting the monophyly of antlered deer in the family Cervidae. In contrast, when only transversions were counted, the same method (i.e., parsimony) led to a solution whereby antlered deer were not monophyletic. The former arrangement favoring antlered deer monophyly is corroborated by morphological data, unlike its alternative solution. Thus, congruence suggests in this case that the use of all mutations provides more reliable results than analysis of transversion differences alone. The failure of transversion parsimony to obtain a reliable result can be attributed to the loss of cladistic information offered by transitions at these lower hierarchical levels.

The primary limitation on obtaining knowledge of phylogenetic history has not been so much with difficulties encountered with any particular method, but often with the actual data (Kraus and Miyamoto, 1991). We either have too few data or the evidence is of such poor quality (because of extensive homoplasy, intraspecific polymorphism, paralogy) that no method can use them effectively. Conversely, when the data are highly structured, the same groupings will very likely emerge no matter what method of phylogenetic inference is employed. The real problem occurs when different methods yield incongruent results for the same data. Is this the result of the methods *per se*, the choice of underlying assumptions used in the approach (e.g., correction factors, weighting schemes), or because the data are ambiguous? By relying on studies of congruence, answers to such questions will continue to accumulate for real-world situations.

THE FUTURE OF MOLECULAR SYSTEMATICS

The growing importance of DNA sequences for phylogenetic inference and for analysis of evolutionary processes at the molecular level requires that investigators strive for complete gene sequences whose accuracy has been verified by sequencing both complements. Such efforts will enhance the overall value of these comparative data to biology as a whole. With regard to systematics, a more detailed understanding of molecular evolution will facilitate the development of improved methods of phylogenetic reconstruction. Because large amounts of sequence data will be needed to address problems about the extent of intraspecific polymorphism, and to resolve many vexing phylogenetic questions, we might expect that cooperative research projects involving different laboratories will be required.

Continued technological developments (with respect to collecting sequence data), coupled with the vast phylogenetic information contained in nuclear and extranuclear genomes, virtually guarantees that nucleotide sequences will become the primary source of systematic data in the near future. Before the full potential of these data is realized, however, many of the problems noted above will need to be resolved. We have proposed that congruence analysis will play an important role in resolving controversies over systematic methods and phylogenetic relationships. Congruence provides the ultimate test of reliability in the absence of revealed truth, and as such, adoption of this methodological criterion can be regarded as crucial to the continued development of the field. Despite the growing importance of sequence data, it cannot be stressed strongly enough that there remains a pressing need to enlarge our nonmolecular database of systematic characters. If we are to gain meaningful insights into the "whats," "hows," and "whys" of the history of life, phylogenetic studies will have to rely on all available comparative information, both molecular and nonmolecular.

ACKNOWLEDGMENTS

We thank M. W. Allard, D. P. Mindell, and M. R. Tennant for their helpful suggestions about the manuscript. This research was supported by grants from the National Science Foundation to MMM (BSR-8857264 and BSR-8918606) and JC (BSR-8805957 and BSR-9007652).

REFERENCES

Allard, M.W., D.L. Ellsworth, and R.L. Honeycutt. (1991) The production of single-stranded DNA suitable for sequencing using the polymerase chain reaction. *BioTechniques* **10:** 24–26.

Doolittle, R.F. (ed.). (1990) *Molecular Evolution: Computer Analysis of Protein and Nucleic Acid Sequences.* Methods in Enzymology, vol. 183. Academic Press, New York.

Eldredge, N., and J. Cracraft. (1980) *Phylogenetic Patterns and the Evolutionary Process.* Columbia University Press, New York.

Erlich, H.A. (ed.). (1989) PCR *Technology. Principles and Applications for DNA Amplification.* Stockton Press, New York.

Farris, J.S. (1981) Distance data in phylogenetic analysis. Pp. 3–23 in *Advances in Cladistics: Proceedings of the First Meeting of the Willi Hennig Society* (V.A. Funk and D.R. Brooks, eds.). New York Botanical Garden, Bronx.

Farris, J.S. (1983) The logical basis of phylogenetic analysis. *Adv. Cladistics* **2:** 7–36.

Farris, J.S. (1985) Distance data revisited. *Cladistics* **1**: 67–85.

Farris, J.S. (1986a) On the boundaries of phylogenetic systematics. *Cladistics* **2:** 14–27.

Farris, J.S. (1986b) Distances and statistics. *Cladistics* **2:** 144–157.

Felsenstein, J. (1981) Numerical methods for inferring evolutionary trees. *Q. Rev. Biol.* **57:** 379–404.

Felsenstein, J. (1984) Distance methods for inferring phylogenies: a justification. *Evolution* **38:** 16–24.

Felsenstein, J. (1986) Distance methods: a reply to Farris. *Cladistics* **2:** 130–143.

Felsenstein, J. (1988) Phylogenies from molecular sequences: inference and reliability. *Ann. Rev. Genet.* **22:** 521–565.

Fickett, J.W., and C. Burks. (1989) Development of a database for nucleotide sequences. Pp. 1–34 in *Mathematical Methods for DNA Sequences* (M.S. Waterman, ed.). CRC Press, Boca Raton.

Field, K.G., G.J. Olsen, D.J. Lane, S.J. Giovannoni, M.T. Ghiselin, E.C. Raff, N.R. Pace, and R.A. Raff. (1988) Molecular phylogeny of the animal kingdom. *Science* **239:** 748–753.

Gautheret, D., F. Major, and R. Cedergren. (1990) Computer modeling and display of RNA secondary and tertiary structures. Pp. 318–330 in *Molecular Evolution: Computer Analysis of Protein and Nucleic Acid Sequences* (R.F.

Doolittle, ed.). Methods in Enzymology, vol. 183. Academic Press, New York.

Gerbi, S.A. (1985) Evolution of ribosomal DNA. Pp. 419–517 in *Molecular Evolutionary Genetics* (R.J. MacIntyre, ed.). Plenum Press, New York.

Gouy, M., and W.-H. Li. (1990) Archaebacterial or eocyte tree? *Nature* **343:** 419.

Greenberg, B.D., J.E. Newbold, and A. Sugino. (1983) Intraspecific nucleotide sequence variability surrounding the origin of replication in human mitochondrial DNA. *Gene* **21:** 33–49.

Gyllensten, U. (1989) Direct sequencing of in vitro amplified DNA. Pp. 45–60 in *PCR Technology. Principles and Applications for DNA Amplification* (H.A. Erlich, ed.). Stockton Press, New York.

Gyllensten, U.B., and H.A. Erlich. (1988) Generation of single-stranded DNA by the polymerase chain reaction and its application to direct sequencing of the HLA-DQA locus. *Proc. Natl. Acad. Sci. USA* **85:** 7652–7656.

Hein, J. (1989) A new method that simultaneously aligns and reconstructs ancestral sequences for any number of homologous sequences, when the phylogeny is given. *Mol. Biol. Evol.* **6:** 649–668.

Hein, J. (1990) Unified approach to alignment and phylogenies. Pp. 626–645 in *Molecular Evolution: Computer Analysis of Protein and Nucleic Acid Sequences* (R.F. Doolittle, ed.). Methods in Enzymology, vol. 183. Academic Press, New York.

Hennig, W. (1966) *Phylogenetic systematics*. University of Illinois Press, Urbana.

Higuchi, R., B. Bowman, M. Freiberger, O.A. Ryder, and A.C. Wilson. (1984) DNA sequences from the quagga, an extinct member of the horse family. *Nature* **312:** 282–284.

Higuchi, R.G., and H. Ochman. (1989) Production of single-stranded DNA templates by exonuclease digestion following the polymerase chain reaction. *Nucleic Acids Res.* **17:** 5865.

Hillis, D.M. (1987) Molecular versus morphological approaches to systematics. *Ann. Rev. Ecol. Syst.* **18:** 23–42.

Hillis, D.M., A. Larson, S.K. Davis, and E.A. Zimmer. (1990) Nucleic acids III: sequencing. Pp. 318–370 in *Molecular Systematics* (D.M. Hillis and C. Moritz, eds.). Sinauer Associates, Sunderland.

Hillis, D.M., and C. Moritz (eds.). (1990) *Molecular Systematics*. Sinauer Associates, Sunderland.

Hull, D.L. (1989) The evolution of phylogenetic systematics. Pp. 3–15 in *The Hierarchy of Life. Molecules and Morphology in Phylogenetic Analysis* (B. Fernholm, K. Bremer, and H. Jornvall, eds.). Elsevier Science Publishers B.V., Amsterdam.

Irwin, D.M., T.D. Kocher, and A.C. Wilson. (1991) Evolution of the cytochrome *b* gene of mammals. *J. Mol. Evol.* **32:** 128–144.

Johnson, M.S., A. Sali, and T.L. Blundell. (1990) Phylogenetic relationships from three-dimensional protein structures. Pp. 670–690 in *Molecular Evolution: Computer Analysis of Protein and Nucleic Acid Sequences* (R.F. Doolittle, ed.) Methods in Enzymology, vol. 183. Academic Press, New York.

Kluge, A.G. (1989) A concern for evidence and a phylogenetic hypothesis of relationships among *Epicrates* (Boidae, Serpentes). *Syst. Zool.* **38:** 7–25.

Kocher, T.D., W.K. Thomas, A. Meyer, S.V. Edwards, S. Pääbo, F.X. Villablanca, and A.C. Wilson. (1989) Dynamics of mitochondrial DNA evolution in animals: amplification and sequencing with conserved primers. *Proc. Natl. Acad. Sci. USA* **86:** 6196–6200.

Kraus, F., and M.M. Miyamoto. (1991) Rapid cladogenesis among the pecoran ruminants: evidence from mitochondrial DNA sequences. *Syst. Zool.* **40:** 117–130.

Lake, J.A. (1988) Origin of the eukaryotic nucleus determined by rate-invariant analysis of rRNA sequences. *Nature* **331:** 184–186.

Lake, J.A. (1990) Archaebacterial or eocyte tree? *Nature* **343:** 418–419.

Li, W.-H., and D. Graur. (1991) *Fundamentals of Molecular Evolution.* Sinauer Associates, Sunderland.

Li, W.-H., C.-C. Luo, and C.-I. Wu. (1985) Evolution of DNA sequences. Pp. 1–94 in *Molecular Evolutionary Genetics* (R.J. MacIntyre, ed.). Plenum Press, New York.

Mickevich, M.F. (1978) Taxonomic congruence. *Syst. Zool.* **27:** 143–158.

Mindell, D.P., and R.L. Honeycutt. (1990). Ribosomal RNA in vertebrates: evolution and phylogenetic applications. *Ann. Rev. Ecol. Syst.* **21:** 541–566.

Mitchell, L., and C.R. Merril. (1989) Affinity generation of single-stranded DNA for dideoxy sequencing following the polymerase chain reaction. *Anal. Biochem.* **178:** 239–242.

Miyamoto, M.M., and M. Goodman. (1990) DNA systematics and evolution of primates. *Ann. Rev. Ecol. Syst.* **21:** 197–220.

Nei, M. (1986) Stochastic errors in DNA evolution and molecular phylogeny. Pp. 133–147 in *Evolutionary Perspectives and the New Genetics* (H. Gershowitz, D.L. Rucknagel, and R.E. Tashian, eds.). Alan R. Liss, New York.

Nei, M. (1987) *Molecular Evolutionary Genetics.* Columbia University Press, New York.

Nelson, G., and N. Platnick. (1981) *Systematics and Biogeography: Cladistics and Vicariance.* Columbia University Press, New York.

Pääbo, S. (1989) Ancient DNA: extraction, characterization, molecular cloning, and enzymatic amplification. *Proc. Natl. Acad. Sci. USA* **86:** 1939–1943.

Pääbo, S., J.A. Gifford, and A.C. Wilson. (1988) Mitochondrial DNA sequences from a 7000-year old brain. *Nucleic Acids Res.* **16:** 9775–9787.

Pääbo, S., R.G. Higuchi, and A.C. Wilson. (1989) Ancient DNA and the polymerase chain reaction. The emerging field of molecular archaeology. *J. Biol. Chem.* **264:** 9709–9712.

Patterson, C. (1987) Introduction. Pp. 1–22 in *Molecules and Morphology in Evolution: Conflict or Compromise*? (C. Patterson, ed.). Cambridge University Press, New York.

Patterson, C. (1988) Homology in classical and molecular biology. *Mol. Biol. Evol.* **5:** 603–625.

Penny, D., L.R. Foulds, and M.D. Hendy. (1982) Testing the theory of evolution by comparing phylogenetic trees constructed from five different protein sequences. *Nature* **297:** 197–200.

Rieppel, O.C. (1988) *Fundamentals of Comparative Biology.* Birkhauser Verlag, Basel.

Saiki, R.K., D.H. Gelfand, S. Stoffel, S.J. Scharf, R. Higuchi, G.T. Horn, K.B. Mullis, and H.A. Erlich. (1988) Primer-directed enzymatic amplification of DNA with a thermostable DNA polymerase. *Science* **239:** 487–491.

Saitou, N., and T. Imanishi. (1989) Relative efficiencies of the Fitch-Margoliash, maximum-parsimony, maximum-likelihood, minimum-evolution, and neighbor-joining methods of phylogenetic tree construction in obtaining the correct tree. *Mol. Biol. Evol.* **6:** 514–525.

Sambrook, J., E. Fritsch, and T. Maniatis. (1990) *Molecular Cloning. A Laboratory Manual*/second edition. Cold Spring Harbor Laboratory, Cold Spring Harbor.

Sanderson, M.J. (1989). Confidence limits on phylogenies: the bootstrap revisited. *Cladistics* **5:** 113–129.

Sanger, F. (1981) Determination of nucleotide sequences in DNA. *Science* **214:** 1205–1210.

Sanger, F., S. Nicklen, and A.R. Coulson. (1977) DNA sequencing with chain-terminating inhibitors. *Proc. Natl. Acad. Sci. USA* **74:** 5463–5467.

Sankoff, D. (1975) Minimal mutation trees of sequences. *SIAM J. Appl. Math.* **28:** 35–42.

Sankoff, D., and R.J. Cedergren. (1983) Simultaneous comparison of three or more sequences related by a tree. Pp. 253–263 in *Time Warps, String Edits, and Macromolecules: The Theory and Practice of Sequence Comparison* (D. Sankoff and J.B. Kruskal, eds.). Addison-Wesley, Reading.

Smith, A.B. (1989) RNA sequence data in phylogenetic reconstruction: testing the limits of its resolution. *Cladistics* **5:** 321–344.

Sober, E.R. (1983) Parsimony methods in systematics. *Adv. Cladistics* **2:** 37–47.

Sober, E.R. (1988) *Reconstructing the Past: Parsimony, Evolution, and Inference*. MIT Press, Cambridge.

Sogin, M.L., U. Edman, and H. Elwood. (1989) A single kingdom of eukaryotes. Pp. 133–143 in *The Hierarchy of Life. Molecules and Morphology in Phylogenetic Analysis* (B. Fernholm, K. Bremer, and H. Jornvall, eds.). Elsevier Science Publishers B.V., Amsterdam.

Sourdis, J., and C. Krimbas. (1987) Accuracy of phylogenetic trees estimated from DNA sequence data. *Mol. Biol. Evol.* **4:** 159–166.

Sourdis, J., and M. Nei. (1988) Relative efficiencies of the maximum parsimony and distance-matrix methods in obtaining the correct phylogenetic tree. *Mol. Biol. Evol.* **5:** 298–311.

Swofford, D.L., and G.J. Olsen. (1990) Phylogeny reconstruction. Pp. 411–501 in *Molecular Systematics* (D.M. Hillis and C. Moritz, eds.). Sinauer Associates, Sunderland.

Thomas, W.K., S. Pääbo, F.X. Villablanca, and A.C. Wilson. (1990) Spatial and temporal continuity of kangaroo rat populations shown by sequencing mitochondrial DNA from museum specimens. *J. Mol. Evol.* **31:** 101–112.

Waterman, M.S. (1984) General methods of sequence comparison. *Bull. Math. Biol.* **46:** 473–500.

Waterman, M.S. (1989) Sequence alignments. Pp. 53–92 in *Mathematical Methods for DNA Sequences* (M.S. Waterman, ed.). CRC Press, Boca Raton.

Wheeler, W.C., and R.L. Honeycutt. (1988) Paired sequence difference in ribosomal RNAs: evolutionary and phylogenetic implications. *Mol. Biol. Evol.* **5:** 90–96.

Wiley, E.O. (1981) *Phylogenetics. The Theory and Practice of Phylogenetic Systematics*. J. Wiley Sons, New York.

2

DNA Sequencing: Strategy and Methods to Directly Sequence Large DNA Molecules

JERRY L. SLIGHTOM, DAVID R. SIEMIENIAK, AND
LEANG C. SIEU

The goal of the Human Genome Initiative, to produce the complete nucleotide sequence of the genome of human and other organisms, will require the development of rapid and efficient methods for obtaining nucleotide sequences. Thus, improvements in present sequencing methods are of great interest and importance if we hope to sequence the 3×10^9 base pair (bp) human genome. Presently, two methods exist for obtaining nucleotide sequences, the enzymatic method developed by Sanger et al. (1977) which uses DNA polymerases and chain termination dideoxy nucleotides for base determinations; and the chemical method developed by Maxam and Gilbert (1977) which uses base-specific chemical reactions for base determinations. Of these two methods, the enzymatic method has been found to be readily adaptable to automation, and thus has received more attention in the form of research on different polymerases and reagents for the utilization of these polymerases. The T7 polymerase described by Tabor and Richardson (1987) and a polymerase from *Thermus aquaticus* (*Taq*) (Saiki et al., 1988) have, for the most part, replaced the use of *Escherichia coli* DNA polymerase I (Klenow fragment) and reverse transcriptase as the enzymes of choice for enzymatic sequencing.

Automation of different aspects of the nucleotide sequencing reactions (DNA purification, enzymatic reactions, and information transfer) has been the subject of numerous papers, and these steps are continuously being improved. Some of the laborious aspects of DNA sequencing have been partially automated, such as the use of robotic workstations to perform enzymatic reactions (Frank et al., 1988; Wilson et al., 1988; Zimmerman et al., 1988) the use of film readers (Elder et al., 1986), and informational transfer using fluorescent tags (Smith et al., 1986; Ansorge et al., 1987; Prober et al., 1987).

Present methods using random sequencing strategies for automating DNA sequencing suffer from several weaknesses: (1) the need to subclone short DNA fragments; (2) the length of reliable nucleotide sequence reads are short, in the range of 500 bp; (3) meaningful analysis cannot be done until a substantial amount of data has been accumulated; and (4) the strategy requires a considerable amount of "over-sequencing." For the random sequencing strategy, it is generally accepted that an insert should be sequenced a minimum of five times (Edwards et al., 1990), after which the individual sequences are assembled into a contiguous sequence with the aid of computer programs known as random sequence handlers (Staden, 1982). These computer programs do have their limitations, especially when repetitive sequences are encountered, such as the highly repetitive *Alu* element. In many cases, even after excessive random sequencing of an insert, gaps remain that must be closed using a directional sequencing approach. It would be much more efficient if large DNA molecules could be directly sequenced, in a directional manner, because coverage could be done by sequencing the insert only 2.5 times, allowing for overlaps between sequencing runs on both strands.

This report describes our attempt to develop a nucleotide sequencing strategy that avoids many of the problems described above, yet remains as simple as possible (relying on proven methods). Our strategy has two major goals. First, to conveniently sequence large DNA molecules [50 kilobase (kb) range]; and second, to obtain the longest sequence reads as possible. We describe here methods for obtaining the nucleotide sequence of cosmid and λ cloned inserts utilizing mostly double-stranded enzymatic sequencing (Sanger et al., 1977; Chen and Seeburg, 1985; Zagursky et al., 1985) using T7 polymerase, and to a smaller degree, the chemical sequencing method (Maxam and Gilbert, 1977).

EXPERIMENTAL AND DISCUSSION

Strategy for Directly Sequencing Cosmid and λ Cloned DNA Inserts

The use of T7 polymerase to obtain nucleotide sequence information from large DNA molecules such as cosmid and λ clones has been described previously (Zagursky et al., 1985; Manfioletti and Schneder, 1988; Levedakou et al., 1989; Kesterson et al., 1989). However, these methods are somewhat limited because sequences are obtained only from the vector cloning junctions or from genetic elements (e.g., genes, repetitive DNA) known to be located within the cloned insert. Having such a small number of sequencing points for a "walk" across a large insert would be too time consuming. This problem can be overcome by obtaining more sequencing points (internal to the insert) from which to initiate double-stranded dideoxy sequencing; these points are referred to as initiation points (IPs). These additional IPs should be obtained using a method that is independent of the cloned DNA insert; that is, it could be used for any cloned insert. Several methods for obtaining IPs are feasible: (1) the use of the chemical sequencing procedure (Maxam and Gilbert, 1977); (2) the use of splinkers to add a known primer sequence to a restriction enzyme site (Kalisch et al., 1986); (3) the more recently developed use of exonuclease III (Sorge and Blinderman, 1989); and (4) the use of short oligomer primers for random priming (Studier, 1989; Siemieniak and Slightom, 1990). Any one of these methods is appropriate for obtaining additional IPs, and extensive sequence reads from IPs are not necessary, as the switch to oligomer primer-directed sequencing method requires an accurate sequence reading of only the length of the oligomer primer. However, after switching to the enzymatic sequencing method, its efficiency can be greatly improved as the length of the sequence reads is increased. Thus, our second goal, to obtain the longest reads possible (see below), allows us to reduce the following: the number of oligomer primers needed, the number of enzymatic reactions, the amount of template DNA, the amount of overlapping sequence, and, most importantly, the number of sequencing gels needed.

Our sequencing strategy for large DNA molecules is outlined in Table 2-1. For this strategy we use the chemical sequencing method to obtain IPs by obtaining the sequence of small (100 to 1,500 bp) fragments from within the cloned insert. These small DNA fragments are obtained by using two different restriction enzymes known to yield, upon sequential diges-

Table 2-1 Strategy for Sequencing Large DNA Inserts

1. Obtain "stable" cosmid or λ DNA clones. The use of overlapping clones is desirable but not necessary.
2. Obtain oligomers specific for the 5′ and 3′ regions flanking the vector cloning site.
3. Analyze insert with various restriction enzymes for the identification of enzyme pairs that can be used to obtain single ^{32}P-end-labeled fragments (100 to 1,500 bp in length). The sequence of these 5′-end-labeled fragments provides internal IPs.
4. Sequence the insert 5′ and 3′ flanking regions using flanking oligomer primers and obtain the sequence of four to five ^{32}P-end-labeled internal fragments (see Fig. 2-1).
5. Design oligomer primers to continue the sequence of the flanking and IPs within the cloned insert.
6. Use PCR amplification to map the orientation and relative positions of sequence contigs with respect to each other and flanking contigs (see Fig. 2-2). Obtain additional IPs if needed.
7. Extend sequence contigs using the enzymatic sequencing method; enter new sequence information into the computer as soon as available, and order new sets of oligomer primers. Alternate among clones (if available) to avoid waiting for the delivery of oligomer primers. Each oligomer set should yield about 5 to 8 kb of new sequence information.
8. Order opposite-strand oligomer primers as sequence information becomes available, and store oligomers until needed or until time allows for their use. This allows for effective use of DNA synthesis instrument, and will accelerate obtaining the sequence of the opposite strand.
9. Continue the sequencing process described in steps 7 and 8 until all contigs merge.

tion, identifiable DNA fragments. The first restriction enzyme digestion product is subjected to 5'-end-labeling with [^{32}P-γ]-adenosine triphosphate (ATP) (see below) followed by digestion with the second enzyme which releases a single ^{32}P-end-labeled DNA fragment. Generally, the sequence of a small number (four to eight) of these unmapped DNA fragments are obtained from a cloned insert. Each IP is extended in both 5′ and 3′ directions, resulting in a growing continuous DNA sequence region that is referred to as a "sequencing contig."

The overall efficiency of this method can be greatly enhanced by sequencing several large cloned inserts at one time; this process helps to eliminate "dead time" between the delivery of oligomer primers. Presently, extensive regions of the genomes of human and many other species have been cloned as sets of overlapping clones; for example, the T-cell receptor variable gene families (Lai et al., 1988). The use of this strategy for sequencing three overlapping cosmid clones is schematically illustrated in Figure 2-1. The use of overlapping cosmid clones is also useful because the overlapping boundary regions provide additional IPs within the adjacent cosmid clones (Fig. 2-1).

The relative location of the randomly generated IPs can be mapped using polymerase chain reaction (PCR) amplification. The 5′ and 3′ oligomer primers from each IP are used in a series of PCR amplifications designed to cross-challenge each primer, including the oligomers from regions flanking the vector cloning site. An example of this type of PCR mapping is shown in Figure 2-2, where the relative location of four contigs was determined; two contigs are separated by about 4.1 kb (lane 1), and another pair of contigs is separated by about 8 kb (lane 6). Mapping by PCR can be used effectively for determining the location of contigs even if they are separated by a considerable distance; knowledge of the relative locations of four to eight contigs would be sufficient to obtain a partial map of a complete 40 kb insert. The method used to amplify the DNA fragments in the range of 8 kb is identical to that described by Saiki et al. (1988), except that 5 mM DTT was used and the extension time was 10 minutes. This mapping information is useful in determining if the IPs are sufficiently spread throughout the insert to allow for efficient sequencing; if not, additional IPs should be obtained before closure of existing (closely spaced) contigs.

With the synthesis of 10 to 16 oligomer primers (for five to eight contigs, including flanking contigs) the first round of oligomer-directed sequencing can be initiated. After sequencing and film exposure, the sequence information near the ends of each read can be used for designing the next set of oligomer primers. The remaining sequence information can be read while waiting for delivery of the new oligomer primer set. This process is continued until the ends of each sequencing contig merge. The primary goal of this initial closure is to obtain a DNA sequence of the insert at a confidence level of about 95% or better. While accumulating the first strand sequence, the oligomer primers needed for sequencing the opposite strand should be ordered.

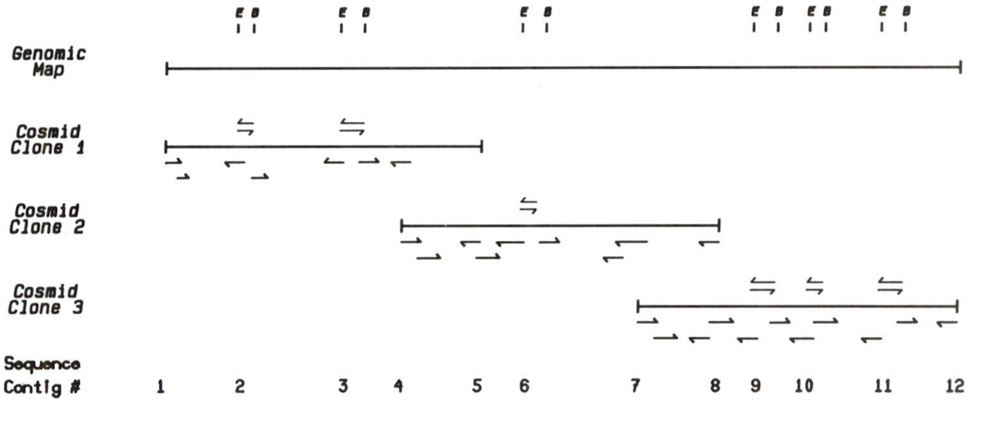

Figure 2-1 Schematic representation of the sequencing strategy outlined in Table 2-1. Here we show how the strategy could be used to obtain the sequence of three overlapping cosmid clones. The region of cloned genomic DNA is represented by the top line and the positions of three overlapping cosmid clones are shown below this line. Internal initiation points (IPs) are obtained by chemically sequencing ^{32}P-end-labeled fragments that are the result of sequential digestion with two restriction enzymes; the use of the restriction enzymes *Eco*RI (E) and *Bam*HI (B) is illustrated. Nucleotide sequences obtained by the chemical sequencing of ^{32}P-end-labeled fragments are shown by the arrows above the cosmid cloned inserts. Additional internal contigs are obtained at the 5′ and 3′ overlapping regions, which in this illustration adds four sequencing contigs. Extensions of the flanking and internal sequencing contigs, using the enzymatic sequencing method, are shown by arrows below the cosmid cloned inserts.

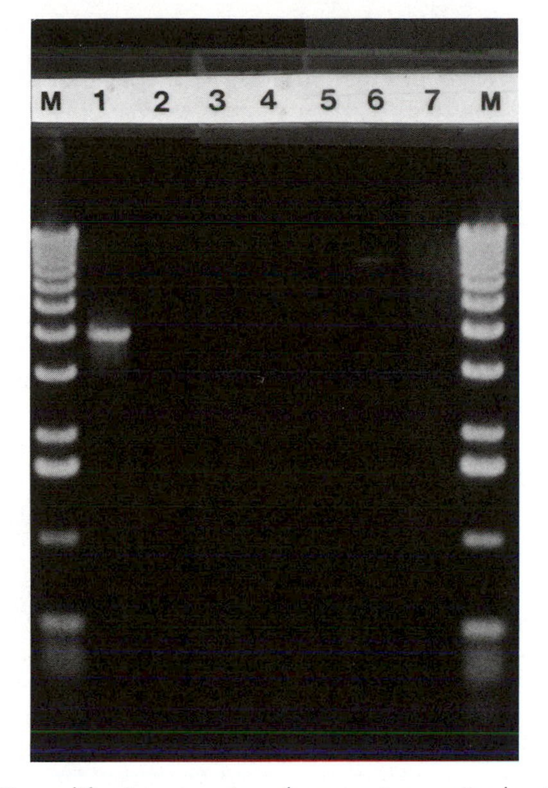

Figure 2-2 PCR amplification mapping of sequencing contigs located within cosmid H7.1, which contains part of the human T-cell receptor β-gene family (Lai et al., 1988). Oligomer primers corresponding to DNA regions which flank (5′ and 3′) the cloning site of the cosmid vector pTL-5 (Lund et al., 1982) were used in various combinations with oligomer primers from internal sequencing contigs. The lanes marked M, contain the kb DNA size standard purchased from Bethesda Research Laboratories (BRL). Lane 1 shows that two internal sequencing contigs are separated by about 4.1 kb. Lanes 2 to 5 and 7 show no amplified DNA fragment indicating that the oligomer primer parts used are either separated by large distances, or that the primer pairs are not orientated in the direction needed for amplification (Saiki et al., 1988). Lane 6 shows that two other sequencing contigs are separated by a distance of about 8 kb.

The strategy described above and in Table 2-1 possesses several distinct advantages, which include:

1. no subcloning of cosmid and λ inserts is needed;
2. limited restriction enzyme site mapping data are needed;
3. the use of overlapping inserts increases efficiency;
4. meaningful analysis of the accumulated sequence information is straightforward;
5. inserts can be completely sequenced by sequencing only 2.5-fold;
6. films provide hard-copy data storage.

Of these advantages, the fourth is particularly interesting, especially to the individual(s) involved in obtaining the sequence information because it allows them the opportunity to immediately evaluate the accumulated data.

In fairness, to those who may want to try our strategy, we must also describe some disadvantages that are associated with the direct sequencing of large DNA molecules:

1. the need to isolate large amounts of cosmid or λ DNAs (about 400 μg for a cosmid and about 200 μg for a λ insert);
2. the need for a large number of oligomer primers (that presently cost about $100 each);
3. insert instability can be a problem (however, this is a problem for any sequencing method);
4. lack of an accurate film reader for transferring sequence information into a computer.

Of these disadvantages, the lack of an efficient method to transfer information into the computer is the most severe. There are many DNA sequencing methods that have the ability to generate sequencing films faster than they can be read. The need to obtain a large amount of DNA template is not an extreme disadvantage, but does require a commitment to isolated cosmid and λ DNAs from several liters of *E. coli* cells. From the standpoint of cost and the need for a large number of oligomer primers, this method is quite expensive, and at our present price it would cost more than $10,000 for only the oligomers to sequence a 40 kb insert. However, because of recent advances in the synthesis of oligomer primers (K. Beattie, personal communication), their cost should be reduced considerably (about one-fifth to one-tenth) in the near future, at which time this disadvantage will be removed from the list.

The Chemical Sequencing Method

The use of specific chemicals for DNA strand breakage to generate a breakage pattern which is consistent with the four DNA bases was developed by Maxam and Gilbert (1977). A more detailed description of these procedures, including trouble-shooting information for the base-specific reactions, has also been described by Maxam and Gilbert (1980). The chemical reactions involve three consecutive steps: base modification, removal of the modified base from its sugar, followed by strand breakage at the exposed sugar molecule. The specific base reactions involve the use of dimethyl sulfate (DMS) to methylate the N-7 of guanine and the use of the nucleophile hydrazine to split thymine or cytosine ring structures. The identification of either a thymine or cytosine ring structure is achieved by performing the reaction in the presence of 1 M NaCl; only cytosine reacts at an appreciable rate. The presence of an adenine can be determined by treatment with formic acid, which weakens both adenine and guanine glycosidic bonds by protonating the purine ring nitrogens. In all cases, piperidine is used to catalyze β-elimination of phosphates from the exposed sugar, which results in breakage of the DNA strand. However, the base-specific chemical reaction is the final reaction step in this DNA sequencing method. Prior to performing these reactions, the DNA fragment must be end-labeled with [^{32}P-γ]-ATP, separated on a preparation gel, extracted from the gel, and then subjected to the base-specific chemical reactions.

Although the chemical sequencing procedure has been extensively used for the past 14 years, its use has declined during the past few years due to the use of modified DNA polymerases, such as T7 and *Taq* polymerases, which are much more reliable than *E. coli* polymerase I (Klenow fragment). In the past, sequences obtained using the chemical procedure were considered to be less subject to errors. However, the reactions required considerable trouble shooting, quality control of the chemicals used, and very pure DNAs. These drawbacks have probably been responsible for much of its continued decline and disfavor. Despite these problems, we have found the chemical sequencing method to be very useful for obtaining sequence information from regions internal to large DNA inserts (see Table 2-1). We feel that the advantage of obtaining additional IPs compensates for difficulties encountered in using the chemical sequencing procedure. Because we have had extensive experience with the chemical sequencing procedure, we do not consider its use to be difficult. However, we do realize that this could be a problem for others who do not have such experience. We have included this section to detail some of the procedures and precautions that we use to eliminate many of the problems which beginners have with chemical sequencing. The following descriptions highlight several problem areas associated with chemical DNA sequencing.

DNA Purification

The use of impure plasmid, cosmid, or λ DNAs can be responsible for many of the problems common to both chemical and enzymatic sequencing procedures. It is important to use a DNA purification method that consistently yields DNA which can be sequenced. The use of crude lysate procedures is not appropriate for chemical sequencing due to the presence of large amounts of *E. coli* DNA and RNAs that interfere with the 5′-end-labeling procedure (see below). For the purification of λ phage we have found the best method involves the use of a CsCl step gradient followed by a CsCl equilibrium gradient (Slightom and Drong, 1988).

The isolation of plasmid and cosmid DNAs can be accomplished by any number of methods, and many of these methods are modifications of the alkaline-extraction procedure described by Birnboim and Doly (1979). One particular derivation of this method involves the use of 7 M NH_4OAc for the neutralization step (Morelle, 1989). Table 2-2 describes a modification of this plasmid isolation procedure for the large-scale preparation of plasmid or cosmid DNAs. The most important part of the procedure is the use of two CsCl-ethidium bromide gradient bandings, using a fixed-angle

Table 2-2 Preparation of Plasmid or Cosmid DNAs for Sequencing by Chemical or Enzymatic Methods

1. Grow bacteria overnight culture (50 mL) in appropriate antibiotic.
2. Add supplements to M9 media (Maniatis et al., 1982): 1 mL of 0.1 M $CaCl_2$; 1 mL of 1.0 M $MgSO_4$; 10 mL of 40% glucose.
3. Add appropriate antibiotic and 10 mL of bacterial overnight growth to 1 L of M9 media. Shake at 37°C, 150 rpm until $O.D._{650}$ reaches about 0.5.
4. If plasmid can be amplified, add 150 mg of chloramphenicol (3 mL of stock 50 mg/mL in EtOH) per L of M9. Amplify overnight at 37°C shaking at 120 rpm.
5. After overnight growth, chill cells on ice followed by centrifugation in 1 L bottles at 4,200 rpm (Sorval RC-3B) for 15 minutes.
6. Resuspend cells in 25 mL of 0.9% NaCl, 10 mM Tris-HCl, 10 mM EDTA (pH 7.0). Transfer into a 50 mL polypropylene Oak Ridge tube and centrifuge 4,200 rpm for 10 minutes.
7. Resuspend cells in 5 mL of 50 mM glucose, 25 mM Tris-HCl, 10 mM EDTA (pH 8.0) and 8 mg/mL lysozyme. Incubate at room temperature for 10 minutes.
8. Add 10 mL of freshly prepared alkaline solution (0.2 N NaOH and 1% SDS). Mix by gently inverting the tube 3 to 6 times. The suspension should become somewhat clear and viscous. Place on ice for about 10 minutes.
9. Add 7.5 mL of 7.5 M ammonium acetate solution and mix the contents by gently mixing the tube for about 10 seconds (the pH should be about 7.8 and thus not require any adjustment). Maintain on ice for about 10 minutes to allow most of the protein, high molecular weight RNA, and chromosomal DNA to precipitate.
10. Centrifuge tube for 10 minutes at 15,000 rpm, 4°C. Remove the clear supernate and transfer to a clean tube. Add an equal volume of phenol extraction solution (see Table 2-4) and gently mix. Separate phases by centrifuging at 5,000 rpm for 5 minutes. Remove the lower phenol solution and repeat the extraction using a solution of chloroform and isoamyl alcohol (1:0.04); repeat phase separation centrifugation.
11. Transfer upper DNA solution to a clean tube and add 0.6 volumes of isopropanol and incubate at −70°C for 10 minutes.
12. Centrifuge at 15,000 rpm for 10 minutes, then remove supernate. Wash the DNA pellet by adding 5 mL of 100% EtOH, gently mixing, followed by incubation at −70°C for 10 minutes.
13. Centrifuge at 15,000 rpm for 10 minutes, then remove supernate and excess EtOH using a cotton swab. Add 8 mL of DNA dialysis buffer [10 mM Tris-HCl (pH 8.0), 10 mM NaCl, 1 mM EDTA] and gently resuspend the DNA pellet.
14. Measure volume and add 1.05 g of CsCl/mL. Gently mix until all of the CsCl is in solution; then add 100 μL of ethidium-bromide solution [50 mg/mL in dimethylsulfoxide (DMSO)]. Avoid exposure to light.
15. Prespin the mixture at 19,000 rpm, 15°C, for 30 minutes to remove large RNAs (ribosomes) and denatured proteins.
16. Transfer DNA solution to either a Beckman quick-seal tube or clear Oak Ridge tube. Centrifuge in a fixed-angle rotor according to the specifications necessary to establish a CsCl gradient. Generally, quick-seal tubes can be spun at speeds in the range of 60,000 rpm for 24 hours, while Oak Ridge tubes can be spun at about 38,000 rpm for 48 hours.
17. After centrifugation, remove all DNA bands from the top and not the side (DNA removal from the side will mix the small RNAs into the DNA band). First, remove the upper bacterial chromosomal band if present. Then carefully remove the lower supercoil plasmid or cosmid DNA band and add to a new centrifuge tube.
18. The supercoil plasmid or cosmid DNA band should be in a volume of about 2 mL. Fill the remaining portion of the centrifuge tube with 1.58 ϱ CsCl (1.05 g CsCl/mL of DNA dialysis buffer) and add an additional 25 μL of ethidium-bromide solution (50 mg/mL in DMSO). Gently mix and centrifuge again as described in step 16.

Table 2-2 Continued

19. Remove supercoiled plasmid or cosmid DNA band from the top and transfer to a 15 mL conical tube (Falcon). Extract ethidium-bromide by adding an equal volume of isopropanol (which has been saturated with a NaCl solution to ensure phase separation). Isopropanol extract 2 to 3 times or until no "pink" ethidium-bromide color remains.
20. Transfer DNA solution into a dialysis tube and dialyze against 1 L of DNA dialysis buffer; 3 changes.
21. After dialysis, determine DNA concentration by measuring its $O.D._{260}$ (1 O.D. = 50 μg/mL).

rotor such as the Beckman type Ti 70.1. Although it is tempting to reduce or eliminate the time needed for these CsCl bandings by using only one banding or by using vertical angle rotors, we have found that shortcuts can be the cause of many problems encountered in DNA sequencing reactions.

We strictly avoid the use of vertical rotors for the preparation of plasmid and cosmid DNAs which are to be 5'-end-labeled. The vertical rotors do not offer an effective purification of the plasmid or cosmid DNAs from contaminating small RNA molecules (in the range of 10 to 50 nucleotides). It is essential that the concentration of these RNAs be reduced as much as possible because their 5'-OH groups effectively compete with the targeted DNA during the ^{32}P-end-labeling reaction. In fact, because of their small size and large numbers, these small RNA molecules are the preferred substrate of the end-labeling reaction. Removal of these small RNAs is very difficult, and the use of ribonuclease (RNase) A only results in increasing their concentration because the products of RNase A treatment are small ribonucleotides. These RNAs are generally not detected on acrylamide gel electrophoreses because they are rapidly eluted.

5'-End-Labeling with [^{32}P-γ]-ATP

It is well known that restriction enzyme sites that yield protruding 5'-OH groups, end-label much more efficiently than flush-ended 5'-OH groups, and that recessed 5'-OH groups are the most difficult to 5'-end-label. However, we have found that if the DNA is free of these small RNA molecules, all restriction enzyme sites can be end-labeled regardless of the location of its 5'-OH group (Slightom et al., 1980; Slightom et al., 1987; Bosma et al., 1988; Siemieniak et al., 1991). We have 5'-end-labeled sites such as *Pst*I, *Apa*I, *Sac*I, *Kpn*I, and others, and have obtained ^{32}P-specific activities similar to that obtained for sites such as *Bam*HI, *Bgl*II, *Eco*RI, and others that have protruding 5'-OH groups. Having the ability to end-label most restriction enzyme sites greatly increases the usefulness of the chemical sequencing procedure.

End-labeling reactions are generally done using about 10 μg of plasmid, 20 μg of cosmid, or 30 μg of λ DNAs. These DNAs are first subjected to digestion by the restriction enzyme of choice. Digestions are done with five to tenfold enzyme excess and completeness of the restriction enzyme

Table 2-3 5′-End-Labeling of DNA Fragments

1. Obtain restriction enzyme site map of insert DNA to identify which restriction enzymes can be sequentially used to obtain a single ^{32}P-end-labeled DNA fragment.
2. Digest 10 μg of plasmid, 20 μg of cosmid, or 30 μg of λ DNA with the first restriction enzyme. Digest in a total volume of about 50 μL using salts recommended by the enzyme vendor. Digest the DNA using between five- to ten-fold excess enzyme for 1 hour. Digest at the optimal temperature recommended for the enzyme.
3. Check completeness of digest by removing 250 to 500 ng of DNA from the digestion mix and electrophoresing in a mini-agarose gel. During this time, allow the sample to continue to digest.
4. After the DNA sample is completely digested with the first enzyme, add one-tenth volume 1 M Tris-HCl (pH 8.4) to raise the pH to between 8.0 and 8.4. Add 5 to 10 U of calf intestinal alkaline phosphatase (CIP) (from Boehringer-Mannheim, see Table 2-4 for preparation of CIP).
5. Incubate reaction at 55°C for 30 minutes, then remove all enzymes (CIP and restriction enzyme) by 2 extractions with phenol:chloroform:isoamyl alcohol (1:1:0.04) followed by 1 extraction with chloroform:isoamyl alcohol (1:0.04).
6. Add one-tenth volume of 3 M NaOAc and 2.5 volumes of EtOH and incubate at −70°C for 10 minutes. Spin in an Eppendorf centrifuge for 7 to 10 minutes; however, if λ DNA is being end-labeled, this spin should be reduced to about 1 minute so as not to pack the DNA pellet. Wash the DNA pellet once with 100% EtOH and repeat the Eppendorf spin.
7. Remove excess EtOH by briefly drying in a Savant Speedvac; be sure not to dry λ samples completely as they will become difficult to resuspend. Resuspend DNA in 15 μL of H_2O and 15 μL of denaturation buffer (see Table 2-4). Use 25 μL of each liquid for λ samples. Carefully check samples to ensure that all of the DNA pellet has been resuspended.
8. Heat samples to 65°C for 5 minutes, cool on ice, and then add 4 μL polynucleotide kinase buffer (see Table 2-4). Mix the sample briefly, and then add 100 μCi of [^{32}P-γ]-ATP followed by the addition of 10 U of polynucleotide kinase (US Biochemical). Briefly mix the sample, spin down the solution in an Eppendorf centrifuge, and incubate at 37°C for 30 minutes.
9. Heat-kill the polynucleotide kinase by incubating at 65°C for 5 minutes then add one-tenth volume 3 M NaOAc and 2.5 volumes of EtOH and precipitate the labeled DNA away from most of the unreacted [^{32}P-γ]-ATP.
10. Dry the DNA pellet in a Savant Speedvac (do not completely dry λ DNA samples) and then resuspend DNA pellet in 30 μL H_2O followed by the addition of the appropriate second restriction enzyme buffer Check to ensure that the DNA pellet has been resuspended.
11. Add the second restriction enzyme (five- to ten-fold excess) and digest at the appropriate temperature for 1 hour. Following digestion, remove the restriction enzyme by 1 extraction with the phenol extraction solution (see Table 2-4), and 1 extraction with chloroform:isoamyl alcohol (1:0.04). Removal of the restriction enzyme is done to ensure that it [or added bovine serum albumin (BSA)] does not interfere with acrylamide gel separation of the labeled DNA fragments.
12. Add 3 to 5 μL of sucrose loading dye (see Table 2-4) and load DNA sample onto a glycerol acrylamide gel. Electrophorese DNA sample to allow separation of the expected DNA fragments.
13. Locate ^{32}P-labeled DNA fragments by exposing to film (see text). The size of DNA fragments can be determined by using a ^{32}P-end-labeled-size standard mixture such as the 1 kb ladder available from BRL (treat with CIP before ^{32}P-end-labeling, see steps 4 and 5). Bands can be located by the attachment of side strips marked with chemiluminescent paint, followed by a short exposure to film. The exposed film provides an effective guide for localizing the end-labeled fragments.
14. Remove the ^{32}P-end-labeled DNA fragments and place in a 1.5 mL Eppendorf tube. Crush the acrylamide gel matrix using a pointed spatula and then wash the acrylamide

Table 2-3 Continued

fragments remaining on the spatula into the tube using 400 μL of elution solution (see Table 2-4). Incubate the tube at 37°C overnight with shaking in an Eppendorf vortex.

15. Separate the DNA-containing elution solution from the acrylamide gel matrix by passing through a 1 mL plastic Eppendorf pipet tip that has been fitted with a small plug of siliconized glass wool. A 10 × 70 mm borosilicated glass culture tube is used to collect the eluted sample.
16. The elution process can be enhanced by spinning the tip and tube at 3,000 rpm. After the spin, add 100 μL more elution solution to the pipet tip and repeat the spin.
17. Transfer the sample to a fresh 1.5 mL Eppendorf tube, add 1 mL 100% EtOH, mix, and incubate at −70°C for 10 minutes. Pellet the DNA by spinning for 7 to 10 minutes in an Eppendorf centrifuge. Remove the supernatant and resuspend in 50 μL H_2O. Add 500 μL EtOH, incubate at −70°C for 10 minutes.
18. Centrifuge sample for 5 minutes in an Eppendorf centrifuge, dry DNA pellet in a Savant Speedvac, and then resuspend DNA in 30 μL H_2O. Some of the excess acrylamide that remains can be removed by freezing the sample at −70°C, then spinning 5 minutes in an Eppendorf centrifuge, followed by transfer to a fresh 1.5 mL Eppendorf tube.
19. Add 3 μL sonicated carrier DNA, 2.5 μg/μL (final concentration of 0.25 μg/μL) (see Table 2-4). The ^{32}P-end-labeled fragment is now ready to be aliquoted among the 4 base-specific reactions used for the chemical sequencing method (Table 2-6).

digest can be checked by electrophoresis on a small agarose gel. The 5′-terminal phosphates are removed by the addition of 5 to 10 U of calf intestinal alkaline phosphatase (CIP) (from Boehringer-Mannheim) followed by replacement of this phosphate with a radioactive (^{32}P) phosphate. The 5′-^{32}P-end-labeling procedure is outlined in Table 2-3.

The ^{32}P-end-labeled DNA fragments can be located by placing paper strips, which have been marked with a unique pattern using chemiluminescent paint, on both sides of the gel. Exposure of these strips to light, followed by a short exposure to X-ray film, will leave the unique pattern as a guide on the exposed film. Using this film guide, along with a ^{32}P-end-labeled internal size standard (we use the one kb ladder obtained from BRL), the location of ^{32}P-end-labeled fragments can be determined and removed by cutting around them with a razor blade. The ^{32}P-end-labeled DNA fragments are removed from the acrylamide by first placing them into a 1.5 mL Eppendorf tube followed by using a pointed spatula to mash the gel matrix and the addition of 400 μL of elution solution (Table 2-4). Removal of the ^{32}P-end-labeled DNA fragment from the acrylamide gel matrix is described in Table 2-3.

Base-Specific Chemical Cleavage Reactions

The chemicals needed for base-specific reactions can be obtained from the vendors recommended by Maxam and Gilbert (1980) or as a DNA chemical sequencing kit from Sigma (SEQ-1). If a small number of chemical sequencing reactions are planned, use of a chemical sequencing kit is recommended, leaving the chemical quality control problems to the kit's vendor. If a large number of reactions are planned, stock chemicals can be

Table 2-4 Solutions Needed for Chemical and Enzymatic Sequencing

General Solutions

Preparation of phenol extraction solution
- Melt bottle of Mallinckrodt AR phenol at 65°C.
- Saturate with DNA dialysis buffer; about 250 mL/500 g phenol.
- Add 2 M Tris-HCl (pH 9.5) until pH of phenol solution is between 7 and 8 (about 65 mL).
- Add 0.2% 8-OH quinoline.
- Add an equal volume of chloroform:isoamyl alcohol (1:0.04); yields about 850 mL of phenol extraction solution.

Glycerol acrylamide gel (7.5% acrylamide, 20% glycerol; 100 mL)
- 40 mL—50% solution of glycerol (use 50% for ease of pouring)
- 10 mL—10X Tris-borate-EDTA (TBE) electrophoresis buffer
- 15 mL—acrylamide stock solution (48% acrylamide, 2% bis)
- 35 mL—H_2O

10X TBE electrophoresis buffer (4 L)

Tris	484.4 g
Boric acid	205.4 g (crystals)
EDTA	14.9 g

Dissolve in a final volume of H_2O, check pH, should be about 8.3.

Sucrose loading dye (100 mL)

Sucrose	60%
10X TBE	0.5 mL
Dyes	0.1% xylene cyanol, bromophenol blue, and orange G

Solutions Used for 5′-End Labeling

Preparation of CIP alkaline phosphatase:
- Obtain CIP from Boehringer Mannheim, (lyophilized) 4,000 U
- Resuspend in 1 mL glycerol buffer (10 mM Tris-HCl, pH 8.0; 1 mM $MgCl_2$; 40% glycerol).
- Carefully mix. Aliquot 100 μL into 1.5 mL Eppendorf tubes and store at −20°C.

Denaturation buffer
- 20 mM Tris-HCl pH 9.5
- 0.1 mM EDTA
- 1.0 mM spermidine

Polynucleotide kinase buffer
- 0.5 M Tris-HCl pH 9.5
- 0.1 M $MgCl_2$
- 50 mM DTT
- 50% glycerol

Elution solution

0.5 M	NH_4OAc
10 mM	$MgOAc_2$
1.0 mM	EDTA
0.1%	SDS

Chemical Sequencing Solutions

DMS reaction buffer

50 mM	sodium cacodylate (pH 8.0) Sigma # C-0250
1.0 mM	EDTA

DMS stop buffer

1.5 M	NaOAc
1.0 M	β-mercaptoethanol
100 μg	tRNA (carrier)

Table 2-4 Continued

Hydrazine stop buffer

0.3 M	NaOAc
0.1 mM	EDTA
25 μg	tRNA (carrier)

Preparation of carrier DNA

Obtain solution of calf thymus DNA, concentration of about 2.5 mg/mL. Sonicate extensively using the large sonication tip. Average size of fragments should be about 1,000 bp (analyze on acrylamide gel). Phenol extract and EtOH precipitate. Resuspend in DNA dialysis buffer and measure concentration ($O.D._{260}$).

DNA sequencing gel loading solution

80%	Formamide
10mM	NaOH
1.0mM	EDTA
0.1%	Xylene cyanol

Enzymatic Sequencing Solutions

Termination mixes (nucleotide concentrations, as supplied in kits)

G mix	A mix	T mix	C mix
dGTP 80 μM	80 μM	80 μM	80 μM
dATP 80 μM	80 μM	80 μM	80 μM
dCTP 80 μM	80 μM	80 μM	80 μM
dTTP 80 μM	80 μM	80 μM	80 μM
ddGTP 8 μM	ddATP 8 μM	ddTTP 8 μM	ddCTP 8 μM
NaCl 50 μM	50 μM	50 μM	50 μM

Nucleotide chase mix (for extensions, not part of kit)

dGTP 200 μM
dATP 125 μM
dCTP 50 μM
dTTP 200 μM
NaCl 50 μM

Modified termination mix for sequence extension

G mix: 2.5 μL G termination mix, 2.5 μL chase mix
A mix: 3.0 μL A termination mix, 3.0 μL chase mix
T mix: 3.0 μL T termination mix, 3.0 μL chase mix
C mix: 2.5 μL C termination mix, 2.5 μL chase mix

obtained; however, purity of these chemicals should not be taken for granted. Table 2-5 lists some of these chemicals used and the vendor from which they can be purchased. Table 2-5 also lists the conditions that we use for storage of these chemicals. Before using chemicals such as DMS, hydrazine, and formic acid, one should carefully read the instructions for safe handling, storage, and disposal. Both DMS and hydrazine are poisonous and volatile; and hydrazine is flammable. Formic acid is caustic and can cause severe burns to exposed skin. All of these chemicals should be dispensed in a fume hood, and plastic or latex gloves should be worn. Discard waste (even supernatants from ethanol precipitations) and pipette tips into solutions of inactivating reagents; 5 M sodium hydroxide for DMS; and 3 M ferric chloride for hydrazine.

Table 2-5 Chemical Vendors and Storage

CHEMICAL	VENDOR	STORAGE
Acrylamide	BDH Chemicals	Room temperature
Bis-acrylamide	BRL	4°C
Dimethyl sulfate	Aldrich (99%)	25 mL brown bottles, 4°C
Formic acid	Aldrich (97%)	4°C
Hydrazine	Eastman Kodak (#902)	1.5 mL tubes, −80°C
Piperidine	Fisher Scientific	4°C
Sodium cacodylate	Sigma (#C-0250)	Room temperature
SDS	BDH Chemicals	Room Temperature
γ-methacryloxy-propyltrimethoxy-silane	Sigma (M-6514)	4°C
Repelcote	Atomergic Chemical Corporation	Room temperature

The quality of DMS and formic acid supplied by the listed vendors appears adequate, thus, problems with these base-specific reactions are generally not due to the chemicals. We have found the storage of small aliquots (in 25 mL brown bottles) of DMS at 4°C to be helpful, and that formic acid can also be stored for extended periods of time (several years) at 4°C without noticeable changes in reaction specificities. Such stability for hydrazine is not the case, however, as its base-specific reactivity is very inconsistent from bottle to bottle—even if the bottles have the same chemical lot number. This is probably because hydrazine is more susceptible to variations in storage and shipping conditions. We have found it necessary to test three to four different bottles of hydrazine before finding one that yields clean base-specific reactions for both C and C+T reactions. Even a sample of hydrazine of adequate purity **cannot** be stored at 4°C for long without a noticeable loss of reaction specificity. For prolonged storage of the hydrazine, it is aliquoted into 1.5 mL Eppendorf tubes (0.8 mL/tube) and stored at −80°C, and once an aliquot is opened and partially used it is discarded (using the approved procedure described above). We have been able to store hydrazine aliquots at −80°C for over 4 years with little loss of base-specific reactivity. Additional information concerning storage, analysis, and safe use of these base-specific chemicals has been described by Maxam and Gilbert (1980).

All base-specific reactions are conveniently done in 1.5 mL Eppendorf conical polypropylene tubes and, for added convenience, different color tubes can be used to identify each of the four base-specific reactions. Solutions used in various sequencing reaction steps are listed in Table 2-4. The DNA sequencing chemistry begins with the addition of the base-specific chemical, and the extent of the reaction depends on the concentration of these chemicals; DMS (G-specific), formic acid (A + G-specific), hydrazine (C+T-specific), and hydrazine in 1 M NaCl (C-specific). These reactions are also dependent on the concentration of each base represented in the ^{32}P-end-labeled DNA fragment. That is, on a weight basis, the concentration of any particular base in a shorter fragment is higher than that for a larger fragment, thus short DNA fragments will appear to react slower than long DNA fragments. This is important to know, because the chemical sequencing method will be done on fragments that range in size from 100 to 1,500 bp. Reaction times for shorter DNA fragments (in the range of 100 to 700 bp) and for larger DNA fragments (in the range of 700 to 1,500 bp) are included in Table 2-6. Because the reaction time listed for the DMS reaction of larger DNA fragments is quite short, the number of G-specific reactions done at one time should be kept small (two to four). If the reaction is somewhat slow (measured in minutes), the number of samples in each reaction set can be increased; it is not difficult to handle six to eight A-, C+T-, or C-specific reactions at one time.

After performing the base-specific reactions listed in Table 2-6, the reacted DNA samples either contain methylated Gs, missing purine bases, or open pyrimidine bases. Upon treatment of these samples with 1 M piperidine (Table 2-6), the 7-methyl G residues are opened and all of these ring-opened bases are displaced from the sugar backbone followed by β-elimination of the phosphates from the empty sugars. The piperidine reaction is accomplished using high-temperature incubation (90°C) for 30 minutes; caution should be taken to ensure that the tops remain on the Eppendorf tubes during this step. Following the piperidine cleavage step, samples are spun briefly and then placed into a Savant Speedvac evaporator for removal of the piperidine as a water-piperidine azeotrope. To ensure complete removal of piperidine, this evaporation process is repeated twice by adding 100 μL of H_2O for each repeat. These base-specific reacted DNA samples are quite stable and can be left under vacuum overnight to ensure complete removal of the piperidine. After the final drying step, the relative amount of radioactivity in each sample is determined by counting Cerenkov radiation; the 1.5 mL Eppendorf tubes are placed in scintillation vials and counted in a scintillation counter using the counting efficiency setting for tritium. Knowledge of the relative amounts of radioactivity in each sample is needed to determine the volume of sequencing gel loading solution (Table 2-4) to add. Samples are resuspended in as small a volume as possible to maximize the amount of radioactivity loaded onto the sequencing gel. The volume of loading solution added is influenced by several factors: the number of times each sample must be loaded; how consistent the level of radioactivity is among the four base-specific reactions; and the level of radioactivity among the different DNA fragments to be loaded on the same sequencing gel. The final step in the sequencing reaction involves heating the sample to 90°C for three minutes and loading them onto the sequencing gel (described below) for electrophoretic separation of the reacted fragments.

Table 2-6 Reaction Conditions for the Chemical DNA Sequencing Method

G	G + A	T + C	C
200 μL DMS buffer 1 μL carrier DNA 5 μL labeled DNA	18 μL H_2O 1 μL carrier DNA 9 μL labeled DNA	18 μL H_2O 1 μL carrier DNA 9 μL labeled DNA	15 μL 5.0 M NaCl 1 μL carrier DNA 5 μL labeled DNA
0.6 μL DMS	20 μL formic acid	20 μL hydrazine	20 μL hydrazine
Mix, spin	Mix, spin	Mix, spin	Mix, spin
Incubate 20°C	Incubate 20°C	Incubate 20°C	Incubate 20°C
30 sec (100–700 bp)	2.5 min	2.5 min	2.5 min
10 sec (700 bp–1.5 kb)	1.0 min	1.0 min	1.0 min
Add 50 μL DMS stop buffer	Add 350 μL hydrazine stop	Add 350 μL hydrazine stop	Add 350 μL hydrazine stop
Add 1 mL EtOH	Mix, spin, then add 1 mL EtOH	Mix, spin, then add 1 mL EtOH	Mix, spin, then add 1 mL EtOH
Mix, incubate −70°C 10 min	Mix, incubate −70°C 10 min	Mix, incubate −70°C 10 min	Mix, incubate −70°C 10 min
Spin, 4°C, 7 min Discard super′ into 5 M NaOH	Spin, 4°C, 7 min Discard super′ into 5 M NaOH	Spin, 4°C, 7 min Discard super′ into 1 M $FeCl_3$	Spin, 4°C, 7 min Discard super′ into 1 M $FeCl_3$
Add 350 μL 0.3 M NaOAc, pH 6.0	Add 350 μL 0.3 M NaOAc, pH 6.0	Add 350 μL 0.3 M NaOAc, pH 6.0	Add 350 μL 0.3 M NaOAc, pH 6.0
Add 1 mL EtOH	Add 1 mL EtOH	Add 1 mL EtOH	Add 1 mL EtoH
Mix, incubate −70°C 10 min	Mix, incubate −70°C 10 min	Mix, incubate −70°C 10 min	Mix, incubate −70°C 10 min
Spin, 4°C, 7 min Discard super′ into 5 M NaOH	Spin, 4°C, 7 min Discard super′ into 5 M NaOH	Spin, 4°C, 7 min Discard super′ into 1 M $FeCl_3$	Spin, 4°C, 7 min Discard super′ into 1 M $FeCl_3$
Add 0.8 mL EtOH	Add 0.8 mL EtOH	Add 0.8 mL EtOH	Add 0.8 mL EtOH
Mix, incubate −70°C 10 min	Mix, incubate −70°C 10 min	Mix, incubate −70°C 10 min	Mix, incubate −70°C 10 min
Spin, 4°C, 7 min Discard super′ into 5 M NaOH	Spin, 4°C, 7 min Discard super′ into 5 M NaOH	Spin, 4°C, 7 min Discard super′ into 1 M $FeCl_3$	Spin, 4°C, 7 min Discard super′ into 1 M $FeCl_3$
Speedvac 5 min Add 20 μL H_2O Speedvac dry	Speedvac 5 min Add 20 μL H_2O Speedvac dry	Speedvac 5 min Add 20 μL H_2O Speedvac dry	Speedvac 5 min Add 20 μL H_2O Speedvac dry
Add 100 μL 1 M piperidine, mix Incubate 90°C 30 min Spin, Speedvac dry	Add 100 μL 1 M piperidine, mix Incubate 90°C 30 min Spin, Speedvac dry	Add 100 μL 1 M piperidine, mix Incubate 90°C 30 min Spin, Speedvac dry	Add 100 μL 1 M piperidine, mix Incubate 90°C 30 min Spin, Speedvac dry

Oligomer-Directed Enzymatic Sequencing of Cosmid and λ Inserts

Oligomer-directed enzymatic sequencing of cosmid and λ inserts provides an opportunity to efficiently sequence large DNA molecules in a directed manner. This method is much more efficient than using the chemical sequencing approach described above because it uses much less DNA, and it is not subject to limitations due to the lack of or over-availability of restriction enzyme sites. However, as we have discussed above and outlined in Table 2-1, the combined use of both sequencing methods does provide a strategy for efficient sequencing of large DNA inserts. We have found both T7 and *Taq* polymerase to be effective for enzymatic sequencing; however, because most of our experience has been with T7 polymerase, its use is described in the procedure outlined in Table 2-7. The procedure described in Table 2-7 has been found to be simple, yet the most consistent to yield excellent results with either plasmid, cosmid, or λ DNAs. However, for this procedure to be consistent, we recommend that the DNA templates be as pure as possible (see above and Table 2-3). We have recently used this procedure to obtain over 80 kb of nucleotide sequence information for a set of cosmid clones and we have experienced a very high rate of success for primer-directed sequencing: in excess of 85% of the reactions yield meaningful results (D. R. Siemieniak, unpublished results). Additional information concerning direct sequencing of cosmid DNA has been described by Siemieniak et al. (1991). This high rate of success of our cosmid sequencing procedure is directly attributed to the use of pure cosmid DNAs and the accuracy of the sequence used for the design of the next walking oligomer primer.

Enzymatic Sequencing Reactions

The procedure outlined in Table 2-7 can be used for the enzymatic sequencing of plasmid, cosmid, or λ DNAs; these steps are essentially those described by the T7 vendor (US Biochemical) for sequencing plasmid DNAs. We have found that the commercially available T7 sequencing kits are adequate for obtaining sequence reads in the range of 600 bp using double-stranded template. However, to achieve even longer sequencing reads, in the range of 800 bp or more, we have found it necessary to use a modified termination mix, referred to as the extension mix. The termination mix supplied in the T7 kit contains a dideoxy:deoxy nucleotide ratio of 1:10; while a ratio of 1:30 is used in the extension mix (Table 2-4). Using the procedure outlined in Table 2-7 and the gel system described below, we have routinely obtained nucleotide sequence reads which extend 700 to 800 bp beyond the oligomer primer.

The T7 polymerase reaction is stopped by drying down in a Savant Speedvac, followed by resuspending in 12 μL of sequencing gel loading dye (Table 2-4). We have found the loading solution supplied in the T7 kits to be inadequate for obtaining complete strand denaturation when used as described in the protocol book, possibly because the final concen-

Table 2-7 Procedures for Double-Stranded Sequencing

1. Use 2 μg plasmid DNA, 3 μg cosmid DNA, or 4 μg λ DNA, final concentration of 1 μg/μL, adjusted concentration after denaturation.
2. DNA mix:
   2 to 4 μL of DNA (step 1)
   2 μL 5× reaction buffer (200 mM Tris-HCl, pH 7.5, 100 mM $MgCl_2$, 200 mM NaCl)
   1 μl oligomer primer (50 ng)
   × H_2O (to a total of 10 μL)
   10 μL total volume.
3. Denature DNA by placing tube in boiling H_2O for 5 minutes, cool on ice for 5 minutes, then add to DNA mix (step 2).*
4. Aliquot termination mixes (see Table 2-4); for normal sequencing dispense 2.5 μL of the appropriate termination mix into tubes labeled G, A, T, C. Prewarm tubes in a 37°C heating block. If sequencing reads beyond 600 bp are desired, the nucleotide concentrations in the termination mixes are altered by the addition of a chase mix, changing the ratio of dideoxy:deoxy nucleotides to 1:30 (see Table 2-4).
5. Dilute labeling mix (7.5 μM dGTP, 7.5 μM dCTP, 7.5 μM dTTP, 50 mM NaCl) 1:5. This will allow for reads close to the primer.

Labeling mix	2 μL*
H_2O	8 μL

6. Dilute Sequenase (1:8) in cold enzyme dilution buffer. (10 mM Tris-HCl, pH 7.5, 5 mM DTT, 0.5 mg/mL BSA)
   Enzyme dilution buffer 7 μL
   Sequenase version 2.0 1 μL
7. Reaction mixture is obtained by adding the following:

DNA mix	10 μL (from step 2)
0.1 M DTT	1 μL
Dilute labeling mix	2 μL (from step 5)
[^{32}P-α]-dATP (3,000 Ci/mM)	2 μL
Dilute Sequenase (3.25 U)	2 μL (from step 6)
Total	17 μL

   Incubate at room temperature for 5 minutes.
8. Transfer 3.5 μL of reaction mixture (step 7) to each termination tube (G, A, T, C) (step 4), mix and incubate at 37°C for 5 minutes.
9. Stop the reaction by drying in a Savant Speedvac (5 minutes) and then add 12 μL of loading solution (see Table 2-4). Heat samples to 90°C for 3 minutes, cool on ice, and then load gel.

*See US Biochemical Sequenase protocols.

tration of formamide is only 40%. Use of the sequencing gel loading solution described in Table 2-4 (which also contains 10 mM NaOH) or even the supplied loading solution (if added after the drying step) greatly reduced the number of sequence pauses found in our long gels. This indicates that many sequencing artifacts, attributed to the polymerase, may indeed be the result of inefficient denaturation of the duplex. In addition, the use of the thermostated gels (see below) which allow for a constant gel temperature of 55°C is helpful in reducing the number of gel artifacts that many researchers attribute to polymerase pauses. Thus, inefficient denaturation of the duplex can cause serious problems in the accuracy of the sequence obtained.

Sequencing Gel Technology: Reaching for 1 kb

The second goal of our sequencing strategy is to maximize the amount of information obtained from each sequence reaction. Most commercially available sequencing gel apparatuses allow for the use of gels in the range of 40 to 60 cm in length, and a few apparatuses are capable of supporting 100 cm gels (C.B.S. Scientific Inc., and Fotodyne, Inc.). Sequencing gels in the range of 60 cm can routinely separate DNA fragments in the range of 600 bp and can be effectively used with our sequencing strategy. However, to increase efficiency, we routinely use 100 cm gels. The use of long, thin, thermostated DNA sequencing gels has been previously investigated (Garoff and Ansorge, 1981; Ansorge and Barker, 1984) and much of our gel system is derived from the work of these investigators.

Our sequencing gels are 20 cm wide and 100 cm long and can be used with a wedge spacer system (0.2 mm top and 0.4 or 0.6 mm bottom) (Olsson et al., 1984) or can be used with a continuous 0.2 mm spacer. The wedge gel system is referred to as a "bellbottom" gel (Slightom et al., 1987) as it contains 60 cm long wedge spacers (C.B.S. Scientific Inc.) that start with a 0.4 mm wedge at the bottom of the gel and has a thickness of 0.2 mm at the top (60 cm); a 0.2 mm spacer is used for the remaining 40 cm of the gel. The thermostated plates are made according to the diagram shown in Figure 2-3. The material and measurements shown in Figure 2-3 are for the assembly of a thermostated plate measuring 20 cm × 104 cm. Thermostated plates can be made for any size gel, and our experience suggests that their use will increase the resolving power and accuracy of any sequencing gel system. The thermostated plate has been extremely helpful for sequencing DNAs that have a high G+C composition, regardless of whether the chemical or enzymatic sequencing method is used. If needed, gel temperatures can be raised (above 70°C if thermostable glass is used). We recently found such high-temperature gels to be necessary for sequencing a transposable element (70% G+C) from *Streptomyces fradiae* (Siemieniak et al., 1990).

Thermostated plates are treated with Repelcote (an organic based silane solution available from Atomergic Chemical Corporation) to prevent gel sticking, while the face plate is treated with γ-methacryloxypropyltrimethoxysilane that bonds the gel matrix directly to the glass. Procedures for treating both the thermostated and face plates and for removing the gel matrix from the face plate are given in Table 2-8. These large sequencing gels are filled with acrylamide sequencing solution (4% acrylamide) while they are lying flat (horizontal) on a bench. This is done by fitting a one-eighth-inch hole, which has been drilled through the face plate 1 inch from the bottom and 1 inch from one side of the plate, with a 50 mL syringe (Fig. 2-4). The gel mold can be filled using hydrostatic pressure, but the process can be accelerated by fitting the syringe with the plunger and forcing the solution into the gel mold. After filling the gel mold, the slot-forming comb is put into place and the bottom spacer is removed, followed by

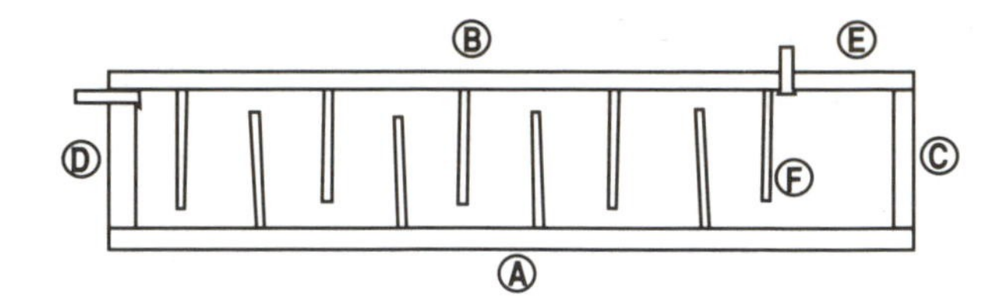

1. **Plate Glass, 2 Sheets, 41 inches (108 cm) x 8 3/4 (22.2 cm)**

2. **Spacers and baffles**

   A. 1 spacer 41 inches (104 cm) x 3/4 inch (2 cm)

   B. 1 spacer 36 inches (91.5 cm) x 3/4 inch (2 cm)

   C. 1 spacer 7 1/4 inches (18.4 cm) x 3/4 inch (2 cm)

   D. 1 spacer 6 3/4 inches (17.25 cm) x 3/4 inch (2 cm)

   E. 1 spacer 4 1/2 inches (11.5 cm) x 3/4 inch (2 cm)

   F. 9 baffles 6 inches (15.25cm) x 1/4 inch

Figure 2-3 Construction of thermostated gel plate. This diagram shows the design of the thermostated gel plate used for the sequencing gels described in the text and shown in Figure 2-4 and 2-5. Measurements are given for the glass parts needed to construct a gel which is 20 cm wide and 104 cm long. The thermostated plates are constructed by first cleaning glass surfaces with acetone (in a fume hood) and the glass parts are glued together using 100% silicone sealer (clear, so as to allow visibility through the fused glass). Apply plenty of silicone sealer and use pressure to remove all air bubbles between glass surfaces. Material used for the inlet (bottom) and outlet (top) connectors can be rigid plastic or metal tubing. Assemble glass parts on a clean, level surface. After assembly, allow plates to set (under the weight of several face plates) for at least 24 hours. Test thermostated plates for leaks prior to using.

excreting a small amount of the acrylamide gel solution from the bottom to ensure that the gel makes uniform contact with the electrophoresis buffer. The hole in the face plate is plugged with a small amount of grease. This method of pouring large gels is rapid (these large gels can be poured in less than a minute) and avoids the introduction of bubbles, which is one of the major obstacles to using long gels.

After pouring, the gels are allowed to set for about one hour, and then placed into the gel stand where the necessary tubing from the circulating bath is attached to the thermostated plate (Fig. 2-5). The connection between the water bath and gel plates uses sealed fittings that, when disconnected, shut off water flow (these connectors are available from Cole-Parmer). After filling the upper buffer chamber, the combs are removed, the slots are washed free of any gel debris, and the circulating water is used to heat the gel to 55°C prior to loading. The 4% bellbottom gel is

Figure 2-4 Horizontal pouring of sequencing gel. Sequencing gel face plates contain a one-eighth-inch hole drilled 1 inch from the bottom and 1 inch from one side. After assembling the gel mold, acrylamide sequencing gel solution is poured into a 50 mL syringe and, following removal of air bubbles, the barrel of the syringe is fitted into the hole and the acrylamide solution is allowed to fill the gel mold. After the solution clears the top region of the syringe, the plunger can be fitted into the syringe and used to force the remaining solution into the gel mold. After the gel mold is filled, the slot former is put in place, the bottom spacer is removed, and the hole is plugged with a small amount of grease.

preelectrophoresed for about one hour at 2,000 volts, but preelectrophoresis of the 4% straight gel is not necessary. Sequenced DNA samples which have been resuspended in the loading solution (Table 2-4) are heated to 90°C for three minutes, placed on ice, and then loaded onto the gels. Both the wedge and straight gels are designed for maximum nucleotide sequencing reads, and they need to be loaded only once. Maximum nucleotide sequencing reads for shorter gels generally require the inconvenience of multiple sample loadings. The amount of sequence sample loaded depends on the size of the slots used; for example, loading volumes of about 0.5 to 1.0 μL can be used with a 5 mm slot; while only about 0.2 μL should be used in a 3 mm slot. Loading of these thin gels is done using plastic micropipet tips that are available from Drummond Scientific Co.

A typical 4% acrylamide-7 M urea bellbottom gel is electrophoresed until the xylene cyanol dye is about 70 cm from the top, while the 4% acrylamide-7 M urea straight gel is electrophoresed for a total of 45,000 to 50,000 volt-hours (no dye markers are visible). After electrophoresis,

Table 2-8 DNA Sequencing Gels: Solutions and Plate Treatments

DNA Sequencing Gel Solutions

Acrylamide sequencing stock solutions (100 mL)

Components	Gel percentage (4%)
Urea	50 g
Bis-acrylamide	0.2 g
Acrylamide	3.8 g
10X TBE	12.5 mL

γ-methacryloxypropyltrimethoxysilane solution

25 mL	EtOH (100%)
0.75 mL	10% acetic acid
75 μL	γ-methacryloxypropyltrimethoxysilane

Silane solution
 2% dichlorodimethylsilane in chloroform or obtain Repelcote from Atomergic Chemicals Corporation

Treatment of Plates

Thermostated plate
- Wash plate with soap and water
- Rinse with water until all the soap residue has been removed
- Wash plate with 100% EtOH, use lint-free wipes (Kimwipes)
- Place thermostated plate in a hood
- Apply Repelcote and allow plate surface to dry
- Buff plate surface with Kimwipes

Treatment of face plate
- Wash plate with soap and water
- Rinse with water until all the soap residue has been removed
- Wash plate with 100% EtOH using Kimwipes
- Apply γ-methacryloxypropyltrimethoxysilane solution with Kimwipe
- Allow plate to dry
- Remove excess γ-methacryloxypropyltrimethoxysilane by rinsing with 100% EtOH and buffing with a Kimwipe
- Repeat rinsing with 100% EtOH—generally, this buffing is repeated 2 to 3 times
- If too much γ-methacryloxypropyltrimethoxysilane remains on the face plate it could cause the acrylamide gel to also adhere to the thermostated plate (even if it has been silane treated). If this happens it could cause the acrylamide gel to rip upon plate separation.

Removal of gel matrix from face plate
- Soak free plates in a solution of 2% KOH for at least 2 hours, or until needed

gels are removed from the stand, separated, and the DNA fixed and the urea removed by soaking the face plate in a 10% acetic acid bath for ten minutes. Excess acetic acid is removed by an extensive water rinse. Then the gels are allowed to dry at room temperature, or, if available, in a warm room (37°C). If a warm room is used, gels dry in about three hours, and once dry, the gel matrix has a thickness of about 10 μ which increases the

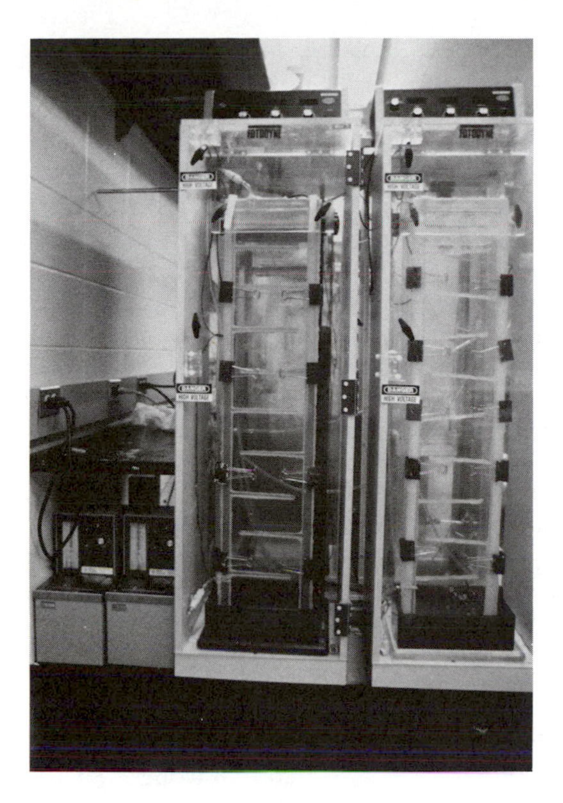

Figure 2-5 Use of thermostated DNA sequencing gel system including safety cabinets. The DNA gel strands and safety cabinets were obtained from Fotodyne, Inc. and fitted with tubing that allows heated water to circulate from the water bath (baths, shown on the lower left, contain about 6 L) into the thermostated plate. Electrophoresis power supplies are located on top of the safety cabinets and contain ground-default safety switches. Safety cabinets also contain shutoff switches that disconnect the power from the gel when open.

efficiency of the film exposure and band resolution. Gels are exposed to X-ray roll film (Eastman Kodak XAR351 available from Fotodyne, Inc.) by forming a sandwich with the film being placed between the gel plate and a clean gel plate followed by wrapping with aluminum foil and enclosing within a large (custom made) black plastic bag. Films are generally exposed at room temperature, except for chemical sequencing reactions, which are exposed at −70°C with the aid of an intensifying screen (Quanta III, 35 cm × 1 m, available from DuPont). Nucleotide sequence reads from a typical 4% bellbottom gel generally start about 50 bp from the end-labeled restriction enzyme site (chemical method) or oligomer primer (enzymatic method) and extend out to about 400 bp; while a typical 4% straight gel can be read starting from about 350 bp and extend out to better than 800 bp.

CONCLUDING REMARKS

Recent advances in nucleotide sequencing technologies have lead us to believe that alternative strategies should be investigated for obtaining nucleotide sequence information. We believe that these new strategies will involve the direct sequencing of larger and larger DNA molecules, and that this information will be collected in a directional manner that will allow for rapid analysis. Toward these goals we have been developing strategies and methodologies for directly sequencing large DNA molecules. The results of our test suggest that a directional sequence strategy can be improved to a level where it competes favorably with the random sequencing strategy. We feel that the major contribution of this work is to show that large DNA molecules can be conveniently sequenced directly and directionally. We realize that additional technological improvements are needed before the direct sequencing procedure and strategies described here will be widely applied. Additional improvements such as automated sequencing reactions, the use of nonradioactive labels, the use of automatic film readers, and reduction in the price of oligomer primers, would greatly benefit our strategy. The use of short primers has recently been suggested by Studier (1989), and the use of a small library of oligomers would eliminate most of the expenses and time involved in oligomer synthesis. The effectiveness of using a limited library of oligomers (nonamers) for genome sequencing was simulated using computer-aided techniques by Siemieniak and Slightom (1990). Thus, because many of these technologies are already in use, or are in the advanced stages of development, their incorporation into a strategy for directly sequencing large DNA molecules is possible and most likely inevitable.

ACKNOWLEDGMENTS

We thank Eric Lai and Lee Hood for the use of cosmid clone H7.1 (described in the legend of Fig. 2-2) and Roger Drong for his help in critically reading this manuscript.

REFERENCES

Ansorge, W., and R. Barker. (1984) System for DNA sequencing with resolution of up to 600 base pairs. *J. Biochem. Biophys. Meth.* **9**:33–47.

Ansorge, W., B. Sproat, J. Stegemann, C. Schwager, and M. Zenke. (1987) Automated DNA sequencing: ultrasensitive detection of fluorescent bands during electrophoresis. *Nucleic Acid Res.* **11**:4593–4602.

Birnboim, H.D., and J. Doly. (1979) A rapid alkaline extraction procedure for screening recombinant plasmid DNA. *Nucleic Acids Res.* **7**:1513–1523.

Bosma, P.J., E. Van den Berg, T. Kooistra, D.R. Siemieniak, and J.L. Slightom. (1988) Human plasminogen activator inhibitor-1 gene: promoter and structural gene nucleotide sequences. *J. Biol. Chem.* **263**:9129–9141.

Chen, E.Y., and P.H. Seeburg. (1985) Supercoil sequencing: a fast and simple method for sequencing plasmid DNA. *DNA* **4**:165–170.

Edwards, A., H. Voss, P. Rice, A. Civitello, J. Stegemann, C. Schwager, J. Zimmerman, H. Erfle, C.T. Caskey, and W. Ansorge. (1990) Automated DNA sequencing of the human HPRT locus. *Genomics* **6**:593–608.

Elder, J.K., D.K. Green, and E.M. Southern. (1986) Automatic reading of DNA sequencing gel autoradiographs using a large format digital scanner. *Nucleic Acids Res.* **14**:417–424.

Frank, R., A. Bosserhoff, C. Boulin, A. Epstein, H. Gausepohl, and K. Ashman.

(1988) Automation of DNA sequencing reactions and related techniques: a workstation for micromanipulation of liquids. *BioTechniques* **6:**1211–1213.

Garoff, H., and W. Ansorge. (1981) Improvements of DNA sequencing gels. *Anal. Biochem.* **115:**450–457.

Kalisch, B.W., S.A. Krawetz, K.-H. Schenwaelder, and J.H. Van de Sande. (1986) Covalently linked sequencing primer linkers (splinkers) for sequence analysis of restriction fragments. *Gene* **44:**263–270.

Kesterson, R.A., S.A. Kerner, and J.W. Pike. (1989) Cosmid sequencing using Sequenase: a novel approach for determining genomic structure. *US Biochemical Comments*: 17–18.

Lai, E., P. Concannon, and L. Hood. (1988) Conserved organization of the human and mouse T-cell receptor β-gene families. *Nature* **331:**543–546.

Levedakou, E.N., U. Langegren, and L. Hood. (1989) A strategy to study gene polymorphism by direct sequence analysis of cosmid clones and amplified genomic DNA. *BioTechniques* **7:**438–442.

Lund, T., F.G. Grosveld, and R.A. Flavell. (1982) Isolation of transforming DNA by cosmid rescue. *Proc. Natl. Acad. Sci. USA* **79:**520–524.

Manfioletti, G., and C. Schneider. (1988) A new and fast method for preparing high quality lambda DNA suitable for sequencing. *Nucleic Acids Res.* **16:**2873–2884.

Maniatis, T., E. Fritsch, and J. Sambrook. (1982) *Molecular Cloning*. Cold Spring Harbor Laboratory, Cold Spring Harbor.

Maxam, A., and W. Gilbert. (1977) A new method for sequencing DNA. *Proc. Natl. Acad. Sci. USA* **74:**560–564.

Maxam, A., and W. Gilbert. (1980) Sequencing end-labeled DNA with base specific chemical cleavage. *Methods Enzymol.* **65:**499–560.

Morelle, G. (1989) A plasmid extraction procedure on a miniprep scale. *BRL Focus* **11:**7–8.

Olssen, A., T. Moks, M. Uhlen, and A.B. Gaal. (1984) Uniformly spaced banding patterns in DNA sequencing gels by the use of field-strength gradients. *J. Biochem. Biophys. Methods* **10:**83–90.

Prober, J.M., G.L. Trainor, R.J. Dam, F.W. Hobbs, C.W. Robertson, R.J. Zagursky, A.J. Cocuzza, M.A. Jensen, and K. Baumeister. (1987) A system for rapid DNA sequencing with fluorescent chain-terminating dideoxynucleotides. *Science* **238:**336–341.

Saiki, R.K., D.H. Gelfand, S. Stoffel, S. Scharf, R. Higuchi, G.T. Horn, K.B. Mullis, and H.A. Erlich. (1988) Primer-directed enzymatic amplification of DNA with a thermostable DNA polymerase. *Science* **239:**487–491.

Sanger, F., S. Nicklen, and A.R. Coulson. (1977) DNA sequencing with chain-terminating inhibitors. *Proc. Natl. Acad. Sci. USA* **74:**5463–5467.

Siemieniak, D.R., L.C. Sieu, and J.L. Slightom. (1991) Strategy and methods for directly sequencing cosmid clones. *Anal. Biochem.* **192:**441–448.

Siemieniak, D.R., and J.L. Slightom. (1990) A library of 3342 useful nonamer primers for genome sequencing. *Gene* **96:**121–124.

Siemieniak, D.R., J.L. Slightom, and S.-T. Chung. (1990) Nucleotide sequence of *Streptomyces fradiae* transposable element Tn*4556*: a class-II transposon related to Tn*3*. *Gene* **86:**1–9.

Slightom, J.L., A.E. Blechl, and O. Smithies. (1980) Human fetal $^{G}\gamma$- and $^{A}\gamma$-globin genes: complete nucleotide sequences suggest that DNA can be exchanged between these duplicated genes. *Cell* **21:**627–638.

Slightom, J.L., T.W. Theisen, B.F. Koop, and M. Goodman. (1987) Orangutan fetal globin genes: nucleotide sequences reveal multiple gene conversions during hominid phylogeny. *J. Biol. Chem.* **262:**7472–7483.

Slightom, J.L., and R. Drong. (1988) Procedures for constructing genomic clone banks. Section A8, pp. 1–42 in *Plant Molecular Biology Manual* (S.B. Gelvin and R.A. Schilperoort, eds.). Martinus Nijhoff, Dordecht.

Smith, L.M., J.Z. Sanders, R.J. Kaiser, P. Hughes, C. Dodd, C.R. Connell, C. Heiner, S.B.H. Kent, and L. Hood. (1986) Fluorescence detection in automated DNA sequence analysis. *Nature* **321:**674–679.

Sorge, J.A., and L. Blinderman. (1989) ExoMeth sequencing of DNA: eliminating the need for subcloning and oligonucleotide primers. *Proc. Natl. Acad. Sci. USA* **86:**9208–9212.

Staden, R. (1982) Automation of the computer handling of gel reading data produced by the shotgun method of DNA sequencing. *Nucleic Acids Res.* **10:**4731–4751.

Studier, F.W. (1989) A strategy for high-volume sequencing of cosmid DNAs: random and directed priming with a library of oligonucleotides. *Proc. Natl. Acad. Sci. USA* **86:**6917–6921.

Tabor, S., and C.C. Richardson. (1987) DNA sequence analysis with a modified bacteriophage T7 DNA polymerase. *Proc. Natl. Acad. Sci. USA* **84:**4767–4771.

Wilson, R.K., A.S. Yuen, S.M. Clark, C. Spence, P. Arakelian, and L.E. Hood. (1988) Automation of dideoxynucleotide DNA sequencing reactions using a robotic workstation. *BioTechniques* **6:**776–787.

Zagursky, R.J., K. Baumeister, N. Lomax, and M.L. Berman. (1985) Rapid and easy sequencing of large linear double-stranded DNA and supercoiled plasmid DNA. *Gene Anal. Tech.* **2:**89–94.

Zimmerman, J., H. Voss, C. Schwager, J. Stegemann, and W. Ansorge. (1988) Automated Sanger dideoxy sequencing reaction protocol. *FEBS Lett.* **233:**432–436.

3

The Application of Automated DNA Sequence Analysis to Phylogenetic Studies

ROBERT J. FERL, C. JOSEPH NAIRN, JIN-XIONG SHE, EDWARD WAKELAND, AND ERNESTO ALMIRA

Automated DNA sequence analysis has developed to the point that many individuals and research organizations have considered or will soon consider investing in this technology. We discuss here some of the factors involved in automated DNA sequence analysis, particularly as performed by a service facility and as applied to phylogenetic studies. We also show, by comparative studies involving ribosomal RNA (rRNA) genes and polymerase chain reaction (PCR)-generated templates, that automated DNA sequencing saves tremendous time and can be every bit as accurate or inaccurate as manual methods.

The purpose of this chapter is to discuss the current state of automated DNA sequencing and its application to the generation of sequence data for use in phylogenetic studies. We address the subject from a purely end-user point of view. That is, we are not on the cutting edge of automated methods development, nor do we have available a wide array of different automated sequencers and large amounts of comparative performance data. Rather, as part of an active, readily accessible service facility (University of Florida Interdisciplinary Center for Biotechnology Research, Gainesville, FL), we have explored the practical aspects of automated DNA sequence analysis as currently available to the general scientific public and practiced by the average user.

We begin with a general introduction to the notion of DNA sequencing by a core service facility. Does such a facility really work? Can service sequencing, particularly automated service sequencing, provide quality data? Can any investigator (the hard-core molecular biologists as well as those with less molecular expertise) have access to DNA sequence information?

The answer to these questions is yes, especially if given the appropriate funding and administrative atmosphere.

We then move on to a brief description of our sequencing facility and the particular automated sequencing instrument upon which the facility is based. After an abbreviated coverage of the methods involved and examples of actual DNA sequencing projects, we present a comparison of automated versus manual examples of the kinds of sequencing projects likely to be undertaken by a user interested in generating sequence information from a large number of alleles or homologous genes.

AUTOMATED DNA SEQUENCING

Automated DNA Sequencing as a Service

The Interdisciplinary Center for Biotechnology Research at the University of Florida, like the biotechnology centers at many universities, participates in a wide range of activities, and maintains a set of core service laboratories.[1] These service laboratories consist of high-end analytical equipment that is purchased, maintained, and staffed from Center funds. Thus, faculty members have access to equipment that would otherwise be beyond their individual funding capabilities. The budget from the Center generally covers all expenses other than the expendables actually used, allowing these service laboratories to provide state-of-the-art technology at a contract price that encourages regular use.

The DNA Sequencing Core Laboratory, for example, is set up to provide DNA sequencing data from any fragment of DNA. The cost (keeping in mind that only expendables are recovered by this fee) is $50 per reaction, and from that reaction the average user will receive approximately 300 bases of DNA sequence information.

Typically, a potential user will contact us for general information on vector and primer choices, DNA cloning and purification procedures, and other hints as to how to develop their sequencing project. For each user, we provide a set of protocols, as well as advice on overall strategy for the project. Then the sequencing and analysis cycle starts, with the user providing the DNA (and primers if necessary) and the Sequencing Core Laboratory providing the data. The information from each sequencing run is stored in computer files, and also supplied to the user as hard-copy output of the raw data together with an interpretation of the data as a sequence of bases. Users are encouraged to become aware of the nuances of data interpretation and to make use of all of the raw data, not just the string of machine-interpreted bases. Thus the user still maintains the primary

[1]For further information on the DNA Sequencing Core Laboratory at the University of Florida, feel free to contact Dr. Ernesto Almira or Dr. Robert Ferl. Potential purchasers of automated sequencers are encouraged to make full use of the demonstrations and literature available from the manufacturers. The manufacturers will also provide a wealth of specific technical literature on DNA purification and sequencing reactions.

responsibility for sequence accuracy, including final decisions on base calling, number of repetitive runs to obtain the final sequence, and the production of sequence from both complementary strands.

By using the Sequencing Core Laboratory, the user is freed from the purchase of expensive sequencing apparatuses, radiolabeled nucleotides, X-ray film, and all of the other minutiae associated with manual DNA sequencing. Instead, funds and effort can be directed to producing the clones or fragments for sequencing and analyzing the subsequent data. Indeed, most of the faculty that currently use the Sequencing Core Laboratory are new to DNA sequencing, though faculty already established in DNA sequencing also use the Laboratory for occasional or particularly difficult sequencing projects. We have had users come to us for very short-term projects, such as to simply confirm the sequence of a probe or insert. We have also had users generate sequences of tens of kilobases (kb) over longer periods of time.

From our experience as a service sequencing facility, it has become clear that automated DNA sequencing through such a facility is a viable alternative to manual sequencing. The cost, at least within the center concept as developed at the University of Florida, is minimal, and the automated nature of the process means a short turnaround time. Thus, many investigators who would not otherwise have done so are in the business of generating and analyzing DNA sequence information.

A Brief History of Automated DNA Sequencing and Factors Involved in Choosing an Automated Sequencer

While a full review of automated DNA sequencing is beyond the needs of this discussion, familiarity with certain aspects of the technology will provide a useful perspective for the data comparisons presented in the next sections. It should also be noted that automated DNA sequencer research and production is a rapidly evolving field in its own right, and some of the views presented here (and indeed some of the factors that influence the decision of which sequencer to purchase) might well change in the near future.

Automated DNA sequencing can be a misleading phrase. What is really meant is automated analysis of DNA sequencing reactions. Other than some modifications that we will discuss below, the enzymatic protocol to produce the DNA sequence is currently the same, time-honored Sanger dideoxy terminator system. What has become automated is the electrophoretic separation and detection of the products of the Sanger method (Sanger et al., 1977).

Manual sequencing utilizes fairly large gels and electrophoretic separation for a specified time, followed by autoradiography. The sequence is read from the bottom of the gel to the top, because in a given period of time the smaller fragments will have migrated farther than the larger ones. The autoradiogram presents a detailed view of the separation achieved at a certain time-point.

All automated sequencers also utilize electrophoresis, but in a fundamentally different way. Automated sequencers, instead of looking at the whole gel at one point in time (as in an autoradiogram), look at one point of the gel over time. They measure the time it takes for a band to traverse a specified distance in the gel. Smaller bands traverse the gel more rapidly than larger ones, and arrive at the detection window in a shorter period of time. Thus the output is very different from the "ladder" observed in the manual sequencing autoradiogram. Instead, an electrophoretogram is produced, presenting the detected bands as peaks on the Y axis, and time of electrophoresis on the X axis. Each peak is then identified as an A, T, G, or C depending upon the detection system (Fig. 3-1).

The main difference among automated sequencers is the mechanism by which the bands are detected. One type of machine detects radioactively-labeled DNA, while the other detects fluorescently-labeled DNA. As a service facility, we did not consider the machines designed to read radioactive bands as they elute from the bottom of the gel for two reasons. First, we did not want to handle the large amounts of radioactive label that would be involved in large-scale service sequencing. Second, all of these machines require four reaction tubes and four gel lanes per sequence template. In machines designed to detect fluorescently-labeled DNA, only one gel lane is required to analyze the sequence from a template, and the number of reaction tubes per template can be reduced as well.

The two companies currently marketing automated sequencers that operate by detection of fluorescently-labeled DNA fragments are DuPont and ABI (Applied Biosystems). Sequencers from both companies use a laser to scan the gel to detect the DNA fragments. Both use different fluorescent tags to label the DNA from the different reactions, so that a single lane is all that is needed to produce sequence from a template. (The slightly different color of each tag allows the machine to not only detect when a band is passing the detector, but also whether it represents a fragment terminated at A, T, G, or C.) Both employ essentially standard sequencing gel technology. Both also provide a similar type of sequence hard-copy output. The main difference between the machines is in the chemistry of the flourescent tags, and this leads to differences in performance and overall approach.

Except for some very recent developments, the chemistry of the system from ABI employs a set of four differently labeled primers, one primer for each of the sequencing reactions, A, T, G, and C (Smith et al., 1986). After the annealings and extension reactions in the presence of the standard dideoxy chain terminators, the reactions are stopped, combined, and run on a single lane of the sequencing gel.

DuPont employs a set of four differently labeled dideoxy terminators (Prober et al., 1987). After annealing with a standard primer in a single tube, the extension reaction takes place in the presence of the four labeled dideoxy chain terminators. After the reaction is stopped, it is loaded in a single lane on the sequencing gel. In the output shown in Figure 3-1, each

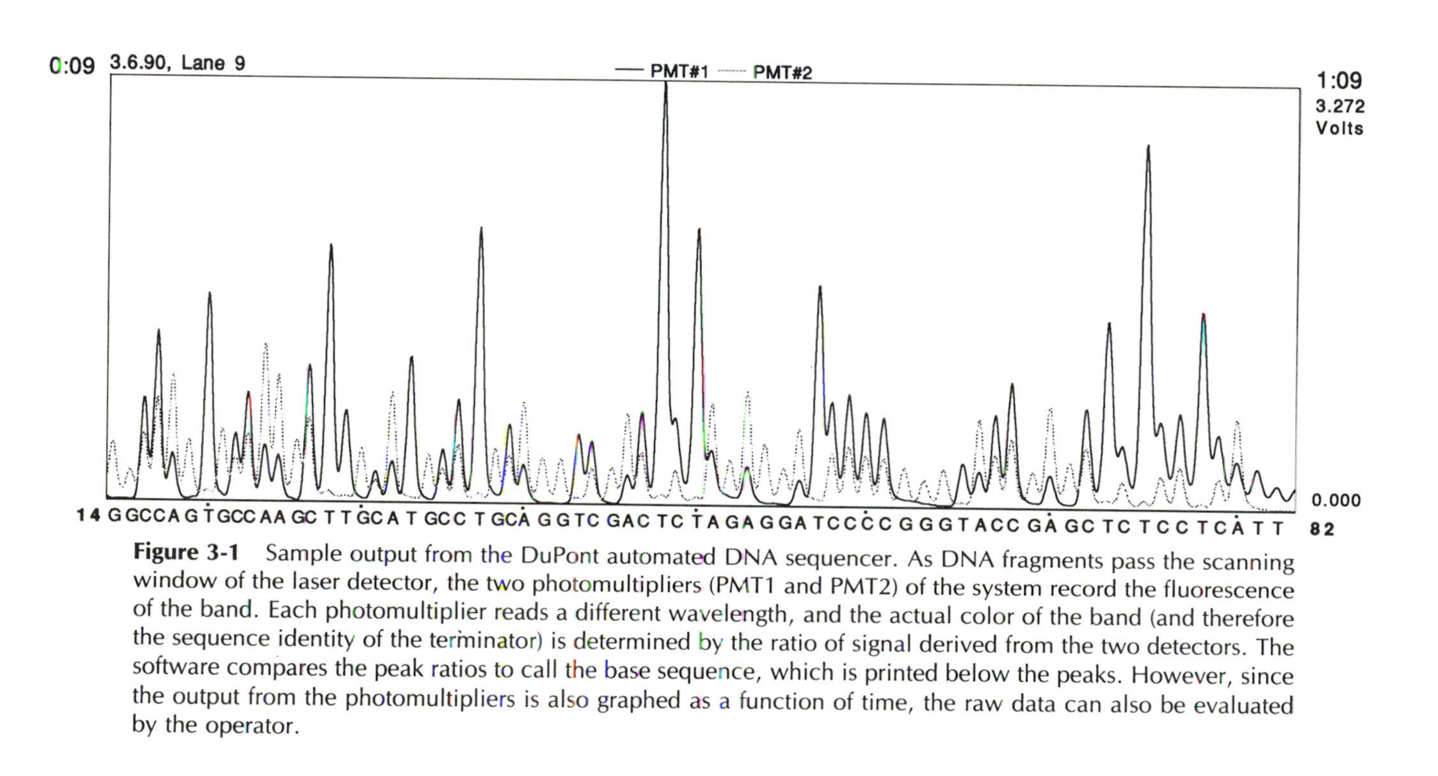

Figure 3-1 Sample output from the DuPont automated DNA sequencer. As DNA fragments pass the scanning window of the laser detector, the two photomultipliers (PMT1 and PMT2) of the system record the fluorescence of the band. Each photomultiplier reads a different wavelength, and the actual color of the band (and therefore the sequence identity of the terminator) is determined by the ratio of signal derived from the two detectors. The software compares the peak ratios to call the base sequence, which is printed below the peaks. However, since the output from the photomultipliers is also graphed as a function of time, the raw data can also be evaluated by the operator.

band (or peak) is then identified as an A, T, C, or G depending upon its color, which is determined by the ratio of fluorescence detected by two photomultiplier tubes (PMT1 and PMT2).

The choice between machines becomes a choice between chemistries and the limitations and benefits therein. In 1987 we chose the DuPont machine for the following three reasons. First, we (correctly) anticipated that a large proportion of our sequencing would involve custom primers and the DuPont system allows their simple and direct use. The ABI system did have the capability to use custom primers, but at the cost of the extra steps to add the fluorescent labels to the primers. From a service facility perspective, we wanted to minimize our efforts for each step, while still maximizing our ability to use any primer. (It should be noted that ABI now offers labeled terminator chemistry on their latest system.) Second, the labeled dideoxy terminator chemistry allows the extension reactions to be carried out in a single tube. Thus, by using the DuPont system we could reduce our manipulations per sequence template by 75%. Again, this was desirable from a service facility perspective. Third, as we will demonstrate below, the labeled terminator chemistry removes many types of artifacts that can impact a sequencing analysis.

These reasons were our own, and firmly based on the availabile data at the time. Any person or group currently considering the purchase of an automated sequencer should carefully reexamine the performance characteristics, chemistries, and flexibilities of all machines now available. This discussion should not be taken as an endorsement of the DuPont automated sequencer, but merely reflects our views at the time of our purchase.

In order to test the effectiveness of the automated sequencing approach for phylogenetic analysis, we chose two examples for comparison with manual methods. First, we chose a relatively large gene, the small-subunit rRNA gene from *Gingko biloba* (Nairn et al., 1991). With a panel of 12 primers for conserved regions, we conducted a standard set of manual sequencing experiments to produce a confident sequence, and we submitted the same template DNA and primers to the DNA Sequencing Core Laboratory for automated analysis. Second, we compared manual and automated sequencing of a PCR generated fragment produced from mouse genomic DNA (exon 3 of MHC Class II Aβ gene).

RESULTS AND DISCUSSION

Automated Sequencing of a Cloned Gene: the 18*S* rRNA Gene from *Gingko biloba*

The small-subunit rRNA gene from *Gingko biloba* was recovered from a genomic library and subcloned into pUC 19 for sequencing. The pUC subclone was used directly as a double-stranded template for both the manual and automated runs. The coding region of the rRNA gene is approximately 1,800 base pairs (bp). The overall setup of the 12 primers is shown in Figure 3-2. The average distance between adjacent primers is approximately 300 bases. The locations for the primers were chosen from within those regions exhibiting the most conservation over wide taxa. For the purposes of this comparison, we have not included the price of the primers, since that cost would be the same for both the automated and manual methods.

The manual sequencing strategy and resulting read lengths are presented in Figure 3-2. The method for sequencing was according to the KiloBase Sequencing System instruction manual as delivered by Bethesda Research Laboratories, Life Technologies Inc. (Chen and Seeburg, 1985). Initially, three separate gel runs of up to 14 hours each were required to generate the primary data set. This resulted in an average read length of over 500 bases of sequence per run. However, as is common with rRNA gene sequencing experiments, several areas of severe collapse or compression (where multiple bands appear in all lanes) were encountered that required additional reactions and gels for final resolution (see further discussion below). We have found that each collapse requires a separate approach for resolution, and no single modified reaction type was sufficient to remove all collapse regions. In all, four additional gels, with several modified reaction sets (such as ITP, deaza-7-GTP, and Sequenase) were required to produce the final sequence.

Thus, after eight sequencing gels and over 60 reactions, a sequence was generated that contained no ambiguities. Because of the length of the readable sequence, each area was sequenced on both strands and duplicated on at least one strand. All collapsed regions were fully resolved.

Automated sequencing utilizing the protocols suggested by the manufacturer required only 12 sequencing reaction tubes and one five-hour run on the DuPont automated sequencer. The general results of the automated sequencing run are shown in Figure 3-3. The distinguishing feature of the automated analysis is that the effective reading lengths per primer were about half that derived from manual methods. In fact, in several cases of nonoverlap, sequence information was therefore obtained from only one strand. (The consequences of these shorter reads will be considered in the next section.)

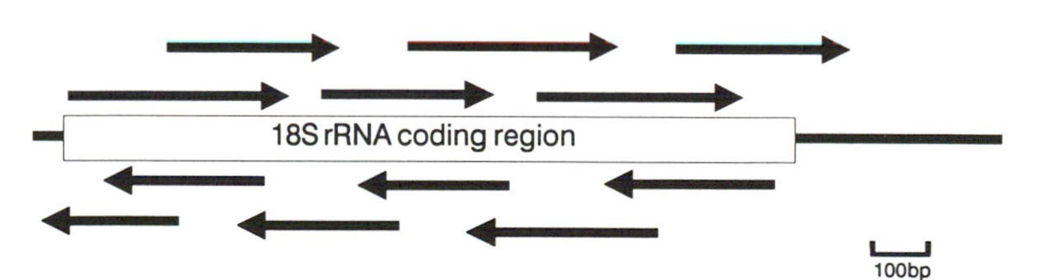

Figure 3-2 Manual sequencing strategy and results for the *Gingko biloba* 18*S* rRNA gene. Note that the read lengths of the individual sequence runs allow a complete overlap of data from both strands.

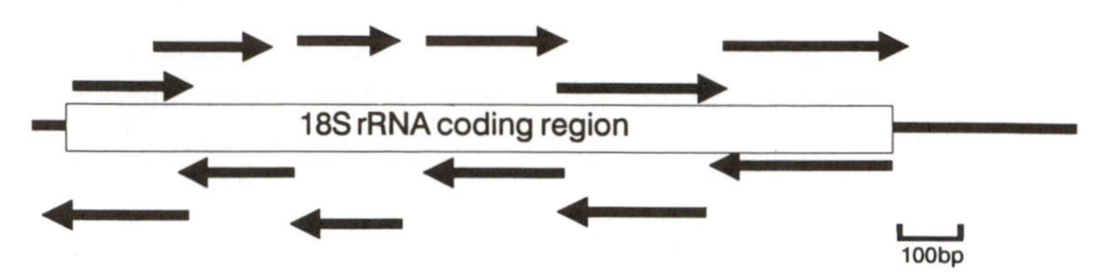

Figure 3-3 Automated sequencing strategy and results for the *Gingko biloba* 18*S* rRNA gene. Note that the shorter read lengths resulted in several areas of the gene that were *not* sequenced on both strands.

Table 3-1 *Ginkgo biloba* rRNA Gene Analysis: Automated versus Manual Sequencing

Sequence length	1,811 base pairs
Base mismatches	10 positions
Percent sequence agreement	99.5%
Percent sequence agreement minus regions representing "overlap failure" by automated sequencing	100%

Table 3-1 presents the comparison of the data obtained from an exhaustive manual sequencing approach and a single automated analysis. In this table, the data are strictly compared, and any mismatches in the alignment are counted as machine errors. From this data set, an error rate for the DuPont automated sequencer of 0.5% would be calculated. However, a very key point is that this comparison is based on the data as derived from the single reads as shown in Figure 3-4. That is, it includes sequence derived from some areas that were covered only on one strand.

Therein lies the most important lesson in automated sequence analysis. The same rules for responsible sequencing must apply to both manual and automated methods. In other words, only those data that are confirmed with opposite-strand data should be given confidence.

Since the readable runs on the DuPont sequencer failed to extend beyond 300 bases from each primer, the overlap failures created gaps in the se-

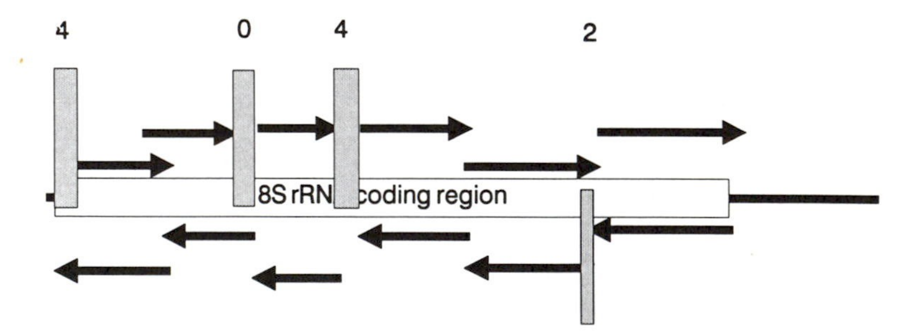

Figure 3-4 Location of sequence mismatches by the automated method. Note that all cases of miscalled bases in the automated analysis occurred in areas of the sequence that were not covered by complementary data from both strands. *N*, number of errors in single-stranded regions; □, overlap failures.

quence data where data were available from only one strand (see Fig. 3-4). In every case of mismatch between the automated and the manual analyses, the error in the automated run occurred in an *area of failed overlap*. Indeed, when the errors within the overlap gaps are eliminated from the analysis, the agreement between automated and manual methods is 100% (Table 3-1).

At the time of this comparison, read length on the DuPont sequencer was a significant limitation. However, as mentioned earlier, protocols and equipment continue to evolve, and our average run length now exceeds 300 bases. Thus, it is likely that a repeat of the automated analysis would achieve complete overlap.

There was one remarkable benefit from the automated analysis. The regions of collapse that plagued the manual sequencing gels were totally absent from the automated run. Figure 3-5 shows one example in which a dramatic collapse region exists on the regular sequencing gel. The collapse was subsequently resolved by modified reaction and gel conditions. However, it can be clearly seen that the DuPont sequencer read through the region the first time and without modification of the standard reaction conditions. Collapse regions in a manual gel are most likely the result of premature chain termination in the sequencing reaction, or hairpins in the fragments. In the DuPont system, any premature terminations are not detected (because the label is on the dideoxy terminators) and thus have no adverse influence on the read. In addition, because of the sensitive computer-controlled electrophoresis, precise temperature control is always realized and fold-back collapses are essentially eliminated.

Thus, using cloned DNA as template, automated sequence analysis is extremely accurate and much faster than traditional manual methods. The only drawback in the comparison was that the automated sequencer produced runs of shorter readable length. If the shorter read length prevents overlap, then errors can occur because of the inability to cross-check with the complementary strand. However, as automated protocols continue to develop, this will become less and less of a problem.

Automated Sequencing of PCR-Generated Templates

Clearly, one of the main approaches in future phylogenetic analysis of DNA sequences is in the application of PCR technology (Erlich, 1989). Direct sequencing of PCR-generated products is the fastest way to gain DNA sequence information from any DNA source that is not already subcloned into a convenient vector.

In order to evaluate the efficacy of combining automated sequence analysis and PCR technology, we compared manual and automated sequences derived from alleles of exon 3 of the mouse MHC Class II Aβ gene. PCR was performed according to the method described by She et al. (1991). PCR amplification of 1 μg of genomic DNA produced a DNA fragment which was subsequently purified by agarose gel electrophoresis. The pur-

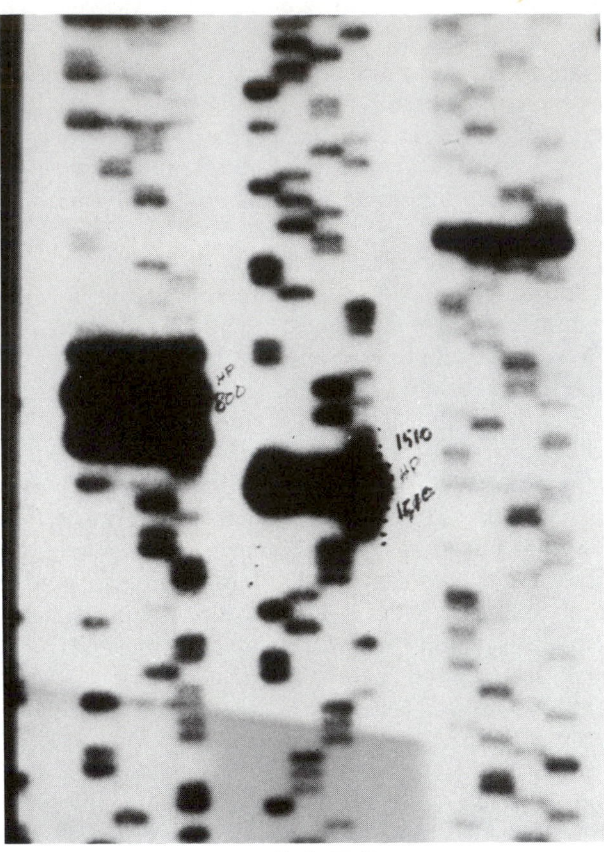
HP
800
1510
HP
1510

GATC

A

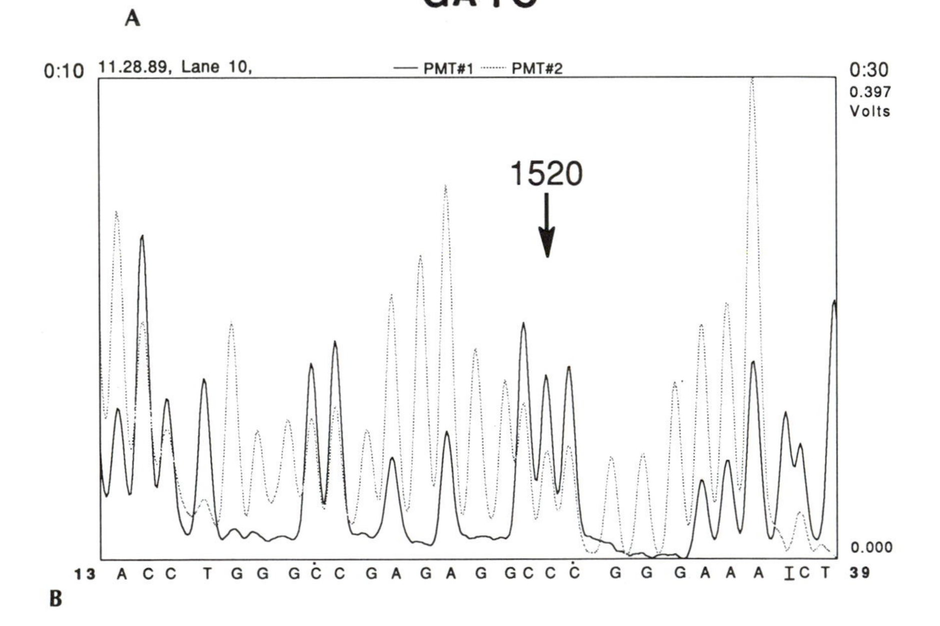
0:10
11.28.89, Lane 10,
PMT#1
PMT#2
0:30
0.397
Volts
1520
0.000
13 A C C T G G G C C G A G A G G C C C G G G A A A I C T 39

B

ified DNA fragment and one of the primers was then used to perform an asymmetrical amplification of one strand. The amplified single-stranded DNA product was divided and subjected to both manual and automated sequence analyses using the opposite member of the primer pair to prime the sequencing reactions.

The results clearly indicated that PCR-derived templates can be directly sequenced by both manual and automated methods. In a comparison of the manually derived sequence aligned with the sequence produced from the automated analysis, over the effective region of the analysis (approximately 220 bases), there is 100% agreement between the two methods. Again, it must be remembered that confident sequence can only be arrived at if sequences from both strands are compared and analyzed.

On-Screen Data Analysis

The comparison utilities present in the computer control system of the automated sequencer allow direct, on-screen comparison of different data sets (Amorese and Hochberg, 1989). Thus, in a search for sequence polymorphisms, the actual data as well as the derived sequences for several runs can be loaded into windows on the screen. Presumptive polymorphisms can be visually confirmed by this direct comparison method, rather than by the tedious rereading of autoradiograms. Another example from the mouse MHC II locus (Fig. 3-6) shows a comparison of the output data generated from analysis of the sequences of PCR-generated templates for two different alleles. Visual confirmation of a G/A transition is easy and obvious.

In a similar fashion, the direct comparison utility of the automated sequencer also allows the direct, on-screen alignment of the two complementary reads for any given area of the sequence. If the sequences derived from the complementary strands show any mismatches or ambiguities, the raw sequence data can be reevaluated in paired windows from each sequence. From these raw data the operator can make an immediate informed interpretation based on the quality of the peaks in each sequence, and can correct one of the sequences or can decide to rerun the reaction(s).

Figure 3-5 Comparison of one dramatic collapse region affecting the manual but not the automated analysis. (Top) The center four lanes of this sequencing gel demonstrate a common sequencing artifact that is especially prevalent in rRNA gene sequencing projects. This particular collapse region required several additional attempts to finally resolve the collapse and derive the correct sequence (the complement to the automated run is shown here). (Bottom) The automated run through this region, however, shows no artifact. The collapse region is indicated on the output and the correct sequence is displayed.

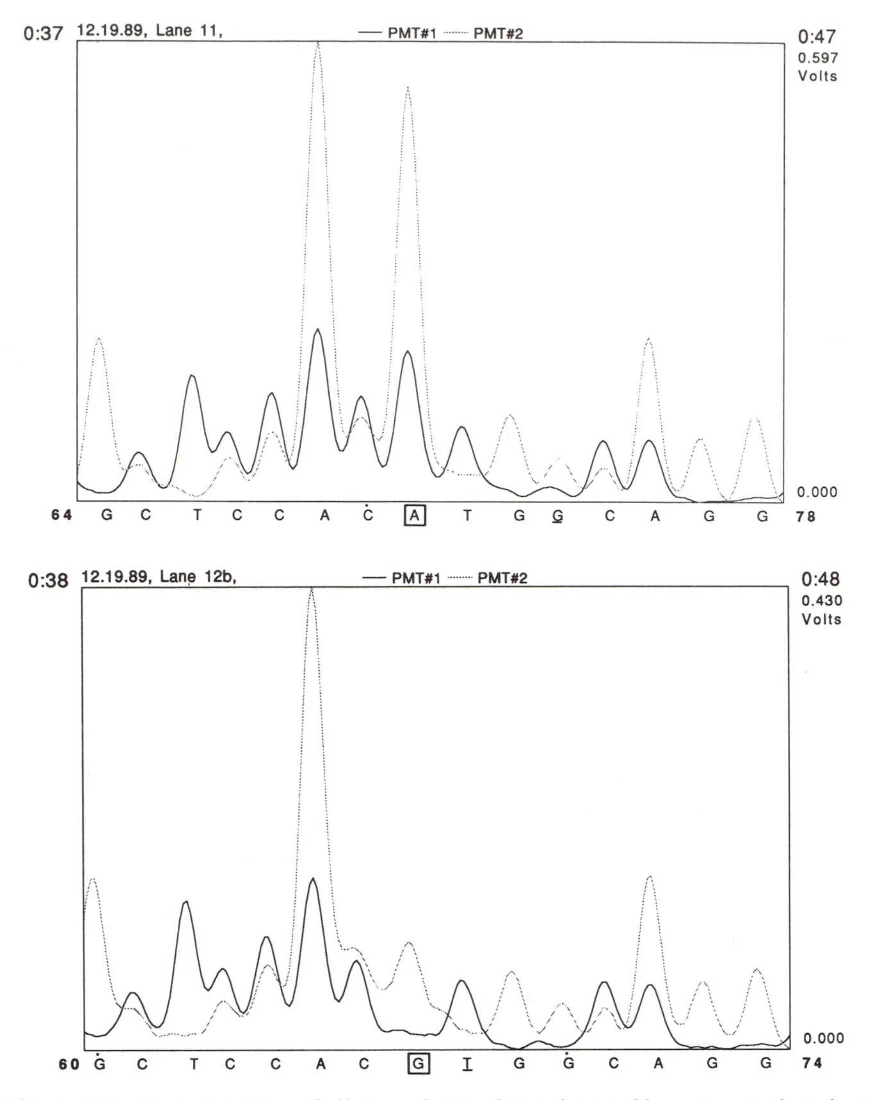

Figure 3-6 Demonstration of allelic polymorphism detected by automated analysis of PCR-generated templates. These two panels represent output from automated sequencer analysis of PCR templates derived from two distinct alleles of the mouse MHC II locus. The G/A transition (highlighted in boxes) is clear as is the similarity of the remainder of the sequence output.

CONCLUSIONS

We hope that this presentation has shown that automated sequencing, even by a central sequencing facility, is a viable, perhaps even a desirable, alternative to traditional DNA sequencing methods.

The concept of a central sequencing service facility that services a wide array of researchers is appropriate for several reasons. First, the automated systems can handle a fairly large throughput of sequencing reactions and can thereby service a number of users simultaneously. Second, as a group service facility, maximum use of the sequencer allows the cost of machine purchase and/or maintenance to be amortized among the users (or, under the Center concept, to be absorbed by the organization providing the service). Finally, use of an automated sequencing facility relieves the individual researcher from a large amount of the effort and other capital outlays involved in producing sequence information. This allows more time to be spent in data analysis or production of the templates for analysis.

We have also shown that automated sequence analysis can be every bit as accurate as manual sequencing, but only if the same rules of responsible sequencing are employed. That is, all sequences should be confirmed on both DNA strands and any ambiguities should be resequenced. Given one run on one template, the machines are capable of what might seem to be high error rates, but a similar error-rate observation must also be admitted for any single manual sequencing run as well. Responsible sequencing protocol demands complementary data in each case. Our data indicate that for both cloned genes and PCR-generated templates, automated sequencers equal the accuracy of manual analyses.

The one limitation of automated sequencers is the length of readable sequence derived from a single reaction. In our hands, the DuPont sequencer produces very accurate sequence for 250 to 400 bases, depending on the quality of the template. While machines from other manufacturers (such as ABI) claim higher run lengths, it seems that automated sequencers are only beginning to be able to match the 600 to 900 bases that can be read in many laboratories using manual techniques. This can seriously effect experimental design, as we saw in our analysis of the rRNA gene from *Gingko biloba*. While the manual sequencing provided large amounts of redundant overlap, the automated sequencer could not provide overlapping sequence in every case, leading to the potential loss of information. However, even though run lengths continue to increase, the shortfall in overlaps could be avoided by simply reducing the space between adjacent primers.

The benefits of automated analysis include the ability to sequence a large number of templates at one time, with a relatively short gel run that does not require constant attention or numerous loadings. For the production of sequence information from a large number of alleles or species, this approach is ideal. This is especially true when combined with PCR to generate the templates. In some cases, such as with the rRNA genes, gel and enzymatic artifacts are greatly reduced or absent in automated runs, thus adding to the overall accuracy. Finally, because all the data are computer-resident, entry errors are eliminated and comparisons can be made on-screen and directly from the raw data, thus enhancing the proofreading of ambiguities in the sequence and eliminating the storage, shuffling, and rereading of autoradiograms.

ACKNOWLEDGMENT

The DNA sequencer was purchased through a Biomedical Research Grant through the Division of Sponsored Research at the University of Florida, Gainesville, FL.

REFERENCES

Amorese, D.A., and A.M. Hochberg. (1989) A fluorescence-labeled DNA analysis system for sequencing and mapping. *Am. Biotech. Lab.* **7**:38–46.

Chen, E.J., and P.H. Seeburg. (1985) Supercoil sequencing: a fast and simple method for sequencing plasmid DNA. *DNA* **4**:165–170.

Erlich, H.A. (ed.). (1989) *PCR Technology. Principles and Applications for DNA Amplification*. Stockton Press, New York.

Nairn, C.J., M. Miyamoto, and R.J. Ferl. (1991) Phylogenetic implication of the DNA sequence of the 18s rRNA gene of *Gingko biloba*. *Mol. Biol. Evol.*, submitted.

Prober, J.M., G.L. Trainor, R.J. Dam, F.W. Hobbs, C.W. Robertson, R.J. Zagursky, A.J. Cocuzza, M.A. Jensen, and K. Baumeister. (1987) A system for rapid DNA sequencing with fluorescent chain-terminating dideoxynucleotides. *Science* **238**:336–341.

Sanger, F., S. Nicklen, and A.R. Coulson. (1977) DNA sequencing with chain terminating inhibitors. *Proc Natl. Acad. Sci. USA* **74**:5463–5467.

She, J.X., F. Bonhomme, E. Desmarais, and E.K. Wakeland. (1991) Mitochondrial DNA sequence evolution and phylogeny in the genus *Mus*. *Mol. Biol. Evol.*, submitted.

Smith, L.M., J.Z. Sanders, R.J. Kaiser, P. Hughes, C. Dodd, C.R. Connell, C. Heiner, S.B.H. Kent, and L. Hood. (1986) Fluorescence detection in automated DNA sequence analysis. *Nature* **321**:674–679.

4

Computer Alignment of Sequences

MICHAEL S. WATERMAN, JANA JOYCE, AND MARK EGGERT

Sequencing of DNA has precipitated a revolution in biology. Rapid sequencing techniques were introduced a little more than a decade ago, and the rate of sequencing continues to accelerate. As of spring 1991 the international databanks (EMBL, GenBank, and DDBJ) contained approximately 50×10^6 base pairs (bp) of DNA sequences with an average sequence length of about 1,000 bp. Improvements in sequencing technology continue to be made, and the associated discoveries in biology are staggering.

It is instructive to look at volumes of Dayhoff's Atlas (1979) and consider the data available in the early 1970s. Hemoglobin and cytochrome *c* sequences had been determined from a variety of organisms, and efforts were being made to infer the evolutionary relationships from macromolecular sequences. See Fitch and Margoliash (1967) for a pioneering paper in this area. Today, in contrast, there are families of sequences for many genes. Journals such as *Journal of Molecular Evolution* and *Molecular Biology and Evolution* are largely devoted to the study of sequence evolution.

Recently, there has been an exciting new initiative in molecular biology, the Human Genome Initiative. This project has as its objective the characterization of an entire human genome of 3×10^9 bp. Even though the basic goal of the genome project is very important and will take several years to accomplish, the project has much broader implications and could be the basis of biology for the next century. First of all, the human genome cannot be considered in isolation from those of other organisms. For example, the genomes of *Escherichia coli*, nematode, and yeast are being characterized. With a fairly detailed restriction map of *E. coli* already in existence (Kohara et al., 1987), both the experimental techniques and the basic biology learned from work on these organisms are essential to the Human Genome Initiative. From our point of view, genomic analysis is an important new emphasis in biology. Genomes from different organisms can be studied from a comparative viewpoint, and variation within a species is becoming accessible at the molecular level. When large amounts of sequence are produced, it will be an enormous task to detect functional and evolutionary relationships within and between the DNAs of organisms.

In this chapter we will consider sequence alignment, one of the most utilized computational tools for the study of molecular sequences. See Bell and Marr (1989) for related discussions of these topics. In the section on sequence alignment, we discuss the alignment problem, including a maximum-likelihood heuristic to motivate alignments that maximize a scoring function. The computer algorithms used to align full sequences and to locate maximum segment alignments are given. Alignments from 5*S* ribosomal RNA (rRNA) gene sequences are presented to illustrate the algorithms. The statistical significance of alignments is also discussed. Finally, there is a brief survey of multiple sequence alignments.

Sequence Alignment

Several different questions might be asked about a sequence, one of which concerns unexpected relationships with other sequences. These discoveries are sometimes made by screening a database with the sequence. Wilbur and Lipman (1983), and Lipman and Pearson (1985), devised a search algorithm based on the computer science technique of hashing, and theirs is the method of choice for such questions. Several groups have implemented versions of their technique. Essentially, the search looks for exact matching k-mers (usually $k = 4$ to 6 for DNA, and $k = 2$ for protein) and reports regions where the test sequence has a high density of matches with the database. Then an optimized alignment region is presented. There is no further discussion of database searches in this chapter, but they should not be overlooked.

Sequence alignment is a popular computer activity. The alignments are often based on some explicit optimization function, rewarding matches and penalizing mismatches, insertions, and deletions. Sequence alignment can give useful information about the evolutionary or functional relationships between sequences.

The generic two-sequence alignment problem can be stated as follows: $\mathbf{a} = a_1a_2 \ldots a_n$ and $\mathbf{b} = b_1b_2 \ldots b_m$ are two sequences over a finite alphabet. For DNA and RNA, the alphabet has four letters, and for proteins, the alphabet has 20 letters. Molecular evolution events include substitutions of one letter for another as well as insertions and deletions (indels) of letters. What is the minimum number of these events required to transform **a** to **b**? The relationships between **a** and **b** implied by the minimal set of events are usually displayed as alignments where letters of **a** are written directly over their corresponding letters in **b**. It is necessary to weight the transformations in order to come closer to biology, so the problem has a more general statement.

Let us look at a simple example where **a** = AAGTTC and **b** = AGCCC.

An alignment that might be suggested by the sequences is

AAGTTC

A– GCCC

where there are three matches, two mismatches, and one indel. This alignment represents a specific hypothesis about the evolution of the sequences; three of the nucleotides have not changed since the common ancestor, there have been (at least) two substitutions, and one nucleotide has been either inserted or deleted. Sequences are aligned by computer to find good alignments. A computer is necessary because of the exponential number of possible alignments. For example, two sequences of length 1,000 have over 10^{600} possible alignments.

There remains the question of how to score an alignment. We present a simple heuristic derivation of "alignment score." If p is the probability of an identity, q the probability of a substitution, and r the probability of an indel, the above alignment has probability

$$\Pr = p^3 q^2 r^1.$$

Define score S' by the log likelihood:

$$S' = \log \Pr = 3(\log p) + 2(\log q) + 1(\log r).$$

Define $S = S' - n \log s = S' - 6 \log s$, where s is a constant satisfying $\log(p/s) = 1$. We have simply subtracted a constant from S'. S becomes

$$S = 3 - 2\mu - 1\delta,$$

where $\mu = \log(q/s)$ and $\delta = \log(r/s)$. Fortunately, score defined by

$$S = \#\text{identities} - \mu\#\text{substitutions} - \delta\#\text{indels}$$

can be efficiently computed as we will see in the next section. It also has the simple maximum likelihood interpretation we have presented. We caution the reader that this simple reasoning is only a heuristic. See Tavaré (1986) for a related mathematically rigorous discussion.

We distinguish two types of alignments: (1) alignments of full sequences; and (2) finding segments of sequence that can be well aligned. Full sequence alignment should only be attempted when it is believed that the sequences are related, from one end to the other. If this is not the case, the sequences can be forced into incorrect relationships due to the necessity of matching the less similar segments. A more conservative approach is to run a maximum segments analysis that only finds those segments of the sequences matching at or above some preset level.

Consensus sequence analysis is usually done by the biologists "eye" and by experiment. Of course, we only believe a protein-binding site when it is verified by experiment, but analysis by inspection can be biased. So it is useful to have computer methods that can find consensus patterns best fitting explicitly stated criteria. Computer algorithms for consensus can be applied to align sets of multiple sequences.

DYNAMIC PROGRAMMING ALIGNMENT OF TWO SEQUENCES

In 1970 Needleman and Wunsch wrote a landmark paper that approached sequence comparison (alignment) with an algorithm that is one of a class of optimization techniques known as dynamic programming. Their algorithm finds maximum similarity between two sequences, where matches receive positive weight, and mismatches, insertions, and deletions receive nonpositive weight. Some mathematicians began to attempt to define a distance between sequences and so to construct a metric space. This search culminated with Sellers (1974), where the desired results were obtained for single insertions and deletions, and with later workers who extended this work to multiple insertions and deletions (Waterman et al., 1976). While it has been proven that similarity and distance are equivalent concepts when matching full sequences (Smith et al., 1981), for certain other problems similarity is superior, and we presented similarity here.

Modern DNA sequencing has revealed introns and exons. This suggests the problem of finding the best matching pieces (segments) between two sequences. While this problem was first approached with distance methods, a much simpler approach through similarity was taken by Smith and Waterman (1981), and an extension of that technique is presented below (Waterman and Eggert, 1987).

Matching or aligning entire sequences should be attempted when the sequences are known or suspected to be closely related. Even when this is the case, an extraordinary number of best alignments can result; many of these will differ only slightly from one another. We illustrate this below with some recommendations on how to deal with the situation. Most sequence comparisons will, however, involve sequences only significantly related in pieces, if at all. In those cases a full alignment is not informative, and the maximum segments algorithm is the most appropriate. These algorithms can produce segment matchings which are best, second best, and so on; this is shown in the examples below.

Aligning Full Sequences

As explained above, what is to be presented is a similarity method for sequence alignment. The function $s(a,b)$ is to define similarity between the letters a and b. In the examples below, matches receive weight 1 and mismatches receive -1, so that $s(a,b) = 1$ if $a = b$ and $s(a,b) = -1$ if $a \neq b$. Deletions of length k receive weight $-w_k$ and below, only w_1 is used where $w_k = kw_1$. The algorithm is based on recursively computing a matrix S.

First

$$S_{0,0} = 0,\ S_{i,0} = -w_{i,}\ 1 \leq i \leq n\ S_{0,j} = -w_{j,}\ 1 \leq j \leq m.$$

Then

$$S_{i,j} = \max\{S_{i-1,j-1} + s(a_{i,}\ b_j),\ \max_{k\geq 1}\{S_{i,j-k} - w_k\},\ \max_{k\geq 1}\{S_{i-k,j} - w_k\}\}.$$

For single insertions and deletions only,

$$S_{i,j} = \max \{S_{i-1,j-i} + s(a_{i,}b_j),\ S_{i-1,j} - w_1, S_{i,j-1} - w_1\}.$$

The idea of these calculations is that $S_{i,j}$ is the maximum similarity of $a_1a_2 \ldots a_i$ and $b_1b_2 \ldots b_j$. This is why for example, $S_{0,j} = w_j$. The recursion is based on the ways an alignment can end:

$$\begin{matrix} \ldots a_i \\ \ldots b_j \end{matrix} \text{ corresponds to } S_{i-1,j-1} + s(a_i,b_j)$$

and

$$\begin{matrix} \ldots - \\ \ldots b_j \end{matrix} \text{ corresponds to } S_{i,j-1} - w_1$$

and so on.

When $n = m$, the multiple insertions and deletions program runs in time proportional to n^3, while the single insertion and deletion program runs in the much preferred time n^2. Fortunately, when the function w_k is linear in k, $w_k = \alpha + \beta^*k$, the running time can still be made n^2 (Gotoh, 1982). Multiple insertions and deletions are important, as adjacent bases are deleted or inserted by one event and should be weighted accordingly.

Alignments can be produced from the matrix by two methods. One is accomplished by saving information at each matrix entry that indicates which events were necessary to calculate $S_{i,j}$. Then, beginning at $S_{n,m}$, this information is used to retrace our steps to $S_{0,0}$, thus producing the optimal alignments. The second technique is recomputing to see which events produced the value of each matrix entry beginning at $S_{n,m}$. Both methods take little time in comparison with the matrix construction.

For the examples, we first align genes of 5*S* rRNA *E. coli* (*rrnB* operon) with the closely related 5*S* rRNA *Beneckia harveyi*. The 5*S* rRNA gene sequences used in this chapter are taken from Erdmann et al. (1983). As mentioned above, $s(a,b) = +1$, if $a = b$, and $s(a,b) = -0.9$ if $a \neq b$ and $w_1 = 2$. ($w_k = \infty$, $k \geqslant 2$, so that only single insertions and deletions are allowed.) The two sequences have similarity of 78.1. There are two optimal alignments, which result from simply changing which of the two Ts is mismatched with the C. See Figure 4-1 for these two alignments.

```
TGCCTGGCGGCAGTAGCGCGGTGGTCCCACCTGAC (C-) CCATGCCGAACTCAGAAGTGAAA/
||| ||||| |  |||||   ||| ||||||||||      |||| ||||||||||||||||||\
TGCTTGGCGACCATAGCGATTTGGACCCACCTGAC (TT) CCATTCCGAACTCAGAAGTGAAA/

 /CGCCGTAGCGCCGATGGTAGTGTGGGGTCTCCCCATGCGAGAGTAGGGAACTGCCAGGCAT
 \||   ||||||||||||||||||||||  |||||||| |||||||||  |  ||||||| |
 /CGAATTAGCGCCGATGGTAGTGTGGGGCTTCCCCATGTGAGAGTAGGACATCGCCAGGCTT
```

Figure 4-1 The optimal full sequence alignments of the genes for 5*S* rRNA *E. coli* and 5*S* rRNA *B. harveyi* (top and bottom sequences, respectively). The maximum segment alignments of the same sequences are identical to the full alignments.

For a second example, we align genes of 5S rRNA *E. coli* with the more distantly related 5S *Mycoplasma capricolum*. Here the similarity is 18.3, with 175 different but optimal alignments. The 175 alignments are represented in the four alignments of Figure 4-2 (Top). The representation is similar to that of Figure 4-1, so that the first alignment actually represents a total of $2 \times 5 \times 3 = 30$ alignments, the second represents $2 \times 5 \times 2 = 20$, the third $5 \times 5 \times 3 = 75$, and the fourth $5 \times 5 \times 2 = 50$. This

```
TGCCTGGCGGCAGTAGCGCGGTGGTCCCACCTGACCCCATGCCGAACTCAGAAGTGAAACGCC/
|   ||| || | ||||    | |||| ||||||  ||||||||||| ||||||| || | | \
T---TGGTGGTA-TAGCATAGAGGTCACACCTGTTCCCATGCCGAACACAGAAGTTAAGCTCT/

   /GTAGCGCCGATGGTAGTGTGGGG(TC)T(CCCCA)TGCGAGAGTAGGGAACTGCCAG(GCA)T
   \ |  ||  || | || | |    |     |     || || | |||  | |||||||     |
   /ATTACGGTGAAGATA-T-TACTG(A-)T(G----)TGAGAAAATAGCAAGCTGCCAG(T--)T

TGCCTGGCGGCAGTAGCGCGGTGGTCCCACCTGACCCCATGCCGAACTCAGAAGTGAAACGCC/
|   ||| || | ||||    | |||| ||||||  ||||||||||| ||||||| || | | \
T---TGGTGGTA-TAGCATAGAGGTCACACCTGTTCCCATGCCGAACACAGAAGTTAAGCTCT/

   /GTAGCGCCGATGGTAGTGTGGGG(TC)T(CCCCA)TGCGAGAGTAGGGAACTGCCAGG(CA)T
   \ |  ||  || | || | |    |     |     || || | |||  | |||||| |    |
   /ATTACGGTGAAGATA-T-TACTG(A-)T(G----)TGAGAAAATAGCAAGCTGCCA-G(T-)T

TGCCTGGCGGCAGTAGCGCGGTGGTCCCACCTGACCCCATGCCGAACTCAGAAGTGAAACGCC/
|   ||| || | ||||    | |||| ||||||  ||||||||||| ||||||| || | | \
T---TGGTGGTA-TAGCATAGAGGTCACACCTGTTCCCATGCCGAACACAGAAGTTAAGCTCT/

   /GTAGCGCCGATGGTAGTGT(GGGGT)CT(CCCCA)TGCGAGAGTAGGGAACTGCCAG(GCA)T
   \ |  ||  || | || | |        ||      || || | |||  | |||||||     |
   /ATTACGGTGAAGATA-T-T(A----)CT(GATG-)TGAGAAAATAGCAAGCTGCCAG(T--)T

TGCCTGGCGGCAGTAGCGCGGTGGTCCCACCTGACCCCATGCCGAACTCAGAAGTGAAACGCC/
|   ||| || | ||||    | |||| ||||||  ||||||||||| ||||||| || | | \
T---TGGTGGTA-TAGCATAGAGGTCACACCTGTTCCCATGCCGAACACAGAAGTTAAGCTCT/

   /GTAGCGCCGATGGTAGTGT(GGGGT)CT(CCCCA)TGCGAGAGTAGGGAACTGCCAGG(CA)T
   \ |  ||  || | || | |        ||      || || | |||  | |||||| |    |
   /ATTACGGTGAAGATA-T-T(A----)CT(GATG-)TGAGAAAATAGCAAGCTGCCA-G(T-)T

TGGCGGCAGTAGCGCGGTGGTCCCACCTGACCCCATGCCGAACTCAGAAGT/
||| || | ||||    | |||| ||||||  ||||||||||| |||||||\
TGGTGGTA-TAGCATAGAGGTCACACCTGTTCCCATGCCGAACACAGAAGT/

                                  /GAAACGCCGTAGCGCCGATGGTAGT
                                  \ || | |   |  ||  || | || |
                                  /TAAGCTCTATTACGGTGAAGATATT

TGCGAGAGTAGGGAACTGCCAG
|| || | |||  | |||||||
TGAGAAAATAGCAAGCTGCCAG
```

Figure 4-2 (Top) The four alignments represent the 175 optimal alignments of the genes for 5S rRNA *E. coli* and 5S rRNA *M. capricolum* (top and bottom sequences, respectively). (Bottom) The two alignments represent the two highest scoring segment alignments of the same sequences.

makes a total of 30 + 20 + 75 + 50 = 175 alignments. These alignments have a great deal in common. As will be seen in the next section, a maximum segments algorithm will produce almost all of the common alignment.

Maximum Segments

This is a preferred method for exploring sequence relationships if computer time is available. See Smith et al. (1985) for database searches with the algorithm. It consists of a simple alteration of the preceding method due to Smith and Waterman (1981). The recursively defined matrix here is H and recursion begins with $H_{0,0} = H_{i,0} = 0, 1 \le i \le n$ $H_{0,j} = 0, 1 \le j \le m$ and

$$H_{i,j} = \max\{H_{i-1,j-1} + s(a_i,\mathrm{b}_j), \max_{k \ge 1}\{H_{i,j-k} - w_k\}, \max_{k \ge 1}\{H_{i-k,j} - w_k\}, 0\}.$$

For single insertions and deletions only,

$$H_{i,j} = \max\,\{H_{i-1,j-1} + s(a_i,b_j),\ H_{i-1,j} - w_1, H_{i,j-1} - w_1,\ 0\}.$$

Notice that a simple addition of zeros in the boundary conditions and recursion is the only change from the definition of S. These simple changes have the pleasant effect of causing $H_{i,j}$ to be the maximum similarity of *all possible* segments ending in a_i and b_j. Thus $\max_{ij}H_{ij}$ is the score of the best aligned segments in **a** and **b**.

After the 1981 Smith-Waterman algorithm, much work has gone into finding second, third, . . . best segment matches. Fortunately, there is also a simple, useful solution to this question. There is no problem with the observation that the best segment matching is associated with the entries where $\max_{ij}H_{ij}$ is achieved. Since large entries influence the entries nearby, it is not clear whether or not the second-best matching is near the first or not. One way to make certain of finding the second best is to recalculate the matrix, only this time no match, mismatch, insertion, or deletion from the maximum segment can be used. With single or linear insertion and deletion functions, only a small part of the matrix need be recomputed.

To illustrate the algorithm, we compare the gene sequences of 5*S* rRNA *E. coli* and 5*S* rRNA *M. capricolum* as above. There the full similarity was with 175 different but optimal alignments. They are presented in the four collections of alignments in Figure 4-2 (Top). Asking for the two best nonintersecting segment alignments gives the alignments of Figure 4-2 (Bottom). Notice that the regions of alignment common to all 175 alignments in Figure 4-2 (Top) are almost entirely given in Figure 4-2 (Bottom). A portion of the first matrix calculation is given in Table 4-1. No recalculation has been done so that the reader can check understanding of the method.

STATISTICAL DISTRIBUTION OF ALIGNMENT SCORES

Spurious matches are a problem in interpretation of the sequence alignments. In this section, we present some results for the distribution of scores $\max H_{ij}$ from the maximum segments algorithm.

Table 4-1 Maximum Segments Matrix for the Genes of 5*S* rRNA *E. coli* and 5*S* rRNA *M. capricolum*

	T	G	C	T	T	G	G	C	G	A	C	C	A	T	A	G	C	G	A	T	T
T	10	0	0	10	10	0	0	0	0	0	0	0	0	10	0	0	0	0	0	10	10
G	0	20	0	0	1	20	10	0	10	0	0	0	0	0	1	10	0	10	0	0	1
C	0	0	30	10	0	0	11	20	0	1	10	10	0	0	0	0	20	0	1	0	0
C	0	0	10	21	1	0	0	21	11	0	11	20	1	0	0	0	10	11	0	0	0
T	10	0	0	20	31	11	0	1	12	2	0	2	11	11	0	0	0	1	2	10	10
G	0	20	0	0	11	41	21	1	11	3	0	0	0	2	2	10	0	10	0	0	1
G	0	10	11	0	0	21	51	31	11	2	0	0	0	0	0	12	1	10	1	0	0
C	0	0	20	2	0	1	31	61	41	21	12	10	0	0	0	0	22	2	1	0	0
G	0	10	0	11	0	10	11	41	71	51	31	11	1	0	0	10	2	32	12	0	0
G	0	10	1	0	2	10	20	21	51	62	42	22	2	0	0	10	1	12	23	3	0
C	0	0	20	0	0	0	1	30	31	42	72	52	32	12	0	0	20	0	3	14	0
A	0	0	0	11	0	0	0	10	21	41	52	63	62	42	22	2	0	11	10	0	5
G	0	10	0	0	2	10	10	0	20	21	32	43	54	53	33	32	12	10	2	1	0
T	10	0	1	10	10	0	1	1	0	11	12	23	34	64	44	24	23	3	1	12	11
A	0	1	0	0	1	1	0	0	0	10	2	3	33	44	74	54	34	14	13	0	3
G	0	10	0	0	0	11	11	0	10	0	1	0	13	24	54	84	64	44	24	4	0
C	0	0	20	0	0	0	2	21	1	1	10	11	0	4	34	64	94	74	54	34	14
G	0	10	0	11	0	10	10	1	31	11	0	1	2	0	14	44	74	104	84	64	44
C	0	0	20	0	2	0	1	20	11	22	21	10	0	0	0	24	54	84	95	75	55
G	0	10	0	11	0	12	10	0	30	10	13	12	1	0	0	10	34	64	75	86	66
G	0	10	1	0	2	10	22	2	10	21	1	4	3	0	0	10	14	44	55	66	77
T	10	0	1	11	10	0	2	13	0	1	12	0	0	13	0	0	1	24	35	65	76
G	0	20	0	0	2	20	10	0	23	3	0	3	0	0	4	10	0	11	15	45	56
G	0	10	11	0	0	12	30	10	10	14	0	0	0	0	0	14	1	10	2	25	36
T	10	0	1	21	10	0	10	21	1	1	5	0	0	10	0	0	5	0	1	12	35
C	0	1	10	1	12	1	0	20	12	0	11	15	0	0	1	0	10	0	0	0	15
C	0	0	11	1	0	3	0	10	11	3	10	21	6	0	0	0	10	1	0	0	0
C	0	0	10	2	0	0	0	10	1	2	13	20	12	0	0	0	10	1	0	0	0

The average sequence in GenBank or EMBL is 1,000 bases long. Figure 4-3 shows best segment alignments for independent simulations of ten independent pairs of length 1,000 DNA sequences with $P(A) = P(C) = P(G) = P(T) = 0.25$. The algorithm parameters are $s(a,b) = 1$ if $a = b$, $s(a,b) = -\mu$ for $a \neq b$, and $w(k) = \delta k$, where $\mu = 1.5$ and $\delta = 2.1$. It is remarkable that these segmental matchings from random sequences are so long and score so well. Simulations such as this suggest that understanding the distribution of score ($\max H_{ij}$) under the null hypothesis of independence is an important goal. If the analysis of "interesting" alignments proceeds on an *ad hoc* basis, it is easy to be misled by statistically insignificant alignments. As the genome projects get underway and megabases of sequence are produced, these statistical considerations will assume increasing importance. The examples of this chapter are of DNA sequences, but the general theory allows analysis of protein and other sequences.

When two random sequences of length n(**a** and **b**) are written in a fixed alignment, the resulting sequence of matches and mismatches can be iden-

```
                                                                    mis-
Alignment                             H           matches   matches  indels

CCGATAAT-TTGTGGGCCA                   10.9        16        2        1
||||||||  ||||| |||
CCGATAATGATGTGGTCCA

TTCTGCGAAGTCGGATATAACATCAGAAC         12.2        23        3        3
|| ||||||| | ||| ||||| | ||||
TTGTGCGAAGGC-GATTTAACA-C-GAAC

ACTCATGTATCGTGACC                     14.5        16        1        0
||||||| |||||||||
ACTCATGCATCGTGACC

TATAGCTACTGTAT-G-ATTCCCTTTC           15.2        21        1        3
|| ||||||| ||| | ||||||||||
TAAAGCTACT-TATGGAATTCCCTTTC

AAATATAGTTTAGCGTGTTTCGTTGCGATGTGT     13.7        26        4        3
|||| ||||  ||| | |||| ||| |||||||
AAAT-TAGTCGAGC-TTTTTC-TTGTGATGTGT

AGGG-CTATCACGTGC-GATTC                10.8        18        2        2
|||| |||| | |||| |||||
AGGGCCTATGATGTGCAGATTC

ACACTCCCGAGGCT-CTAA                   10.9        16        2        1
|||||| |||| || ||||
ACACTCGCGAGCCTGCTAA

GAGCAGATATCGGGATTTCCGAT               12.4        19        3        1
||||| || | ||||||| ||||
GAGCAAATGT-GGGATTTTCGAT

AGGGCCG-CTCTTAAGCTATCG                10.2        18        1        3
||||||| ||| |||| | |||
AGGGCCGACTC-TAAG-TGTCG

CCTTTATTATCGTGGGTAGCC                 10.4        17        3        1
||||| | ||||||  |||||
CCTTT-TCATCGTGATTAGCC
```

Figure 4-3 Simulation results from the maximum segments algorithm applied to ten pairs of independent identically distributed DNA sequences of length $n = 1{,}000$ with equally likely letters. The algorithm parameters are $\mu = 1.5$ and $w(k) = 2.1k$.

tified as a sequence of coin tosses. The probability that the kth toss is a head is $P(x_i = y_i)$. Our object in this section is to study long runs of matches, which in this case are long head runs. The celebrated Erdös-Rényi law (Erdös and Rényi, 1970) gave order magnitude behavior of the longest head run containing $(1 - \alpha) \times 100\%$ tails where $\alpha > P(H) = p$. For length R_n of pure head runs ($\alpha = 1.0$) their result is

$$\frac{R_n}{\log_{1/p}(n)} \rightarrow 1 \text{ with probability one,}$$

while for general $\alpha > p$ their result is

$$\frac{R_N}{\dfrac{\log (n)}{G(\alpha,p)}} \rightarrow 1 \text{ with probability one,}$$

where $G(\alpha,p) = \alpha \log(\alpha/p) + (1 - \alpha)\log((1 - \alpha)/(1 - p))$ is relative entropy. For $\alpha = 1$, $G(\alpha,p) = (\log 1/p)$ and the two results are consistent. Other work (Arratia et al., 1986) gives more precise results for this law and gives

$$E\{R_n\} = \log_{1/p}(n) + \frac{(0.577\ldots)}{\theta} - \frac{1}{2} + r_1(n) \tag{1}$$

and

$$\text{Var}\{R_n\} = \frac{\pi^2}{6\theta^2} + \frac{1}{12} + r_2(n), \tag{2}$$

where $0.577\ldots$ is the Euler-Mascheroni constant, $\theta = \ln(1/p)$, and the remainders $r_1(n)$ and $r_2(n)$ are of very small, but nonzero magnitude. For 512 fair coin tosses, the mean is approximately 9.33, while the standard deviation is about 1.93.

The probability results quoted here are for independent and identically distributed coin tosses. To carry the results over to sequence matching, the two sequences of length n are assumed to have bases chosen independently and identically with $p = P\{\text{two bases match}\} = p_A^2 + p_C^2 + p_G^2 + p_T^2$. The formulae (1) and (2) hold with n replaced by n^2.

To consider approximate matching, we allow a fixed number of mismatches k. The mean length of the longest match with k mismatches becomes, from Arratia et al. (1986), approximately

$$\log(qn^2) + k\log\log(qn^2) + k\log\left(\frac{q}{p}\right) - \log(k!) + k + \frac{0.577\ldots}{\theta} - \frac{1}{2} \tag{3}$$

where $q = 1 - p$ and all logs are taken to base $1/p$. The variance remains approximately $\pi^2/(6\theta^2) + 1/12$.

The Erdös-Rényi law for the length R_n of the longest 100% head run of n coin tosses then extends to a law for the length M_n of the longest match between two sequences. We have recently shown (Arratia and Waterman,

1989) that the length of the longest run of matches containing $(1 - \alpha) \times 100\%$ mismatches satisfies

$$\frac{M_n}{\frac{\log (n^2)}{G(\alpha,p)}} \rightarrow 1 \text{ with probability one,}$$

which is the Erdös-Rényi law with n replaced by n^2. This last theorem has been obtained by use of the theory of large deviations.

Considering these results, it is not surprising that log(n) laws hold far beyond the longest exact head run or match. The expected behavior of max H_{ij} is of importance in evaluating sequence comparisons. If a located match is at or below that expected from random sequences of similar composition, then the match should not be further considered without additional biological information. These distributions have been shown to fit biological sequences quite well for algorithm parameters not covered by the theorems above. We have also proven that max H_{ij} undergoes a phase transition (Waterman et al., 1987). For larger values of (μ,δ), max H_{ij} grows proportional to log(n) and for smaller values max H_{ij} grows linearly with the sequence length. There are only two modes of behavior at this precision. The logarithmic and linear regions of this two-dimensional parameter space have been determined numerically in a Monte Carlo study.

MULTIPLE SEQUENCE ALIGNMENT

Many problems of interest in biology involve more than two sequences. For example, the analysis of *E. coli* promoter sequences for regions of the sequences critical to polymerase binding to DNA involves all sequenced promoters. See Galas et al. (1985) for such an analysis. While analysis of a set of sequences for evolutionary relationships can be performed by pairwise sequence analysis, it is often essential to simultaneously consider the full set of sequences.

Alignment of $r > 2$ sequences by dynamic programming is possible in principle, but serious computational difficulties must be overcome. In Sankoff (1975) a technique is presented to optimally align r sequences given a tree that relates them. This method uses dynamic programming and Fitch's (1971) parsimony algorithm. The method is very computationally intensive. Another method is that of Waterman et al. (1976) in which a tree is not explicitly assumed. All sequences are treated symmetrically so that a so-called star phylogeny is implicitly assumed. These methods take at least time proportional to n^r where n is the common sequence length.

Recently, some significant progress has been made in the area of multiple alignment by dynamic programming (see Pearson and Lipman, 1988). Lipman and Carrillo have made it feasible to find optimal alignments of up to six sequences by greatly reducing the number of cells the dynamic programming algorithm must consider. Carrillo and Lipman (1988) give the basic algorithm. In Altschul and Lipman (1989), it is extended to an algorithm that aligns the sequences given a tree that relates them. In Lipman et al. (1989) up to eight sequences the length of an average protein can be aligned.

It is often important to analyze many sequences. A consensus algorithm for multiple DNA sequence alignment is given that matches words of length and degree of mismatch chosen by the user (Waterman, 1986). As above, all sequences are treated symmetrically. The alignment maximizes an alignment scoring function. This method can easily align 100 sequences of length 1,000. In Waterman and Jones (1990) the method is extended to protein sequences.

A heuristic approach that has proven very useful is to infer the evolutionary tree while making a multiple alignment. Usually, the various methods proceed by performing pairwise alignments. In Johnson and Doolittle (1986), and Feng and Doolittle (1987), one such development is pursued. Taylor (1986, 1987) has also written algorithms very useful for protein analysis. Many of these methods of multiple sequence analysis are presented in chapters of a comprehensive volume edited by Russell Doolittle (1990). The program of Hein (1989) combines and refines many of the other techniques, and this would be our first choice to align and deduce an evolutionary tree for a set of sequences.

PROGRAM AVAILABILITY

The maximum segments algorithm and consensus alignment algorithms for DNA and protein sequences are all available from the authors. The implementations are written in the C language.

ACKNOWLEDGMENTS

This work was supported by grants from the National Institutes of Health and from the National Science Foundation.

REFERENCES

Altschul, S.F., and D.J. Lipman. (1989) Trees, stars, and multiple biological sequence alignment. *SIAM J. Appl. Math.* **49**:197–209.

Arratia, R., L. Gordon, and M.S. Waterman. (1986) An extreme value theory for sequence matching. *Ann. Statist.* **14**:971–993.

Arratia, R., and M.S. Waterman. (1989) The Erdös-Rényi strong law for pattern matching with a given proportion of mismatches. *Annals Prob.* **17**:1152–1169.

Bell, G.I., and T.G. Marr. (eds.). (1989) *Computers and DNA: The Proceedings of the Interface Between Computation Science and Nucleic Acid Sequencing Workshop*. Santa Fe Institute Studies in the Series of Complexity, vol. 7. Addison-Wesley, Redwood City.

Carrillo, H., and D. Lipman. (1988) The multiple sequence alignment problem in biology. *SIAM J. Appl. Math.* **48:**1073–1082.

Dayhoff, M.O. (ed.). (1979) *Atlas of Protein Sequence and Structure*, vol. 5, suppl. 3. National Biomedical Research Foundation, Washington, D.C.

Doolittle, R.F. (ed.). (1990) *Molecular Evolution: Computer Analysis of Protein and Nucleic Acid Sequences*. Methods in Enzymology, vol. 183. Academic Press, New York.

Erdmann, V.A., E. Huysman, A. Vandenberghe, and R. De Wachter. (1983) Collection of published 5S and 5.8S ribosomal RNA sequences. *Nucleic Acids Res.* **11:**r105–r133.

Erdös, P., and A. Rényi. (1970) On a new law of large numbers. *J. Anal. Math.* **22:**103–111; reprinted in *Selected Papers of Alfred Rényi, vol. 3, 1962–1970*. Akademiai Kiado, Budapest, 1976.

Feng, D., and R.F. Doolittle. (1987) Progressive sequence alignment as a prerequisite to correct phylogenetic trees. *J. Mol. Evol.* **25:**351–360.

Fitch, W.M. (1971). Towards defining the course of evolution: minimum change for a specific tree topology. *Syst. Zool.* **20:**406–416.

Fitch, W.M., and E. Margoliash. (1967) Construction of phylogenetic trees. *Science* **155:**279–284.

Galas, D.J., M. Eggert, and M.S. Waterman. (1985) Rigorous pattern recognition methods for DNA sequences: analysis of promoter sequences from *E. coli*. *J. Mol. Biol.* **186:**117–128.

Gotoh, O. (1982) An improved algorithm for matching biological sequences. *J. Mol. Biol.* **162:**705–708.

Hein, J. (1989) A new method that simultaneously aligns and reconstructs ancestral sequences for any number of homologous sequences, when the phylogeny is given. *Mol. Biol. Evol.* **6:**649–668.

Johnson, M.S., and R.F. Doolittle. (1986) A method for the simultaneous alignment of three or more amino acid sequences. *J. Mol. Evol.* **23:**267–278.

Kohara, Y., K. Akiyama, and K. Isono. (1987) The physical map of the whole *E. coli* chromosome: application of a new strategy for rapid analysis and sorting of a large genomic library. *Cell* **50:**495–508.

Lipman, D.J., S.F. Altschul, and J.D. Keceioglu. (1989) A tool for multiple sequence alignment. *Proc. Natl. Acad. Sci. USA* **86:**4412–4415.

Lipman, D.J., and W.R. Pearson. (1985) Rapid and sensitive protein similarity searches. *Science* **227:**1435–1441.

Needleman, S.B., and C.D. Wunsch. (1970) A general method applicable to the search for similarities in the amino acid sequences of two proteins. *J. Mol. Biol.* **48:**444–453.

Pearson, W.R, and D.J. Lipman. (1988) Improved tools for biological sequence comparison. *Proc. Natl. Acad. Sci. USA* **85:**2444–2448.

Sankoff, D. (1975) Minimal mutation trees of sequences. *SIAM J. Appl. Math.* **28:**35–42.

Sellers, P. (1974) On the theory and computation of evolutionary distances. *SIAM J. Appl. Math.* **26:**787–793.

Smith, T.F., and M.S. Waterman. (1981) Identification of common molecular subsequences. *J. Mol. Biol.* **147:**195–197.

Smith, T.F., M.S. Waterman, and C. Burks. (1985) The statistical distribution of nucleic acid similarities. *Nucleic Acids Res.* **13:**645–656.

Smith, T.F., M.S. Waterman, and W.M. Fitch. (1981) Comparative biosequence metrics. *J. Mol. Evol.* **18:**38–46.

Tavaré, S. (1986) Some probabilistic and statistical problems in the analysis of DNA sequences. *Lec. Math. Life. Sci., Am. Math. Soc.* **17:**57–86.

Taylor, W.R. (1986) Identification of protein sequence homology by consensus template alignment. *J. Mol. Biol.* **188:**233–258.

Taylor, W.R. (1987) Multiple sequence alignment by a pairwise algorithm. *CABIOS* **3:**81–87.

Waterman, M.S. (1986) Multiple sequence alignment by consensus. *Nucleic Acids Res.* **14:**9095–9102.

Waterman, M.S., and M. Eggert. (1987) A new algorithm for best subsequence alignments with application to tRNA-rRNA comparisons. *J. Mol. Biol.* **197:**723–728.

Waterman, M.S., L. Gordon, and R. Arratia. (1987) Phase transitions in sequence matches and nucleic acid structure. *Proc. Natl. Acad. Sci. USA* **84:**1239–1243.

Waterman, M.S., and R. Jones. (1990) Consensus methods for DNA and protein sequence alignment. Pp. 221–237 in *Molecular Evolution: Computer Analysis of Protein and Nucleic Acid Sequences* (R.F. Doolittle, ed.). Methods in Enzymology, vol. 183. Academic Press, New York.

Waterman, M.S., T.F. Smith, and W.A. Beyer. (1976) Some biological sequence metrics. *Advances in Math.* **20:**367–387.

Wilber, W.J., and D.J. Lipman. (1983) Rapid similarity searches of nucleic acid and protein data banks. *Proc. Natl. Acad. Sci. USA* **80:**726–730.

5

Aligning DNA Sequences: Homology and Phylogenetic Weighting

DAVID P. MINDELL

Ever since the term homology was defined as "the same organ in different animals under every variety of form and function" by Owen (1848), it has frequently been both used and debated. Homology is a concept representing an hypothesis of correspondence between features, and is used at different levels of biological organization (e.g., phenotypic, genotypic). Most evolutionary biologists, particularly systematists, would now agree that homology describes the relationship between features that are similar due to common descent among organisms, as opposed to those that are similar due to convergent or parallel evolution. Thus, identification of homologous features is a crucial step in determining phylogenetic relationships. Demonstrating this, homology and synapomorphy have been presented by some as equivalent relationships (Hennig, 1966; Platnick, 1979), both being used to identify monophyletic groups.

Phylogenetic analysis of DNA sequences involves their alignment in such a way that homologous nucleotide base positions within homologous genes are compared. Sequences of DNA offer the largest set of characters and, hence, potential homologies for phylogenetic studies. By logical extension, however, they also offer the largest set of potentially convergent and, hence, phylogenetically misleading characters, indicating the importance of homology assessment. Sequences differ from morphological characters in their linear organization, and in their greater (although highly variable) susceptibility to mutational change. These differences have required different methods in assessment of molecular sequence, as opposed to morphological, homology relationships. The objective of this chapter is to examine the practice of DNA sequence alignment in the context of homology assessment. I point out that species sequences should be aligned in descending order of phylogenetic relationship (phylogenetic weighting of alignments) to maintain the continuity of information which forms the basis of relationships of homology. Using mitochondrial ribosomal RNA (rRNA) sequences, I also show how shuffling the order of input for sequences in multiple alignments may be used to help determine phylogenetic relationships among taxa whose divergences occurred relatively close together in time.

TYPES OF HOMOLOGY AND "RECOGNITION CRITERIA"

A useful distinction has been made between features that are homologous *between* different individuals and those that are homologous *within* a single individual organism. These two types, as recognized in a morphological context, have been called phylogenetic and repetitive homology, respectively (Van Valen, 1982; Roth, 1984). Repetitive homology includes both serial homology (or repetition), as in the legs of a millipede; and nonserial repetitions, as in a snake's scales or leaves of a tree. This distinction is useful as it facilitates the practice of systematics; phylogenetic homologies are appropriate for phylogenetic analysis of organisms, repetitive homologies are not.

The same and additional homology distinctions have been made for molecular features, in accord with increased understanding of mechanisms of molecular change. Sequences of DNA or amino acids that are homologous between individuals have been termed orthologous, whereas duplicate genes or proteins within an individual are paralogous (Fitch, 1970). Chromosomal mutations mixing various pieces of genes in new combinations (Gilbert, 1978, 1985) can cause "partial paralogy" among genes and proteins, and horizontal transfer of DNA sequences between species can mix "native" and "foreign" sequences (termed 'xenology") within an individual (Gray and Fitch, 1983; see review by Patterson, 1987). For molecular data, orthologous sequences are appropriate for phylogenetic analysis of organisms, whereas paralogous sequences are appropriate for phylogenetic analysis of genes.

In its original (morphological) context the main criteria for recognizing homologous features have been: (1) an "essential sameness," as determined by properties of similarity in function, structural position, composition, or ontogenetic origin (Remane, 1956; Jardine, 1967; Bock, 1974; Riedl, 1978); and/or (2) congruence with other postulated homologies, as determined by their agreement in supporting the same phylogenetic relationships (Wiley, 1975; Eldredge and Cracraft, 1980). These same criteria of similarity (excluding the ontogenetic origin property) and congruence are applicable to DNA sequence characters; however, the similarity criterion has required a "new" means of assessment, known as sequence alignment.

DNA SEQUENCE ALIGNMENTS

Homology For Sequences Is Not "A Statistical Concept"

The aligning of DNA sequences is the hypothesizing of a homology relationship for each nucleotide base position. Alignments are relatively easy for protein-coding genes, due to the presence of structured reading frames with predictable features of more frequent change at third base positions within codons, and recognizable start and stop codons. Nonprotein-coding sequences, such as rRNA genes and mitochondrial D-loop regions, are difficult to align because they lack these readily distinguishable features. This is due to the relative abundance of nucleotide base insertion and deletion events that can occur when there are no selective constraints to maintain reading frames coding for specific amino acids. Most of the following discussion will pertain to nonprotein-coding DNAs. In phylogenetic analyses, nucleotide base positions are treated as characters, and each of the four possible bases represent alternative states. Insertion and deletion of nucleotide bases connote gain or loss of nucleotide positions; however, they also may be considered as additional character-states.

A variety of approaches have been taken in aligning sequences, all of which share a fundamental similarity criterion, in seeking to maximize the "property" of identical nucleotide bases at aligned positions, and thereby minimize differences between sequences. In computer-based alignments, maximizing identical bases is accomplished by assigning various costs (or penalties) for invoking insertion/deletion (gap) events and nucleotide base substitutions. This is followed by scoring of different potential alignments, and selection of the least costly one. The *a priori* assignment of costs is crucial, as the number of potential unconstrained alignments for DNA sequences from even two species is large, and, given the existence of only four different nucleotide bases and unlimited insertions/deletions, gaps could be used to achieve "perfect alignment" for any set of sequences.

For example, in the sequence alignment program by Myers and Miller (1988) that I have used in analyses reported below, the solution involves finding a set of operations that converts one sequence to another at a minimum cost. The allowed operations are substitution, deletion, and insertion of nucleotides. Costs for each type of substitution are specified in a matrix, and the cost of inserting a gap n bases long is $g + nh$, where g is the cost, assigned by the user, for opening a gap and h is the user-assigned cost for increasing gap size by one base. Assignment of different costs can significantly alter alignments (Fitch and Smith, 1983), and should be carefully considered. Thus, homology assignments are based on a probabilistic assessment, in that preference for a particular alignment is based on preassigned costs and comparison of multiple, overall alignment scores.

This quantitative aspect of determining homologies contributed to an earlier terminological confusion (i.e., use of the term "percent homologous" to describe the relationship between two sequences), which is now resolved (Reeck et al., 1987). Choice among possible alignments is probabilistic; however, once an alignment is adopted, those (hypothesized) homologous nucleotide positions are qualitatively related. Recognition of this distinction is important, as it enables us to state clearly that homology hypotheses based on sequence alignments are no different than other (e.g., morphological) homology hypotheses, and that homology in the case of DNA sequence data has not become a statistical concept as Patterson (1987:9) has inferred.

Confusing the means of determining homology relationships with the nature of that relationship may lead to discomfort among some systematists when they hear that homologies are quantitatively determined in molecular sequence alignments. However, a quantitative basis for assessing proposed homologies is nothing new. It is inherent in the homology recognition criterion (discussed above) of congruence with other postulated homologies, as determined by their agreement in supporting the same phylogeny. Phylogenetic congruence among proposed homologies is quantitative, in that if two or more sets of postulated homologies (synapomorphies) conflict (support different relationships among taxa), the larger set is preferred. This is the basis of a most-parsimonious phylogeny.

Alignment Sensitivity to Two Classes of Variation

The fact that we discriminate among possible alignments based on quantitative scores makes those alignments, and subsequent phylogenies, sensitive to two classes of variation: (1) variation in rates of DNA sequence change; and (2) variation in the inclusion or exclusion of taxa from different alignment bouts. Alignment sensitivity is related to an inherent assumption of rate constancy in many alignment procedures. In the absence of *a priori* weighting schemes for different types of substitutions, or for order of input of sequences into an alignment (of more than two sequences), alignment algorithms seek maximal matching at all nucleotide positions. Any differences in rate of change between sequences can, potentially, introduce differences in the amount of nonhomologous nucleotide matches accumulating per time-unit since species' divergences, and such nonhomologous nucleotide matches, or homoplasy, might then contribute to an "inaccurate" phylogeny. Nonhomologous matches can arise from multiple substitutions ("multiple hits") at a single nucleotide position occurring since two taxa last shared a common ancestor.

Regarding the first class of variation, rates of DNA sequence change are known to vary: (1) by type of base substitution, as seen in the bias toward transitions in both mitochondrial (Brown et al., 1982) and nuclear (Fitch, 1967; Li et al., 1984) DNA; (2) by gene region depending on local functional constraints, as seen in faster rates of change in introns relative to exons (Li et al., 1985); (3) by gene, as seen in greater sequence conservation in functionally important genes; and (4) across taxa, as seen in faster rates of change in rodents compared to humans (Wu and Li, 1985; Britten, 1986). Rate heterogeneity has been a primary concern among phylogeneticists, hence the diversity of algorithms for phylogeny reconstruction with different underlying assumptions regarding rate equality or the lack of it.

Accommodating rate heterogeneity at the level of DNA sequence alignments, though important, has received less attention. Initial procedures that can be taken include matching phylogenetic questions with sequences from appropriate (informative) genes or gene regions, and preferentially maintaining alignment within conserved regions over known variable regions, as done by Gray et al. (1984). Further alignment procedures that could be taken correspond to character weighting procedures in phylogenetic analyses. Alignment, in determining nucleotide character homologies, is an integral step in phylogenetic analysis, and as such, weighting based on variable rates of character change may be reasonably applied to alignments as well. Although this presumes knowledge of evolutionary rate heterogeneity across the taxa and sequences involved, examples of this approach include assigning higher alignment costs to transversion over transition substitutions, to substitutions and gaps in functionally constrained over less constrained regions, and in species with relatively slower rates of change over those with faster rates.

The second class of variation to which alignments are sensitive, differences in specific taxa included in any given alignment, raises issues no less complex than those raised by the first class; however, a prescriptive generalization is more easily made. Namely, DNA sequences from various taxa should be included (input) in a multiple alignment bout in decreasing order of phylogenetic relationship, an approach which may be called phylogenetic weighting of alignments. At the beginning of this chapter I discussed the widespread agreement that relationships of homology denote similarity due to shared inheritance (of a feature) from a common ancestor. Thus, homology may also be defined on the basis of continuity of information (Van Valen, 1982). If so, that information is primarily genetic, and the "continuity" is provided by genealogy (Roth, 1988). Alignment of sequences from distant relatives, prior to alignment of more closely related taxa, would misrepresent the continuity of information in those DNA sequences. Similarly, equal treatment (via simultaneous alignment) of sequences from all taxa would also remove the genealogical basis of the homology assessment (unless all the taxa involved experienced simultaneous origins from a common ancestor). Thus, phylogenetic weighting of sequence alignments avoids the problem of confounding similarity and relationship.

Use of phylogeny to inform alignments may, at first, appear circular because alignments themselves are used to infer phylogeny. However, this is an artifact of viewing alignment and phylogenetic analysis as independent procedures with different objectives when, in fact, they are mutually dependent procedures in assessment of evolutionary history based on nucleotide sequences. Few systematists would seek to determine character homology (as in sequence alignment) without considering hypothesized relationships among taxa.

	(A)	(B)
	12345	12345
Species 1	GT-AC	GT-AC
Species 2	G-CAG	GC-AG
Species 3	GTCAC	GTCAC

Figure 5-1 Two alternative alignments for a string of nucleotide bases from each of three species. (A) is preferred on the basis of maximal nucleotide matching, with one substitution and two gaps invoked. (B) requires two substitutions and one gap, and would be preferred given a sister relationship between species 1 and 2, relative to species 3, to preserve continuity of information.

In Figure 5-1, two alternative alignments for an hypothetical sequence of nucleotide bases from each of three species are presented to demonstrate: (1) that optimal matching can mask true homology and continuity of information, in the absence of phylogenetic weighting of alignments; and (2) that the relative weight, or cost, applied to gaps (insertion/deletion events) can effect alignment preference. Given a decision to assign less weight (or cost) to gaps, the first alignment (Fig. 5-1A) is preferred on the basis of maximal nucleotide matching, because only one substitution and two gaps are invoked, as opposed to two substitutions and one gap in the second alignment (Fig. 5-1B). However, if loss of nucleotide position 3 is uniquely shared by species 1 and 2, which are mammals, relative to species 3, which is a bird, the second alignment (Fig. 5-1B) is preferred on the basis of maintaining continuity of information. Specifically, the homologous relationship of the second nucleotide position in species 1 and 2 is retained. Of course, the first alignment is still feasible, even given the "known" sister relationship of species 1 and 2. The priority of continuity of information and genealogy over optimal matching, however, is based on the fact that homologies are genealogical in origin, whereas similarity measures, such as optimal matching, are used to assess homology rather than to define it. The effect of differential weighting of gaps is seen in the preference for the first alignment (Fig. 5-1A) if gaps are discounted and the lack of a clear preference for one alignment over the other if gaps are given the same weight as substitutions.

The early computerized techniques for aligning molecular sequences either maximized matches or minimized the number of substitution, insertion, or deletion events required to make two sequences equivalent (Needleman and Wunsch, 1970; Sankoff, 1972; Sellers, 1974). In generalizing this approach to more than two sequences, Sankoff and colleagues have used "given" phylogenies as the basis for input order of sequence data in multiple species alignments (Sankoff and Rousseau, 1975), clearly recognizing the need for phylogenetic weighting of alignments (see Sankoff et al., 1976:136). Multiple sequence alignment algorithms such as the consensus method of Waterman and Jones (1990), which find sequence patterns that are conserved across taxa but which make no attempt to use phylogenetic information are inappropriate for the reconstruction of phylogeny. Such algorithms are well-suited, however, for similarity searches used in identifying unknown sequences (by scanning databanks such as GenBank), RNA secondary structure, palindromes within sequences, and so on.

Recently, several computer programs have been developed integrating alignment and phylogeny analyses that are essentially phenetic in approach. The "progressive alignment" programs of Feng and Doolittle (1987, 1990) for amino acid sequences proceed by: (1) determining an initial phenogram using modified Needleman and Wunsch (1970) alignment scores and the method of Fitch and Margoliash (1967) to determine branching order; (2) progressively aligning the sequences in order of their inferred relatedness from the phenogram; (3) filling gaps with neutral elements; and (4) constructing a new tree based on distances calculated from the new overall alignment. The same strategy used by Feng and Doolittle has been adapted for nucleotide sequences and microcomputers by Higgins and Sharp (1988, 1989). Konings et al. (1987) (see Hogeweg and Hesper, 1984) also determine an initial tree from nucleotide sequence distance calculations, and align sequences according to inferred relationships. However, if the resulting tree differs from the initial one, they realign the sequences and determine a new tree, repeating this last procedure until the alignment order and inferred phylogenetic relationships match. The "unified approach" programs of Hein (1989a,b, 1990) for nucleic and amino acid sequences also use distances from pairwise alignments to make an initial tree. Rearrangements of the tree are performed to improve fit to the distance data, and sequences are then aligned in order of their relationship inferred from the phenogram. The alignments selected are those that support a parsimony tree having the same topology as the distance tree. In this way, the parsimony criterion of minimizing evolutionary change is applied; however, so-called parsimony analyses of alternative alignment sets are constrained to match the topology of the distance tree.

These approaches have not yet received wide use by systematists, but certainly will as more sequence data become available. It would be informative to compare results of the above approaches with parsimony trees (for the same and alternative sequence alignments) that have not been constrained to fit a distance data set. Where available, independent data sets could be used in parsimony analyses to provide the initial tree used for determining order of sequence input to multiple alignments.

ALIGNMENT PROBLEMS ASSOCIATED WITH SHORT INTERNODES

The integrated nature of sequence alignment and phylogenetic analysis results in a "catch-22" (as coined by novelist Joseph Heller in describing unavailable prerequisites). Phylogenetic weighting of alignments (aligning sequences in decreasing order of phylogenetic relatedness) requires knowledge of phylogeny, which the alignment is intended to provide. An initial tree based on pairwise distances (as in the alignment analyses discussed above) is most likely to be phylogenetically "inaccurate" where species sequences are saturated with change, and where species divergences are close in time (as in radiations of taxa). In the former case, homoplasies due to multiple substitutions may unduly affect the tree, and in the latter case, brevity of uniquely shared ancestry results in few or no synapomorphies, and short internodes within a tree. Cavender (1978) and Felsenstein (1978) have discussed the potential increase in phylogenetically false evidence associated with short internodes and unequal rates of change among taxa. Even given equal rates of change, stochastically evolving molecular characters may be unable to support phylogeny reconstruction where time of independent ancestry for sister taxa is short (see Lanyon, 1988).

Recognizing Short Internodes

The first step in addressing the problem of short internodes is recognizing them. Lack of phylogenetic resolution from variable DNA sequences may stem from: (1) "noise" associated with multiple substitutions outweighing the "signal" from informative differences; and/or (2) brief periods of shared ancestry yielding no "signal" (shared derived characters) at all, as discussed above. In some instances, brief shared ancestry for taxa has been inferred based on fossil or biogeographic evidence and independent phylogenetic analyses failing to find any "signal." Examples of radiations of taxa with relatively short periods of shared ancestry include avian and mammalian orders (see Carroll, 1988 and references therein). Where there is less certainty of brief shared ancestry, the sequences themselves may provide evidence. In theory, percent divergence scores for pairwise comparisons among diverse taxa may be plotted against estimated time since divergence, where divergence times are "known" from fossil evidence, and the resulting line or curve may be used to estimate divergence times in instances where other evidence is lacking. This will, of course, be complicated by heterogeneity in rates among taxa.

As an example, consider a plot of percent sequence divergence scores for species comparisons of mitochondrial small-subunit rRNAs (12*S* in vertebrates) (Table 5-1, Fig. 5-2). The relationship between "% Sequence Divergence" and "Estimated Divergence Time" is roughly linear for species divergences of 85 million years or less. Thereafter, divergence accumulates more slowly over time, as multiple substitutions at individual nucleotide positions lead to increasing homoplastic similarity between sequences. If no fossil evidence was available for the comparisons among representatives of mammalian orders, those sequence divergence values might be secondarily overlain on the curve (Fig. 5-2), and estimated values for divergence times taken from the horizontal axis. Clustering of percent divergence scores suggests brief periods of shared independent ancestry for those taxa, assuming similarity in rates of sequence change. Note, however, the overlap in percent divergence scores for different "known" estimated divergence times (Table 5-1), making this approach, for 12*S* rRNA at least, highly approximate. Conversely, if percent divergence scores were not

Table 5-1 Pairwise Comparisons of Mitochondrial Small-Subunit rRNAs*

Species	1	2	3	4	5	6	7	8	9	10
1. *Mus*		57/20	43/85	41/85	52/85	52/85	52/85	47/300	46/350	44/600
2. *Rattus*	8.6		44/85	46/85	54/85	56/85	54/85	46/300	50/350	42/600
3. *Bos*	21.3	21.8		79/30	64/85	60/85	62/85	49/300	47/350	47/600
4. *Giraffa*	20.4	22.5	11.9		59/85	61/85	60/85	52/300	51/350	48/600
5. *Homo*	24.1	22.9	20.9	22.4		94/8	93/19	50/300	47/350	48/600
6. *Pan*	23.6	22.7	21.1	21.9	4.3		93/16	45/300	49/350	49/600
7. *Pongo*	26.4	24.8	23.1	23.7	9.1	9.3		47/300	48/350	49/600
8. *Gallus*	24.3	30.2	30.0	30.4	27.8	28.8	30.0		60/350	52/600
9. *Xenopus*	28.4	28.7	25.7	27.2	28.2	28.2	28.1	27.3		48/600
10. *Drosophila*	34.7	35.3	37.7	36.2	37.4	38.4	38.8	39.0	40.9	

*Values below the diagonal are percent divergence scores, calculated as: 100{1-[number identical nucleotides/(number shared base positions + total number gaps)]}. Values above the diagonal are: percent of substitutions that are transitions/estimated divergence times in millions of years before present. Species names and data sources are: *Mus musculus* (Van Etten et al., 1980), *Rattus norvegicus* (Kobayashi et al., 1981), *Bos taurus* (Anderson et al., 1982), *Giraffa camelopardalis* (Tanhauser, 1985), *Homo sapiens* (Anderson et al., 1981), *Pan troglodytes* (Hixson and Brown, 1986), *Pongo pygmaeus* (Hixson and Brown, 1986), *Gallus gallus* (Desjardins and Morais, 1990), *Xenopus laevis* (Dunon-Bluteau and Brun, 1986), and *Drosophila yakuba* (Clary and Wolstenholme, 1985). Estimates for divergence times are based on Schopf et al. (1983), Shoshani et al. (1985), Britten (1986), Hixson and Brown (1986), Carroll (1988), and references therein.

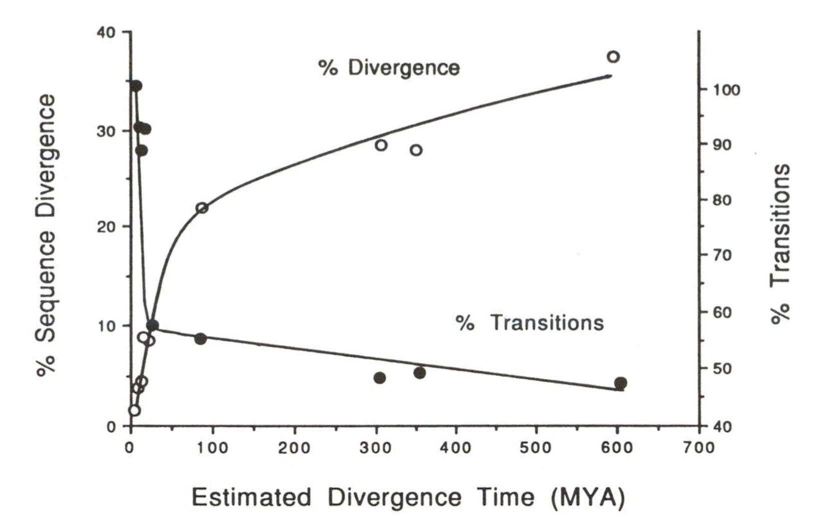

Figure 5-2 Percent sequence divergence for mitochondrial small-subunit rRNA, and percent transitions versus estimated divergence time (open circles and filled circles, respectively) for species pairwise comparisons. Data are from Table 5-1 and Hixson and Brown (1986). Points represent averages for species comparisons having the same estimated divergence times. Percent sequence divergence was calculated as given in Table 5–1.

clustered, the taxa involved in those comparisons would not be inferred as separated by short internodes within a tree. Thus, lack of resolution of their relationships could be more reasonably attributed to overabundance of "noise," rather than an absence of "signal."

Shuffling Input Order for Alignments

Demonstration

Once sets of taxa separated by short internodes have been identified, an obvious procedure in seeking to resolve their relative branching order is to shuffle the input order of their nucleotide sequences in subsequent multiple alignment bouts, and to examine resulting phylogenetic trees. If the resultant trees are the same, order of input of sequences has no major effect, and the relationships may be considered unbiased by alignment input order. I have taken this approach in alignment and phylogenetic analysis of 12*S* and 16*S* rRNA sequences from representatives of three mammalian orders: human (*Homo sapiens*), cow (*Bos taurus*), and mouse (*Mus musculus*), using the chicken (*Gallus gallus*) as an outgroup. The three mammalian (ingroup) species sequences were aligned in all three possible orders: (*Homo, Mus*)*Bos*; (*Homo, Bos*)*Mus*; (*Bos, Mus*)*Homo*. In each case, sequences for the first two species enclosed by parentheses, above, were aligned, and a consensus sequence was made from that alignment in which all variable nucleotide sites were overwritten as "N" (ambiguous). This consensus sequence was then aligned with the third ingroup species. Another consensus sequence was made, as above, from these three sequences combined, and then the outgroup was aligned to this second consensus sequence. Pairwise alignments were done using the algorithm of Myers and Miller (1988). Program parameters for "open gap cost" (cost for opening a gap in sequence) and "unit gap cost" (cost for increasing the size of the gap by one base) were both set to 10. Alignments done with these settings could not readily be improved by visual inspection and adjusting alignments by hand, whereas experimental alignments with various other settings could be improved after visual inspection. Phylogenetic analyses were done using the Phylogenetic Analysis Using Parsimony (PAUP) computer program (Swofford, 1989). This approach is reasonable for small numbers of species; however, as the number of variable sites increases with inclusion of more taxa, ambiguous ("N") designations in the consensus also increase, and too many such ambiguous scores will eventually confound the alignments. Use of "N" scores is conservative, and analyses could also be done in which variable nucleotide sites were overwritten with the appropriate International Union of Pure and Applied Chemistry (IUPAC) ambiguity scores combining the nucleotide variation observed. This would facilitate alignment of more sequences from more species.

To investigate the effects of the order of alignment of species' sequences, I constrained phylogenetic analyses to each of the three possible tree topologies, for each of the three different "alignment order" data sets. I have also employed two weighting strategies for the nucleotide characters, equal weighting for all nucleotide substitutions, and giving zero weight to (omitting) transitions (Fig. 5-3). Omitting transitions, however, is not meant to imply that they contain no phylogenetic information.

Interpretation

The phylogenetic analyses with shuffling of input order of alignments demonstrate that phylogenetic hypotheses can be quite sensitive to alignment order of species sequences. In analyses of 16*S* rRNA sequences, two different most-parsimonious (designated by scores with an asterisk in Figure 5-3) tree topologies were found under conditions of equal weighting for all characters. Considering transversions only, all three possible topologies were found as most parsimonious, varying with alignment order. There is an expected tendency for tree topology to reflect the order of sequence alignments. In 16*S* rRNA analyses of transversions only, branching order within most-parsimonious trees always corresponded with alignment order.

In contrast, 12*S* rRNA analysis under each weighting scheme consistently identifed a single tree topology as most parsimonious, regardless of alignment order. This convergence toward a single topology suggests that hypothesized relationships based on 12*S* rRNA are not an artifact of alignment order. 12*S* rRNA analyses with all nucleotide substitutions equally weighted unite *Bos* and *Mus* as sisters, as found by Cedergren et al. (1988) using a different alignment-phylogeny approach, whereas consideration of

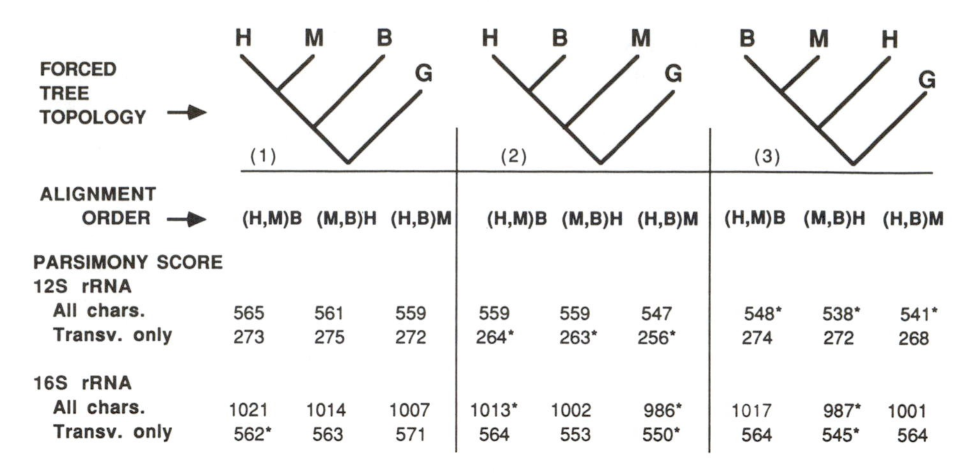

	(1) (H,M)B	(1) (M,B)H	(1) (H,B)M	(2) (H,M)B	(2) (M,B)H	(2) (H,B)M	(3) (H,M)B	(3) (M,B)H	(3) (H,B)M
PARSIMONY SCORE									
12S rRNA									
All chars.	565	561	559	559	559	547	548*	538*	541*
Transv. only	273	275	272	264*	263*	256*	274	272	268
16S rRNA									
All chars.	1021	1014	1007	1013*	1002	986*	1017	987*	1001
Transv. only	562*	563	571	564	553	550*	564	545*	564

Figure 5-3 Lengths for each of three possible phylogenetic trees (parsimony scores) using each of the three possible alignment orders (input order of sequences for multiple alignments) for *Homo sapiens* (H), *Bos taurus* (B), and *Mus musculus* (M) with *Gallus gallus* (G) as an outgroup. For each alignment order, mitochondrial rRNA sequences for two species enclosed by parentheses were aligned (Myers and Miller, 1988), and a consensus sequence was made from that alignment in which all variable nucleotide sites were overwritten as "N" (ambiguous). This consensus sequence was then aligned with the third ingroup species, another consensus sequence was made, and the outgroup sequence was aligned to that second consensus sequence. Parsimony scores are given for analyses under two weighting schemes, equal weight for all substitutions, and zero weight for transitions (transversions only). Asterisks denote the score for the most-parsimonious tree topology under the given alignment order and weighting scheme. Data sources for both small-subunit (12*S*) and large-subunit (16*S*) rRNA analyses are given in Table 5-1.

transversions only place *Homo* and *Bos* as sisters. The bias toward transitions in mitochondrial sequences (Brown et al., 1982; Thomas and Beckenbach, 1989), including rRNAs (Hixson and Brown, 1986), is the basis of the difference in tree topologies. The faster rate of accumulation for transitions makes them more susceptible to multiple substitutions at a given nucleotide position, and less informative for phylogeny as divergence time between species pairs increases. Based on small-subunit mitochondrial rRNA comparisons (Fig. 5-2), percent transitions begins leveling off at about 50%. Transitions begin to saturate with change, and become phylogenetically less informative for species divergences over about 20 million years old. Evolutionary patterns among eutherian orders are not well documented by fossils (Clemens et al., 1979; Novacek, 1982), and estimates for divergence times among the three mammalian orders represented here (Rodentia, Artiodactyla, Primates) vary from 55 to 90 million years ago based on both fossil and biochemical (e.g., Shoshani et al., 1985) evidence, but it is reasonable to suppose that, in the present analyses, transitions are less reliable than transversions. In this light, the best mitochondrial rRNA phylogenetic hypothesis is tree 2 in Figure 5-3, having *Homo* and *Bos* as sisters relative to *Mus*. This variously conflicts (Miyamoto and Goodman, 1986; Novacek, 1986) or agrees (McKenna, 1975; Easteal, 1990) with hypotheses based on other data sets.

Further support for the value of transversions in this case, and the phylogeny inferred from them, is found in the greater relative difference in tree length between the most-parsimonious tree and the other two trees for transversions alone, compared to that for trees based on all characters combined. For analyses presented in Figure 5-3, the sum of additional steps required to support the two less-parsimonious trees constituted 24 and 12% of the mean tree length for transversions only in the 12*S* and 16*S* rRNA genes, respectively, compared to 17 and 9%, in the 12*S* and 16*S* rRNA genes respectively, for all characters combined.

Assigning different weights to different types of nucleotide character change, as above, is unreasonable if it is done after phylogenetic analysis in order to "save" a particular set of relationships. However, it is reasonable where such weighting can be justified beforehand, based on knowledge of relative rates of character change as done here. Weighting could also have been done using a recursive, successive approximations approach (Carpenter, 1988). Because DNA sequence homologies, and hence types of character change (i.e., transitions or transversions), are based on quantitative alignment scores, and because there are only four possible nucleotide character-states, nonhomologous similarity will arise often. Differential weighting of types of molecular character change may be needed in many instances, lest homoplastic similarity overwhelm homologous similarity in phylogenetic analyses.

ACKNOWLEDGMENTS

I am grateful to Jim Carpenter, Joel Cracraft, Brent Mishler, and Michael M. Miyamoto for valuable criticisms of an earlier draft of this paper. Steven Sawchuck kindly assisted in conducting species sequence comparisons.

REFERENCES

Anderson, S., A. T. Bankier, B. G. Barrell, M. H. L. de Bruijn, A. R. Coulson, J. Drouin, I. C. Eperon, D. P. Nierlich, B. A. Roe, F. Sanger, P. H. Schreier, A. J. H. Smith, R. Staden, and I. G. Young. (1981) Sequence and organization of the human mitochondrial genome. *Nature* **290**:457–465.

Anderson, S., M. H. L. de Bruijn, A. R. Coulson, E. C. Eperon, R. Sanger, and I. G. Young. (1982) Complete sequence of bovine mitochondrial DNA: conserved features of the mammalian mitochondrial genome. *J. Mol. Biol.* **156**:683–717.

Bock, W. J. (1974) Philosophical foundations of classical evolutionary classification. *Syst. Zool.* **22**:375–392.

Britten, R. J. (1986) Rates of DNA sequence evolution differ between taxonomic groups. *Science* **231**:1393–1398.

Brown, W. M., E. M. Prager, A. Wang, and A. C. Wilson. (1982) Mitochondrial DNA sequences of primates: tempo and mode of evolution. *J. Mol. Evol.* **18**:225–239.

Carpenter, J. M. (1988) Choosing among multiple equally parsimonious cladograms. *Cladistics* **4**:291–296.

Carroll, R. L. (1988) *Vertebrate Paleontology and Evolution.* W. H. Freeman, New York.

Cavender, J. A. (1978) Taxonomy with confidence. *Math. Biosci.* **40**:271–280.

Cedergren, R., M. W. Gray, Y. Abel, and D. Sankoff. (1988) The evolutionary relationships among known life forms. *J. Mol. Evol.* **28**:98–112.

Clary, D. O., and D. R. Wolstenholme. (1985) The ribosomal RNA genes of *Drosophila* mitochondrial DNA. *Nucleic Acids Res.* **13**:4029–4045.

Clemens, W. A., J. A. Lillegraven, E. H. Lindsay, and G. G. Simpson. (1979) Where, when and what—a survey of known Mesozoic mammal distribution. Pp. 7–58 in *Mesozoic Mammals* (J. A. Lillegraven, Z. Kielan-Jaworowska, and A. Clemens, eds.). University of California Press, Berkeley.

Desjardins, P., and R. Morais. (1990) Sequence and gene organization of the chicken mitochondrial genome. *J. Mol. Biol.* **212**:599–634.

Dunon-Bluteau, D., and G. Brun. (1986) The secondary structures of the *Xenopus laevis* and human mitochondrial small ribosomal subunit RNA are similar. *FEBS Lett.* **198**:333–338.

Easteal, S. (1990) The pattern of mammalian evolution and the relative rate of molecular evolution. *Genetics* **124**:165–173.

Eldredge, N., and J. Cracraft. (1980) *Phylogenetic Patterns and the Evolutionary Process.* Columbia University Press, New York.

Felsenstein, J. (1978) Cases in which parsimony or compatibility methods will be positively misleading. *Syst. Zool.* **27**:401–410.

Feng, D., and R. F. Doolittle. (1987) Progressive sequence alignment as a prerequisite to correct phylogenetic trees. *J. Mol. Evol.* **25**:351–360.

Feng, D., and R. F. Doolittle. (1990) Progressive alignment and phylogenetic tree construction of protein sequences. Pp. 375–387 in *Molecular Evolution: Computer Analysis of Protein and Nucleic Acid Sequences* (R. F. Doolittle, ed.). Methods in Enzymology, vol. 183. Academic Press, New York.

Fitch, W. M. (1967) Evidence suggesting a non-random character to nucleotide replacements in naturally occurring mutations. *J. Mol. Biol.* **26**:499–507.

Fitch, W. M. (1970). Distinguishing homologous from analogous proteins. *Syst. Zool.* **19**:99–113.

Fitch, W. M., and E. Margoliash. (1967) Construction of phylogenetic trees. *Science* **155**:279–284.

Fitch, W. M., and T. F. Smith. (1983) Optimal sequence alignments. *Proc. Natl. Acad. Sci. USA* **80**:1382–1386.

Gilbert, W. (1978) Why genes in pieces? *Nature* **271**:501.

Gilbert, W. (1985) Genes-in-pieces revisited. *Science* **228**:823–824.

Gray, G. S., and W. M. Fitch. (1983) Evolution of antibiotic resistance genes: the DNA sequence of a kanamycin resistance gene from *Staphylococcus aureaus*. *Mol. Biol. Evol.* **1**:57–66.

Gray, M. W., D. Sankoff, and R. J. Cedergren. (1984) On the evolutionary descent of organisms and organelles: a global phylogeny based on a highly conserved structural core in small subunit ribosomal RNA. *Nucleic Acids Res.* **12**:5837–5852.

Hein, J. (1989a) A new method that simultaneously aligns and reconstructs ancestral sequences for any number of homologous sequences, when the phylogeny is given. *Mol. Biol. Evol.* **6**:649–668.

Hein, J. (1989b) A tree reconstruction method that is economical in the number of pairwise comparisons used. *Mol. Biol. Evol.* **6**:669–684.

Hein, J. (1990) Unified approach to alignment and phylogenies. Pp. 626–644 in *Molecular Evolution: Computer Analysis of Protein and Nucleic Acid Sequences* (R. F. Doolittle, ed.). Methods in Enzymology, vol 183. Academic Press, New York.

Hennig, W. (1966) *Phylogenetic Systematics*. University of Illinois Press, Urbana.

Higgins, D. G., and P. M. Sharp. (1988) CLUSTAL: a package for performing multiple sequence alignment on a microcomputer. *Gene* **73**:237–244.

Higgins, D. G., and P. M. Sharp. (1989) Fast and sensitive multiple sequence alignments on a microcomputer. *CABIOS* **5**:151–153.

Hixson, J. E., and W. M. Brown. (1986) A comparison of the small ribosomal RNA genes from the mitochondrial DNA of the great apes and humans: sequence, structure, evolution, and phylogenetic implications. *Mol. Biol. Evol.* **3**:1–18.

Hogeweg, P., and B. Hesper. (1984) The alignment of sets of sequences and the construction of phyletic trees: an integrated method. *J. Mol. Evol.* **20**:175–186.

Jardine, N. (1967) The concept of homology in biology. *Br. J. Philos. Sci.* **18**:125–139.

Kobayashi, M., T. Seki, K. Yaginuma, and K. Koike. (1981) Nucleotide sequences of small ribosomal RNA and adjacent transfer RNA genes in rat mitochondrial DNA. *Gene* **16**:297–307.

Konings, D. A. M., P. Hogeweg, and B. Hesper. (1987). Evolution of the primary and secondary structures of the E1a mRNAs of the adenovirus. *Mol. Biol. Evol.* **4**:300–314.

Lanyon, S. M. (1988) The stochastic mode of molecular evolution: what consequences for systematic investigations? *Auk* **105**:565–573.

Li, W.-H., C.-C. Luo, and C.-I. Wu. (1985) Evolution of DNA sequences. Pp. 1–94 in *Molecular Evolutionary Genetics*. (R. J. MacIntyre, ed.). Plenum Press, New York.

Li, W.-H., C.-I. Wu, and C.-C. Luo. (1984) Nonrandomness of point mutation as reflected in nucleotide substitutions in pseudogenes and its evolutionary implications. *J. Mol. Evol.* **21**:58–71.

McKenna, M. C. (1975) Toward a phylogenetic classification of the Mammalia. Pp. 21–46 in *Phylogeny of the Primates* (W. P. Luckett and F. S. Szalay, eds.). Plenum Press, New York.

Miyamoto, M. M., and M. Goodman. (1986) Biomolecular systematics of eutherian mammals: phylogenetic patterns and classification. *Syst. Zool.* **35**:230–240.

Myers, W. W., and W. Miller. (1988) Optimal alignments in linear space. *CABIOS* **4**:11–17.

Needleman, S. B., and C. D. Wunsch. (1970) A general method applicable to the search for similarities in the amino acid sequence of two proteins. *J. Mol. Biol.* **48**:443–453.

Novacek, M. J. (1982) Information for molecular studies from anatomical and fossil evidence on higher eutherian phylogeny. Pp. 3–41 in *Macromolecular Se-*

quences in Systematics and Evolutionary Biology (M. Goodman, ed.). Plenum Press, New York.

Novacek, M. J. (1986) The skull of leptictid insectivorans and the higher-level classification of eutherian mammals. *Bull. Am. Mus. Nat. Hist.* **183**:1–112.

Owen, R. (1848) *On the Archetype and Homologies of the Vertebrate Skeleton*. R. and J. E. Taylor, London.

Patterson, C. (1987) Introduction. Pp. 1–22 in *Molecules and Morphology in Evolution: Conflict or Compromise*? (C. Patterson, ed.). Cambridge University Press, London.

Platnick, N. I. (1979) Gaps and prediction in classification. *Syst. Zool.* **27**:472–474.

Reeck, G. R., C. de Haën, D. C. Teller, R. F. Doolittle, W. M. Fitch, R. E. Dickerson, P. Chambon, A. D. McLachlan, E. Margoliash, T. H. Jukes, and E. Zuckerkandl. (1987) "Homology" in proteins and nucleic acids: a terminological muddle and a way out of it. *Cell* **50**:667.

Remane, A. (1956) *Die Grundlagen des Naturlichen Systems der Vergleichenden Anatomie und Phylogenetik*. 2. Geest und Portik, Leipzig.

Riedl, R. (1978) *Order in Living Organisms. A Systems Analysis of Evolution*. John Wiley and Sons, New York.

Roth, V. L. (1984) On homology. *Biol. J. Linn. Soc.* **22**:13–29.

Roth, V. L. (1988) The biological basis of homology. Pp. 1–26 in *Ontogeny and Systematics* (C. J. Humphries, ed.). Columbia University Press, New York.

Sankoff, D. (1972) Matching sequences under deletion/insertion constraints. *Proc. Natl. Acad. Sci. USA* **69**:4–6.

Sankoff, D., R. J. Cedergren, and G. Lapalme. (1976) Frequency of insertion-deletion, transversion, and transition in the evolution of 5S ribosomal RNA. *J. Mol. Evol.* **7**:133–149.

Sankoff, D., and P. Rousseau. (1975) Locating the vertices of a Steiner tree in arbitrary space. *Math. Progr.* **9**:240–246.

Schopf, J. W., J. M. Hayes, and M. R. Walter. (1983) Evolution of earth's earliest ecosystems: recent progress and unsolved problems. Pp. 361–384 in *Earth's Earliest Biosphere: Its Origin and Evolution* (J. W. Schopf, ed.). Princeton University Press, Princeton.

Sellers, P. (1974) An algorithm for the distance between two finite sequences. *J. Combination Theory* **16**:253–258.

Shoshani, J., M. Goodman, J. Czelusniak, and G. Braunitzer. (1985) A phylogeny of Rodentia and other eutherian orders: parsimony analysis utilizing amino acid sequences of alpha and beta hemoglobin chains. Pp. 191–210 in *Evolutionary Relationships Among Rodents: A Multidisciplinary Analysis*. (W. Patrick and J.-L. Hartenberger, eds.). Plenum Press, New York.

Swofford, D. L. (1989) *PAUP: Phylogenetic Analysis Using Parsimony*, Version 3.0. University of Illinois, Champaign.

Tanhauser, S. M. (1985) *Evolution of Mitochondrial DNA: Patterns and Rate of Change*. Ph.D. Dissertation, University of Florida, Gainesville.

Thomas, W. K., and A. T. Beckenbach. (1989) Variation in salmonid mitochondrial DNA: evolutionary constraints and mechanisms of substitution. *J. Mol. Evol.* **29**:233–245.

Van Etten, R. A., M. W. Walberg, and D. A. Clayton. (1980) Precise localization and nucleotide sequence of the mouse mitochondrial rRNA genes and three immediately adjacent novel tRNA genes. *Cell* **22**:157–170.

Van Valen, L. M. (1982) Homology and causes. *J. Morphology*. **173**:305–312.

Waterman, M. S., and R. Jones. (1990) Consensus methods for DNA and protein sequence alignment. Pp. 221–237 in *Molecular Evolution: Computer Analysis of Protein and Nucleic Acid Sequences* (R. F. Doolittle, ed.). Methods in Enzymology, vol. 183. Academic Press, New York.

Wiley, E. O. (1975) Karl R. Popper, systematics, and classification: a reply to Walter Bock and other evolutionary taxonomists. *Syst. Zool.* **24**:233–243.

Wu, C.-I., and W.-H. Li. (1985) Evidence for higher rates of nucleotide substitution in rodents than in man. *Proc. Natl. Acad. Sci. USA* **82**:1741–1745.

6

Relative Efficiencies of Different Tree-Making Methods for Molecular Data

MASATOSHI NEI

There are many different tree-making methods that can be used for molecular data (Nei, 1987a; Felsenstein, 1988). Each of these methods has some advantages and disadvantages, and the overall relative efficiencies of the methods in recovering the correct phylogenetic tree are still controversial. The major problem in studying the relative efficiencies is that the true tree is usually unknown for any set of real organisms or any set of real DNA sequences, so that it is difficult to judge which tree is the correct one. However, this problem can be avoided if we use computer simulation. In the case of molecular data, particularly DNA sequences, the pattern of evolutionary change is well-understood so that it is possible to simulate it for any given phylogenetic tree. Sequences of DNA or other molecular data thus generated can be used for reconstructing the tree, and the efficiency of a tree-making method can be studied by examining how often it recovers the correct tree.

This type of study was first conducted by Peacock and Boulter (1975) who considered the evolution of amino acid sequences. However, they compared only two different tree-making methods considering a constant rate of evolution. Furthermore, they did not consider the property of the genetic code in their simulation of amino acid changes. Therefore, their conclusion has limited applicability to actual data. In the late 1970s we initiated a comprehensive study of this problem, considering DNA sequences (Tateno et al., 1982). Initially, we studied relatively simple tree-making methods but later extended our analysis to other methods. Similar studies have also been done by a number of other authors. In this chapter, a summary of the results of these studies is presented. Before the discussion of these results, however, the theoretical basis of each tree-making method that is used for molecular data will be presented.

TREE-MAKING METHODS

As mentioned above, there are many methods for constructing trees from molecular data. According to the type of data used, they can be divided into two categories; that is, distance methods and discrete-character methods. In distance methods, evolutionary distance is computed for all pairs of operational taxonomic units (OTUs; species or populations) or DNA (or amino acid) sequences, and a phylogenetic tree is constructed by considering the relationship among these distances. Once distance values are obtained, there are several ways of obtaining a tree. In discrete-character methods, data with discrete character-states such as nucleotide states in DNA sequences are used, and a tree is constructed by considering the evolutionary relationships of OTUs or DNA sequences at each character or nucleotide position.

It should be noted that some types of molecular data (e.g., DNA hybridization data) exist only as distance data. Therefore, phylogenetic trees for these data can be constructed only by distance methods. By contrast, discrete-character data can usually be converted into distance data, so that they can be analyzed either by distance methods or by discrete-character methods. Some authors (e.g., Farris, 1981; Penny, 1982) have argued that distance methods are inherently inferior to discrete-character methods (e.g., parsimony methods), but their arguments are apparently based on misconceptions of distance methods (Felsenstein, 1986; Nei, 1987b). Actually, some distance methods are often superior to parsimony methods in obtaining the correct tree, as will be mentioned below.

For both distance and discrete-character data, there are several different tree-making methods that are based on different principles. The major ones are as follows.

Distance Data

Average Distance (Linkage) Method [*Unweighted Pair Group Method with Arithmetic Means (UPGMA)*]

The original idea of this method was presented by Sokal and Michener (1958), but their procedure is different from the one currently in use. A formal presentation of this method is given in Sneath and Sokal (1973). To construct a phylogenetic tree by this method, it is necessary to assume a constant rate of evolution.

Transformed Distance (TD) Method

Farris (1977) showed that if a distance matrix satisfies the condition of an additive tree (all substitutions being counted) the UPGMA procedure produces the correct tree topology by using the following transformed distance for species i and j.

$$d_{ij}' = (d_{ij} - d_{ir} - d_{jr})/2 + c, \tag{1}$$

where r stands for a reference species (an outgroup or any given ingroup species) and c is a constant to make d_{ij}' positive. The TD method works well whether the evolutionary rate varies from branch to branch. In practice, of course, the condition of an additive tree is not usually satisfied, so this method may produce a wrong topology. Branch lengths cannot be estimated by this method. However, once a topology is obtained, one can easily estimate branch lengths by using Fitch and Margoliash's (1967) method mentioned below, or some other method. There are several different algorithms for this method (Klotz et al., 1979; Klotz and Blanken, 1981; Li, 1981).

Fitch and Margoliash's (FM) Method

Let d_{ij} be the observed distance between species i and j and e_{ij} be the corresponding patristic distance, which is the sum of the lengths of all branches connecting the two species in a reconstructed tree. Fitch and Margoliash's (1967) method chooses a tree that minimizes the following quantity.

$$s_{\mathrm{FM}} = \left[\frac{2\sum_{i<j}\{(d_{ij} - e_{ij})/d_{ij}\}^2}{n(n-1)}\right]^{1/2} \times 100, \tag{2}$$

where n is the number of species studied. A similar least-squares method was also proposed by Cavalli-Sforza and Edwards (1967). In this method the s_{FM} value must be computed for all or a large number of topologies.

Minimum Evolution (ME) Method

In an unrooted bifurcating tree of n species, there are $2n - 3$ possible branches. Let ℓ_i be the branch length of the i-th branch for a given topology. One can then compute the sum of all branches. That is,

$$L = \sum_i \ell_i. \tag{3}$$

In the ME method, this quantity is computed for all possible topologies, and the topology that shows the minimum L value is chosen as the final tree. This method was originally presented by Cavalli-Sforza and Edwards (1967), but their method of computing ℓ_i was very complicated. Recently, Saitou and Imanishi (1989) simplified the computation considerably by using Fitch and Margoliash's (1967) approach.

One might think that this method is essentially the same as the maximum parsimony method mentioned below. This is not true. Unlike the parsimony method, this method is not affected by parallel or backward mutations, as long as the evolutionary distance is properly measured. It recovers the correct tree if the distance matrix satisfies the condition of an additive tree.

Distance Wagner (DW) Method

This method was originally presented as a distance version of the Wagner parsimony method (Farris, 1972). Farris, therefore, suggested that a measure called "metric" (a distance measure that satisfies the principle of the triangle inequality) be used for this method. However, this requirement does not necessarily improve the performance of this method (Nei et al., 1983; Sourdis and Krimbas, 1987). Tateno et al. (1982) modified this method to make it more appropriate for molecular data. Swofford (1981) and Faith (1985) also proposed modified versions of this method. Comparison of different topologies is built in the method, and the final tree that is supposed to be the best one is automatically produced. This method gives both topology and branch lengths of the final tree.

Neighborliness (ST) Method

The principle of the neighborliness (ST) method (Sattath and Tversky, 1977; Fitch, 1981) is to use Buneman's (1971) four-point condition for an additive tree. Consider a tree with n (≥ 4) OTUs and assume that OTUs 1 and 2 are a pair of neighbors, that is, two OTUs connected through a single interior node in a bifurcating tree (see Saitou and Nei, 1987). Let d_{ij} be the distance between OTUs i and j. We then have the following inequalities for an additive tree.

$$d_{12} + d_{ij} < d_{1i} + d_{2j}, \qquad d_{12} + d_{ij} < d_{1j} + d_{2i}, \tag{4}$$

where i and j are any OTUs ($3 \leq i < j \leq n$). For actual data, the above condition may not hold because of disturbance of additivity of distances. Sattath and Tversky (1977), and Fitch (1981) then proposed algorithms to construct a tree by maximizing the number of cases in which condition (4) holds. In practice, it is easier to use the ST algorithm of Sattath and Tversky (1977).

Neighbor-Joining (NJ) Method

The principle of this method is the same as that of the ME method, but the computational process is much simpler (Saitou and Nei, 1987). Comparison of different topologies is built in the algorithm, and the final tree with both topology and branch lengths is automatically produced. Swofford and Olson (1990) recently stated that this method is for estimating an additive tree, but this statement is incorrect.

Discrete-Character Data

Maximum Parsimony (MP) Method

In this method, the DNA (or amino acid) sequences of ancestral species are inferred from those of extant species, considering a particular tree topology, and the minimum number of evolutionary changes that are required to explain all the observed differences among the sequences is computed. This number is obtained for all possible topologies, and the topology which shows the smallest number of evolutionary changes is chosen as the final tree. This method is used mainly for finding the topology of a tree, but branch lengths can be estimated under certain assumptions (Fitch, 1971).

When the MP method is applied to morphological characters, it is customary to assume that the primitive and derived character-states are known. In the case of molecular data, this assumption generally does not hold, and different character-states are often reversible. It is, therefore, important to use the MP method, which permits reversible mutations (Eck and Dayhoff, 1966; Fitch, 1971). In numerical taxonomy, this type of MP method is sometimes called the Wagner parsimony method (Farris, 1970).

Evolutionary Parsimony (EP) Method

This method is primarily applied to four DNA (or RNA) sequences and utilizes information on the transition/transversion bias in nucleotide substitution (Lake, 1987). The actual procedure is to compute three quantities, X, Y, and Z, which are functions of the numbers of certain nucleotide configurations among the four DNA sequences, and to determine which of the three quantities is significantly different from 0. If only one of them is significant, the tree topology corresponding to the quantity is regarded as the correct one. If two or all of X, Y, and Z are significant, the splitting pattern of DNA sequences is considered unresolvable. A procedure for extending this method to the case of five or more sequences is presented by Lake (1988).

Maximum Likelihood (ML) Method

In this method, the nucleotides of all DNA sequences at each nucleotide site are considered separately, and the log-likelihood of having these nucleotides are computed for a given topology by using a particular probability model (Felsenstein, 1981a). This log-likelihood is added for all nucleotide sites, and the sum of the log-likelihood is maximized to estimate the branch length of the tree. This procedure is repeated for all possible topologies, and the topology that shows the highest likelihood is chosen as the final one.

METHODS OF COMPUTER SIMULATION

Generation of Simulated Sequences

In the case of DNA sequences, we know the basic rules of their evolutionary change (Nei, 1987a: 79–88). There are two types of changes (i.e., nucleotide substitution and deletion/insertion). The latter type of change occurs haphazardly, and it is difficult to quantify the amount of change for a given evolutionary time. Therefore, deletions and insertions are often neglected in the reconstruction of phylogenetic trees. By contrast, the pattern of nucleotide substitution has been studied extensively, and there are several mathematical models that predict the evolutionary change of DNA sequences. These models are, of course, approximations to real changes of DNA sequences, but their predictions are quite accurate unless the number of nucleotide substitutions per site is very high. Therefore, it is possible to simulate the evolutionary change of DNA sequences.

The basic information for modeling the evolutionary change of DNA sequences is the probability of change of a nucleotide (A, T, C, or G) to another nucleotide during a short period of time (e.g., one year, one generation, 1,000 years, etc.). Let λ_{ij} be the probability that the i-th nucleotide changes to the j-th nucleotide during the unit evolutionary time. We then have the following transition matrix:

$$\mathbf{M} = \begin{bmatrix} 1-\lambda_1 & \lambda_{21} & \lambda_{31} & \lambda_{41} \\ \lambda_{12} & 1-\lambda_2 & \lambda_{32} & \lambda_{42} \\ \lambda_{13} & \lambda_{23} & 1-\lambda_3 & \lambda_{43} \\ \lambda_{14} & \lambda_{24} & \lambda_{34} & 1-\lambda_4 \end{bmatrix} \tag{5}$$

where $\lambda_1 = \lambda_{12} + \lambda_{13} + \lambda_{14}$, $\lambda_2 = \lambda_{21} + \lambda_{23} + \lambda_{24}$, $\lambda_3 = \lambda_{31} + \lambda_{32} + \lambda_{34}$, and $\lambda_4 = \lambda_{41} + \lambda_{42} + \lambda_{43}$.

Let g_i be the probability that a given nucleotide site in a DNA sequence is occupied by the i-th nucleotide at a given evolutionary time, and g be the column vector of g_1, g_2, g_3, and g_4. We designate g at time t by g_t. It is then possible to compute g_t by $g_t = \mathbf{M}^t g_0$, where $\mathbf{M}$ is the matrix given by (5) and g_0 is the vector g at time 0. Therefore, if g_0 is given for all nucleotide sites of the DNA sequence under consideration, one can generate a DNA sequence at time t. Note that all nucleotide changes are probabilistic. Therefore, the sequence generated for time t may vary from replication to replication.

When λ_{ij}'s are all the same and are equal to λ, and the initial sequence is a random sequence, the evolutionary change of DNA sequence becomes much simpler. Consider two DNA sequences that diverged t time-units ago. The probability that the two sequences have different nucleotides at a given site is given by

$$P = \frac{3}{4}(1 - e^{-8\lambda t/3}) \tag{6}$$

(Jukes and Cantor, 1969). Similarly, the probability that a descendent sequence is different from the ancestral sequence at a site becomes

$$P_A = \frac{3}{4}(1 - e^{-4\lambda t/3}). \tag{7}$$

Therefore, using pseudorandom numbers, one can easily generate a DNA sequence at any time. We call this the one-parameter model.

Incidentially, equation (6) is useful for deriving an estimator of the total number of nucleotide substitutions per site ($d = 2\lambda t$) between two sequences that diverged t time-units ago. It is given by

$$\hat{d} = -\frac{3}{4}\log_e\left(1 - \frac{4}{3}p\right), \tag{8}$$

where p is the proportion of different nucleotides between the two sequences compared.

Actual data on nucleotide substitution indicate that transitional changes (A $\rightleftarrows$ G and T $\rightleftarrows$ C) are more frequent than transversional changes (all other changes). Considering this difference, Kimura (1980) developed a two-parameter model in which transitional and transversional changes can be treated separately. If we use this model, the probability that two DNA sequences that diverged t time-units ago show a transitional difference at a given site becomes

$$P = \frac{1}{4} - \frac{1}{2}e^{-4(\alpha + \beta)t} + \frac{1}{4}e^{-8\beta t}, \tag{9}$$

whereas the probability that they show a transversional difference is

$$Q = \frac{1}{2} - \frac{1}{2}e^{-8\beta t}. \tag{10}$$

Here α and β denote the rates per unit of evolutionary time of transitional and transversional changes, respectively. These equations can be used for simulating the evolutionary change of DNA sequences. The two-parameter model can also be used for estimating the number of nucleotide substitutions between two sequences. It can be estimated by

$$\begin{aligned} \hat{d} &= 2(\alpha + 2\beta)t \\ &= -\frac{1}{2}\log_e [(1 - 2P - Q)\sqrt{1 - 2Q}] \end{aligned} \tag{11}$$

(Kimura, 1980).

The transition probabilities λ_{ij}'s in equation (5) have been estimated for a number of groups of genes (e.g., Brown et al., 1982; Gojobori et al., 1982; Aquadro and Greenberg, 1983; Graur, 1985; Saitou, 1987). Generally speaking, transitional changes are more frequent than transversional changes, but actual substitution patterns are more complicated than the two-parameter model. For this reason, several more complicated models have been developed (see Nei, 1987a). However, both the one- and two-parameter models are known to be quite robust. For estimating the number of nucleotide subsitutions (d), equation (8) gives a good estimate, provided that d is less than 0.5 and the transition/transversion bias is not extreme, whereas the two-parameter model seems to work even for a higher value of d if the number of nucleotides examined is large. For this reason, the one- or two-parameter model is often used for computer simulation. If this is not satisfactory, the matrix method presented earlier can be used (Jin and Nei, 1990).

Recent data (Britten, 1986; Li et al., 1987b) suggest that the rate of nucleotide substitution varies with evolutionary lineage. This can easily be simulated by changing λ_{ij}'s (or α and β) with evolutionary lineage. Mathematically, however, the same result is obtained by changing t rather than λ_{ij}'s, and this is usually much simpler.

Another important aspect of nucleotide substitution is that the substituion rate varies among nucleotide sites (Shoemaker and Fitch, 1989). The first and second nucleotide positions of codons are usually less variable than the third positions. Nucleotides in the active center regions of genes are also usually less variable (Kimura, 1983). In some exceptional genes such as major histocompatibility complex genes, the active center evolves faster (Nei and Hughes, 1991). This problem can also be resolved by considering spatially varying models of nucleotide substitution (Gojobori et al., 1982; Tateno et al., 1982; Olson, 1987; Jin and Nei, 1990). A commonly used method is to assume that λ_{ij}'s (or α and β) vary according to a gamma-distribution or a lognormal distribution.

Model Trees and Estimation of the Trees

As mentioned earlier, if a model tree is given, we can simulate the evolutionary changes of DNA sequences following the tree and generate the present-day sequences. Once these sequences are obtained, a tree is reconstructed by each tree-making method under consideration. In the case of distance methods, the evolutionary distance between each pair of sequences must be computed. This distance can be computed by various methods. When the number of nucleotide differences per site (p) is small for all pairs of sequences, this number can be used as a distance measure. We call this the p distance. When p is large, however, p is not a good estimate of the number of nucleotide substitutions (d) because of backward and parallel substitutions. In this case, d, is often estimated by equation (8). We call this the d distance in this chapter, whereas the d value estimated by equation (11) will be called the Kimura distance. However, there are many other ways to estimate d, as mentioned earlier. Obviously, the most desirable distance measure is the one that gives the best estimate of the total number of nucleotide substitutions. In the case of discrete-character models, the configuration of nucleotides among all sequences is considered at each nucleotide site.

Since the evolutionary change of nucleotides is stochastic, the DNA sequence generated for a given species varies from replication to replication. Therefore, computer simulation is repeated many times for a given model tree, and the agreement between reconstructed trees and the model tree is statistically determined.

There are two types of deviations of a reconstructed tree from the model tree. One is topological error, and the other is the deviation of estimated branch lengths from the true lengths. Topological error can be evaluated by two measures. One is the proportion of replications in which the correct topology is obtained. This is called the probability of obtaining the correct topology (P_c). The other measure is the topological distance between a reconstructed tree and the model tree. This distance, Tateno et al.'s (1982) distortion index (d_T), is measured by Robinson and Foulds' (1981) index. Roughly speaking, this index is twice the number of branch interchanges that are required to transform the topology of a tree to that of another. When simulation is replicated, the mean topological distance ($\bar{d}_T$) for all replications is used as a criterion. In general, $\bar{d}_T$ is highly negatively correlated to P_c. Therefore, many authors have used P_c only.

Branch-length errors can also be measured by several different methods (Tateno et al., 1982). In this chapter, however, this problem will not be considered.

RESULTS FROM COMPUTER SIMULATION

As mentioned earlier, relative efficiencies of tree-making methods have been studied by many authors. Each author or group of authors has studied a different set of tree-making methods. However, several tree-making methods were examined repeatedly by different groups of authors; in these cases the results obtained were usually consistent. In this review, the general conclusions obtained from these studies (without going into details) are presented.

Computer simulation of the reconstruction (estimation) of phylogenetic trees is quite time consuming, particularly when the number of DNA sequences is large. Therefore, the number of DNA sequences used is usually six to eight, though some authors (e.g., Tateno et al., 1982) examined up to 32 sequences. The shape (topology and branch lengths) of the model tree examined varies among authors, but they can be divided roughly into the two types (A and B) given in Figure 6-1. Some methods show a better performance in topology A, and others in topology B. In general, however,

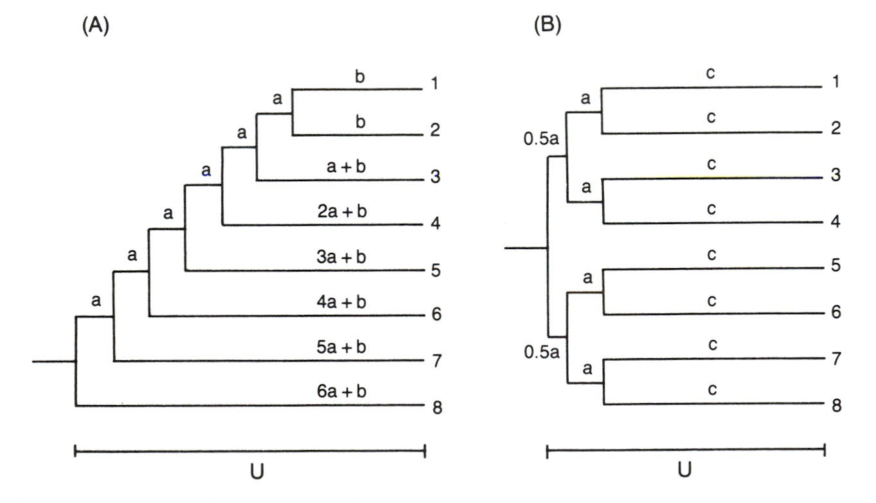

Figure 6-1 Model trees (A) and (B) under the assumption of constant rate of nucleotide substitution. U is the expected number of substitutions per site from the common ancestral sequence to the extant sequences.

the relative merits of different tree-making methods are similar for both topologies (A and B).

In the following discussion, the cases of constant and varying rates of nucleotide substitution will be considered separately.

Constant Rate of Substitution

Number of Nucleotide Substitutions—Effect on Recovery of Correct Topology

When the expected number of nucleotide substitutions per site from the common ancestral sequence to the extant sequence (U) *is small (see Fig. 6-1), the FM, DW, and Tateno et al.'s modified Farris (MF) methods are more efficient than UPGMA in recovering the correct topology. However, when* U *is large and each branch length is large, UPGMA is better than the former methods.*

This was first shown by Tateno et al. (1982), but more general results were obtained by Sourdis and Krimbas (1987). Table 6-1 shows the probabilities of obtaining the correct tree (P_c) for various tree-making methods when model tree A in Figure 6-1 is used. When the number of nucleotides used is small and U is small, any method gives a very small P_c value, as expected. This is true whether rooted or unrooted trees are constructed. However, as long as U is equal to 0.1 or less, the FM, DW, and MF methods are better than UPGMA in obtaining the correct rooted or unrooted tree. (The root for the trees obtained by the FM, DW, and MF methods was given as the midpoint of the longest route connecting two sequences.) The former three methods show more or less the same P_c value. When U is large (e.g., $U = 0.4$), however, UPGMA often shows a higher P_c value than the other methods. Somewhat similar results were also obtained by Blanken et al. (1982). Table 6-1 shows that P_c's for rooted trees are considerably smaller than those for unrooted trees, particularly when the FM, DW, and MF methods are used. This indicates that a substantial amount of error in constructing a rooted tree occurs at the time of rooting.

Comparison of Efficiency of ST and NJ Methods with Other Methods

The ST and NJ methods are almost always better than the UPGMA, DW, MF, and TD methods whether the U *value is small or large.*

Table 6-2 shows the results of another computer simulation for a different set of tree-making methods. This represents the case where U is relatively small ($U = 6a + b = 0.1$ in model tree A, and $U = 1.5a + c = 0.085$ in model tree B of Figure 6-1). In this case, the MF and DW methods are again better than UPGMA in obtaining the correct tree. When tree A is used, they are as efficient as the TD (Li's algorithm), ST, and NJ methods. However, when tree B is used, they are less efficient than the latter methods. Table 6-3 shows the P_c values for the case of a larger U value ($U = 0.52$ in tree A, and $U = 0.465$ in tree B) and a larger number of nucleotides

Table 6-1 Probabilities of Obtaining the Correct Topology ($P_c \times 100$) for Rooted and Unrooted Trees for Four Different Tree-Making Methods in the Case of $a = b$ in Model Tree A of Figure 6-1*

		Rooted Trees				Unrooted Trees			
U	R	UPGMA	FM	DW	MF	UPGMA	FM	DW	MF
				$m = 300$					
0.03	250	4	10	6	10	9	24	21	26
0.05	275	12	19	15	17	23	49	44	48
0.10	275	34	42	34	39	51	70	72	71
0.20	375	56	55	41	52	67	79	82	79
0.40	175	69	53	40	50	83	76	77	73
				$m = 900$					
0.03	275	38	43	42	44	48	82	80	81
0.05	275	60	64	61	64	74	94	95	94
0.10	275	88	84	75	84	92	99	99	99
0.20	250	96	91	85	91	98	98	100	99
0.40	100	99	95	82	95	100	99	99	100
				$m = 1,500$					
0.03	275	60	66	65	66	74	95	96	98
0.05	275	84	84	81	84	88	99	100	100
0.10	275	92	94	90	94	95	100	100	100
0.20	275	99	99	93	99	99	100	100	100
0.40	125	100	95	95	99	100	100	100	100

*The abbreviations used are: U, expected number of nucleotide substitutions per site from the ancestral sequence to the extant sequence; m, number of nucleotides used; R, number of replications; FM, Fitch and Margoliash's method; DW, distance Wagner method; MF, modified Farris method. Trees were constructed by using the d distance. (Adapted from Sourdis and Krimbas, 1987.)

Table 6-2 Probabilities of Obtaining the Correct Unrooted Tree ($P_c \times 100$) and Average Indices ($\bar{d}_T$) of Topological Errors (in Parentheses) for Six Tree-Making Methods for the Case of $a = 0.01$, $b = 0.04$, and $c = 0.07$ in Model Trees in Figure 6-1*

	Model Tree A			Model Tree B		
m	300	600	900	300	600	900
UP: p	14 (3.18)	36 (1.72)	58 (0.98)	14 (4.54)	36 (2.74)	51 (1.68)
UP: d	15 (3.18)	34 (1.74)	56 (1.04)	13 (4.56)	35 (2.70)	52 (1.60)
MF: p	39 (1.76)	73 (0.58)	95 (0.10)	24 (2.86)	51 (1.30)	67 (0.76)
MF: d	38 (1.92)	72 (0.62)	95 (0.10)	19 (2.94)	48 (1.42)	64 (0.86)
DW: p	42 (1.70)	75 (0.54)	96 (0.08)	26 (2.36)	55 (1.12)	79 (0.48)
DW: d	37 (1.74)	74 (0.58)	95 (0.10)	28 (2.36)	58 (1.06)	79 (0.46)
TD: p	41 (1.58)	71 (0.70)	94 (0.12)	40 (2.04)	70 (0.78)	90 (0.22)
TD: d	36 (1.84)	66 (0.82)	89 (0.24)	39 (2.10)	70 (0.78)	90 (0.26)
ST: p	48 (1.26)	75 (0.54)	97 (0.06)	45 (1.66)	75 (0.62)	91 (0.22)
ST: d	44 (1.48)	70 (0.62)	96 (0.08)	43 (1.62)	74 (0.64)	91 (0.22)
NJ: p	48 (1.36)	76 (0.54)	97 (0.06)	46 (1.64)	76 (0.60)	91 (0.20)
NJ: d	41 (1.60)	70 (0.62)	96 (0.08)	45 (1.62)	75 (0.60)	91 (0.20)

*The abbreviations used are: UP, UPGMA; MF, modified Farris method; DW, distance Wagner method; TD, transformed distance method, Li's algorithm was used; ST, Neighborliness, Sattath and Tversky's method; NJ, neighbor-joining method; p, trees reconstructed from p distances; d, trees reconstructed from d distances; m, number of nucleotides used. Number of replications (R) is 100. (Adapted from Saitou and Nei, 1987.)

Table 6-3 Probabilities of Obtaining the Correct Unrooted Tree ($P_c \times 100$) and Average Indices ($\bar{d}_T$) of Topological Errors (in Parentheses) for Six Tree-Making Methods for the Case of $a = 0.03$, $b = 0.34$, and $c = 0.42$ in Model Trees in Figure 1*

	Model Tree A			Model Tree B		
m	500	1,000	2,000	500	1,000	2,000
UP: *p*	9 (3.78)	27 (2.10)	62 (0.86)	10 (5.20)	18 (3.76)	54 (1.32)
UP: *d*	9 (3.78)	27 (2.10)	62 (0.88)	11 (5.30)	18 (3.74)	55 (1.26)
MF: *p*	15 (4.02)	41 (1.82)	62 (0.92)	3 (5.68)	17 (3.64)	28 (2.40)
MF: *d*	13 (4.42)	34 (2.14)	55 (1.14)	3 (5.72)	13 (3.80)	26 (2.48)
DW: *p*	16 (3.78)	46 (1.54)	63 (0.82)	4 (5.42)	18 (3.28)	41 (1.72)
DW: *d*	15 (4.22)	40 (1.96)	58 (0.98)	5 (5.50)	18 (3.48)	35 (1.82)
TD: *p*	3 (4.26)	37 (2.00)	53 (1.18)	15 (4.48)	28 (2.98)	70 (0.90)
TD: *d*	3 (4.84)	25 (2.60)	39 (1.66)	12 (4.72)	27 (3.06)	66 (1.02)
ST: *p*	10 (3.56)	44 (1.62)	68 (0.76)	13 (4.00)	36 (2.34)	74 (0.62)
ST: *d*	6 (4.06)	40 (1.82)	56 (1.04)	10 (4.32)	34 (2.34)	71 (0.72)
NJ: *p*	11 (3.70)	44 (1.68)	67 (0.80)	13 (4.46)	34 (2.38)	75 (0.62)
NJ: *d*	5 (4.24)	38 (2.00)	57 (1.06)	14 (4.44)	32 (2.42)	73 (0.72)

*Abbreviations used are the same as those in Table 6-2. $R = 100$. (Adapted from Saitou and Nei, 1987.)

examined. In this case, the MF and DW methods are as efficient as the ST and NJ methods when tree A is used, but are less efficient than the latter when tree B is used. The TD method is slightly less efficient than the ST and NJ methods for both trees A and B.

It is known that the principle of triangle inequality applies to the p distance but not to the d distance. Therefore it is interesting to compare the P_c values for the trees obtained by using the p and d distances. Tables 6-2 and 6-3 show that there is no real difference in P_c between the p and d distances when UPGMA is used. In other tree-making methods, the p distance tends to give a slightly higher P_c than the d distance. However, the difference between them is generally small, except when the number of nucleotides examined (m) is small. As will be shown later, when the rate of nucleotide substitution varies with evolutionary lineage, and p or d is large, the d distance shows a better performance than the p distance.

Comparison of the MP Method with the ST, NJ, DW, and TD Methods

The MP method generally has a smaller P_c *value than the ST and NJ methods. However, when* U *is small and the number of nucleotides examined is very large, the MP method is as good as or slightly better than the latter.*

As mentioned earlier, the MP method requires examination of all possible topologies, and since the number of possible topologies rapidly increases as the number of OTUs increases, computer simulation has been done for a relatively small number of OTUs. For example, the number of possible topologies is 105 when the number of OTUs is six, whereas it is 10,395 when the number is eight. Table 6-4 shows the P_c values for the model trees A and B in Figure 6-2. When model tree A is used and the number of nucleotide substitutions is small ($U = 0.05$), the NJ and TD methods are better than the MP and DW methods provided that the number of nucleotides examined is ≤ 600. When the number of nucleotides is large

Table 6-4 Probabilities of Obtaining the Correct Tree ($P_c \times 100$) for the Maximum-Parsimony (MP) and Other Tree-Making Methods for Model Trees A and B in Figure 6-2*

	Constant Rate (Tree A)			Varying Rate (Tree B)		
m	300	600	1,200	300	600	1,200
			$U = 0.05$			
MP	46.3	79.0	97.0	16.7	39.3	58.0
NJ	58.7	82.0	95.3	30.0	46.7	63.3
DW	47.7	72.0	92.0	19.2	35.0	52.0
TD	54.7	82.0	95.0	25.3	42.3	62.3
			$U = 0.5$			
MP	44.7	69.3	84.0	16.7	22.3	31.0
NJ	62.7	84.3	97.0	25.0	34.7	49.3
DW	49.0	73.0	91.3	16.0	30.7	40.0
TD	64.3	84.3	96.0	25.0	33.3	48.0

*Abbreviations used are the same as those in Table 6-2. MP, maximum parsimony. $R = 300$. (Adapted from Sourdis and Nei, 1988.)

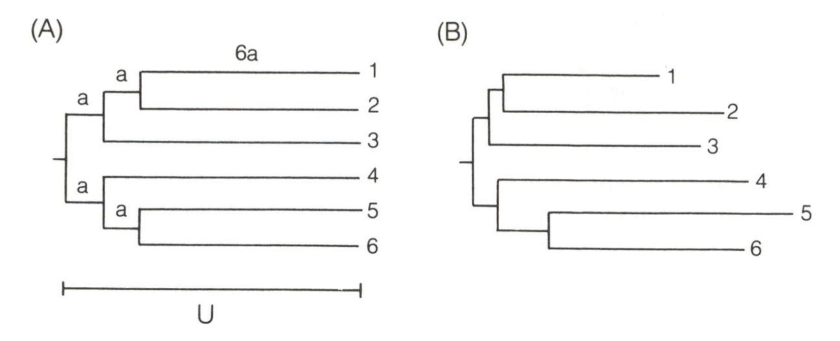

Figure 6-2 Model trees for six DNA sequences. (A) Case of constant rate of nucleotide substitution. (B) Case where the substitution rate varies with evolutionary lineage.

(m = 1,200), however, the MP method seems to be as good as or slightly better than the other methods.

When the number of substitutions per site is large (i.e., $U = 0.5$), the story is different. In this case, even for $m = 1{,}200$, the MP method shows a smaller P_c value than the NJ and DW methods. This is apparently because there are many backward and parallel substitutions in this case. Distance methods are not affected by these substitutions as long as the distances are correctly estimated.

The low performance of the MP method when m and U are both small is due to the fact that in this method only so-called "informative sites" are used, whereas in distance methods information for all variable sites is used. (Note that noninformative variable sites in parsimony methods are actually informative for constructing a tree in distance methods). When m and U are small, the number of "informative sites" is also small. Therefore, the MP method tends to produce many equally parsimonious trees, which often (but not always) include the correct tree (Sourdis and Nei, 1988). In our study, the case where the correct tree and other tied trees were obtained was not included in the probability of obtaining the correct tree. This is part of the reason why P_c became smaller in the MP method than in the NJ and other methods. When m is large, however, the number of "informative sites" becomes large, thus, P_c increases.

In this connection, one might argue that P_c is not an appropriate measure for comparing the MP method with others because many tied trees, including the correct one, are often produced in this method. When there are many tied trees, it is customary to construct a consensus tree with multifurcating branches in parsimony methods. A multifurcating consensus tree can also be constructed with distance methods (e.g., Tajima, 1990). However, a multifurcating tree is clearly wrong in the present case, and it is not clear whether the comparison of consensus trees obtained by parsimony and distance methods is actually meaningful.

When there are many tied trees obtained, a more reasonable way of comparison is to compare $\overline{d}_T$ for the MP and other methods. If many tied trees are close to the correct one, $\bar{d}_T$ is expected to be close to 0; otherwise it will be high. Since Sourdis and Nei did not compute $\bar{d}_T$, Li Jin and I performed a small-scale computer simulation to evaluate this value using model tree A in Figure 6-2. The results obtained for the MP and NJ methods are presented in Table 6-5. The P_c values in this table are essentially the same as those for the corresponding cases of Table 6-4. Table 6-5 also includes the probability of obtaining the correct tree and other tied trees (P_2) and the probability of obtaining only incorrect tree(s) (P_3). As in Sourdis and Nei's study, P_2 for the MP method is large when m is small but decreases as m increases. For the NJ method, it is always 0. This result is slightly different from that of Sourdis and Nei (1988), because in their simulation some numbers with many digits were rounded to save computer time.

Table 6-5 shows that even if we use $\bar{d}_T$ as the criterion of comparison, the MP method is generally inferior to the NJ method. The only exceptions are the cases of m = 600 and m = 1,200 with U = 0.05. Therefore, our conclusion remains the same, whether we use P_c or $\bar{d}_T$ as the criterion.

Comparison of ML Method with the MP, FM, ME, and NJ Methods

The ML method is generally more efficient than the MP and FM methods but is slightly less efficient than the ME and NJ methods.

The P_c values for the ML method for model trees A and B in Figure 6-3 are presented in Table 6-6 in comparison with those for the MP, FM, ME, and NJ methods. The trees for the distance methods were constructed by using both p and d distances. It is clear that the FM method is poorest in obtaining the correct tree among all the methods examined here. The ML method is better than the MP method except for the case of U = 0.5 in tree A, but is slightly less efficient than the ME and NJ methods when m = 300. When m = 600, it has a P_c value similar to that of the latter two methods.

Table 6-5 Probabilities of Obtaining the Correct Unrooted Tree ($P_c \times 100$), the Correct Tree and Other Tied Trees ($P_2 \times 100$) and Incorrect Trees ($P_3 \times 100$), and Average Indices of Topological Errors ($\bar{d}_T$) for the MP and NJ Methods*

	U = 0.05			U = 0.5		
m	300	600	1,200	300	600	1,200
			Maximum Parsimony (MP)			
P_c	44	79	96	40	63	88
P_2	41	14	3	13	8	2
P_3	15	7	1	47	29	10
$\bar{d}_T$	0.96	0.31	0.05	1.30	0.72	0.22
			Neighbor-Joining (NJ)			
P_c	62	86	97	55	79	96
P_2	0	0	0	0	0	0
P_3	38	14	3	45	21	4
$\bar{d}_T$	0.87	0.31	0.07	1.05	0.41	0.07

*P_c is the same as P_1 of Sourdis and Nei (1988). Model tree A in Figure 6-2 was used. R = 300.

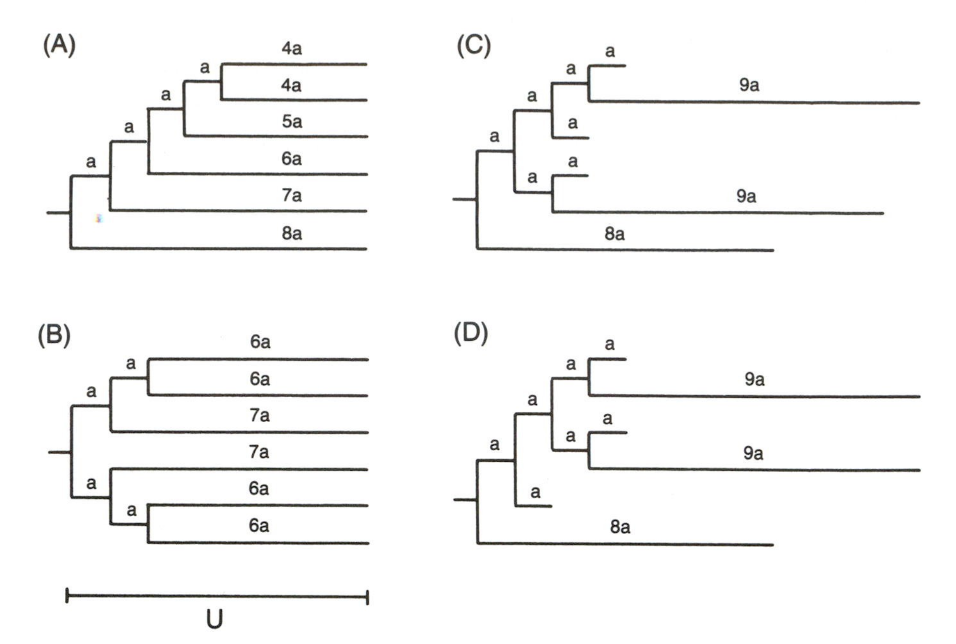

Figure 6-3 Model trees used by Saitou and Imanishi's (1989) computer simulation.

It should be noted that this computer simulation was conducted under the conditions which are favorable for the ML method, but do not completely satisfy the assumptions of this method (Saitou and Imanishi, 1989; N. Saitou, personal communication). Therefore, it seems that a slightly lower performance of the ML method compared with the ME and NJ methods is due to a small violation of the assumptions of this method. This suggests that the ML method is sensitive to violations of its assumptions, at least in this case.

Incidentally, Table 6-6 shows that the ME and NJ methods give essentially the same P_c value, as mentioned earlier. Indeed, Saitou and Imanishi (1989) showed that these two methods usually reconstruct the same tree from a given set of data. This indicates that the NJ method usually finds the minimum evolution tree, though its procedure is very simple.

Recently, Rohlf and Wooten (1988) reported results of their computer simulations on the gene frequency maximum likelihood (GFML) method (Felsenstein, 1981b). They considered situations where the assumptions of the GFML method hold as closely as possible, yet their results showed that UPGMA is generally better than the GFML method for gene-frequency data. Curiously, however, these authors concluded otherwise, considering the possibility that the performance of the ML method would increase as the number of genetic loci used increases. This conclusion is not warranted because the performance of distance methods also increases as the number of loci increases. A similar computer simulation was also conducted by Kim and Burgman (1988). However, since this simulation was conducted under the same assumptions (Gaussian process without

Table 6-6 Probabilities of Obtaining the Correct Tree ($P_c \times 100$) for the Maximum Likelihood (ML) and Other Tree-Making Methods for Model Trees A and B of Figure 6-3*

			p Distance			*d* Distance		
	MP	ML	FM	ME	NJ	FM	ME	NJ
				Model Tree A				
$U = 0.05$								
$m = 300$	34	38	26	36	42	26	40	40
$m = 600$	76	80	58	80	78	58	80	82
$U = 0.50$								
$m = 300$	60	48	30	56	58	22	42	46
$m = 600$	84	70	54	92	92	40	82	82
				Model Tree B				
$U = 0.05$								
$m = 300$	54	62	42	68	68	40	68	70
$m = 600$	84	88	56	82	84	56	80	86
$U = 0.50$								
$m = 300$	48	56	34	70	68	30	60	60
$m = 600$	58	76	36	76	76	36	70	70

*Abbreviations used are the same as those of previous tables; ML, maximum likelihood; ME: minimum evolution method. $R = 100$. (Adapted from Saitou and Imanishi, 1989.)

mutation and selection) as those of the GFML model, the results of the simulation are not very meaningful. Since actual gene-frequency changes almost never follow the assumptions of the GFML model, the simulation should have been conducted by using a more realistic model of gene frequency change due to mutation, selection, and genetic drift (see Nei et al., 1983).

Varying Rate of Substitution

Comparison of UPGMA with Other Tree-Making Methods

When the rate of nucleotide substitution varies with evolutionary lineage, UPGMA is worse than most other tree-making methods.

Since UPGMA requires the assumption of a constant rate of evolution, it is expected to have a poor performance in obtaining the correct tree when the rate varies with evolutionary lineage. Indeed, computer simulations have shown that it is very poor compared with other methods in this case (Table 6-7). The P_c values in this table were obtained by using model trees given in Figure 6-4.

Comparison Among Distance Methods

Among the distance methods available now, the ST, NJ, and ME methods generally show a better performance than other methods. The ST, NJ, and ME methods are nearly equally efficient.

This can be seen from Table 6-7 and Table 6-8. The P_c values in Table 6-8 were obtained by using model trees C and D in Figure 6-3. Note that the total number of nucleotide substitutions between the two most distantly

Table 6-7 Probabilities of Obtaining the Correct Unrooted Trees ($P_c \times 100$) and Average Indices ($\bar{d}_T$) of Topological Errors (in Parentheses) for the Case of Varying Rate of Nucleotide Substitution*

	Model Tree A	Model Tree B
UP:	0 (8.06)	0 (9.74)
MF:	77 (0.50)	57 (1.46)
DW:	69 (0.72)	59 (1.26)
TD:	46 (1.30)	45 (1.68)
ST:	77 (0.50)	69 (0.82)
NJ:	75 (0.56)	72 (0.78)

*Abbreviations used are the same as those in Table 6-2. Model trees A and B are given in Figure 6-4. $m = 600$; $R = 100$. p distances were used. (Adapted from Saitou and Nei, 1987.)

related sequences (d_m) is relatively small in Table 6-7, whereas Table 6-8 includes both cases of a small d_m and a large d_m. When d_m is small, the difference in P_c between the p and d distances is relatively small, though in the case of model trees C and D of Figure 6-3, the d distance gives a higher P_c value than the p distance (Table 6-8). At any rate, Tables 6-7 and 6-8 show that the P_c values for the ST, NJ, and ME methods are generally higher than other distance methods.

Table 6-8 shows that when d_m is large, the d distance gives a higher P_c value than the p distance. This is because the p distance is seriously affected by backward and parallel substitutions, whereas the d distance gives an appropriate estimate of the total number of substitutions. As mentioned earlier, many distance methods are capable of recovering the correct tree as long as the total number of substitutions is correctly estimated.

Comparison of the MP Method with Other Methods

The MP method is worse than several distance methods such as the NJ, ST, and ME methods and the ML method.

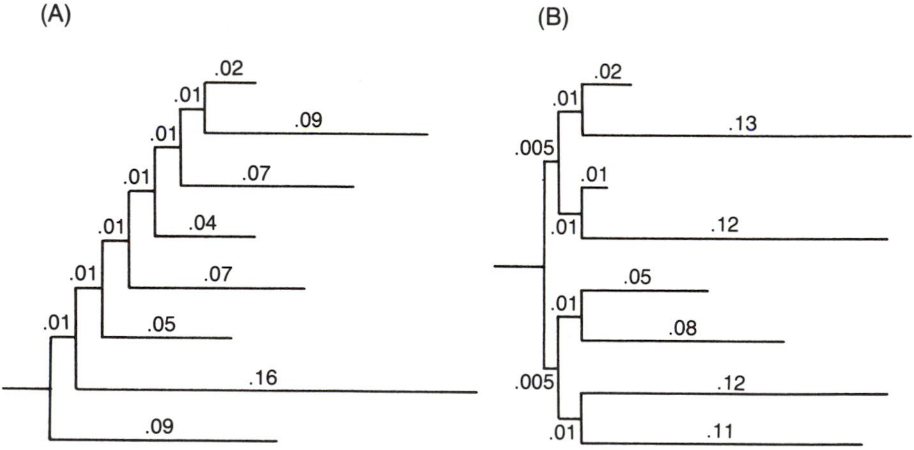

(A) (B)

Figure 6-4 Model trees used for the simulations in Table 6-7. The value given to each branch represents the expected number of nucleotide substitutions per site for that branch.

Table 6-8 Probabilities of Obtaining the Correct Unrooted Tree ($P_c \times 100$) for the MP, ML, and Other Methods in the Case of Varying Substitution Rate*

			p Distance			*d* Distance		
	MP	ML	FM	ME	NJ	FM	ME	NJ
				Model Tree C				
a = 0.01								
300 bp	64	78	26	56	56	34	72	72
600 bp	90	98	42	80	80	68	92	92
a = 0.05								
300 bp	24	92	0	2	2	22	68	68
600 bp	20	100	0	0	0	60	96	96
				Model Tree D				
a = 0.01								
300 bp	68	80	44	64	64	64	74	74
600 bp	94	96	56	90	90	88	92	92
a = 0.05								
300 bp	26	96	0	6	4	30	78	78
600 bp	46	100	0	10	6	64	100	100

*Abbreviations used are the same as those in previous tables; bp, base pairs. R = 50. Model trees C and D are those in Figure 6-3. (Adapted from Saitou and Imanishi, 1989.)

This statement is supported by the "varying rate" part of Table 6-4 and Table 6-8. In all cases examined here, the MP method always gives a smaller P_c value than the NJ, ME, and ML methods. However, it is better than the FM method when d_m is small (a = 0.01). The inferiority of the MP method is explained by Felsenstein's (1978) finding that when evolutionary rate varies with lineage, two OTUs that have long branches tend to form a cluster in the MP method even if this is not the correct branching pattern. This does not happen in the NJ and ME methods as long as the distances between sequences are correctly estimated. This also generally does not occur with the ML method (Hasegawa and Yano, 1984; Saitou, 1988; Saitou and Imanishi, 1989).

Some authors (e.g., Sober, 1985) claimed that the MP method is better than other methods because no assumption is necessary about the process of evolutionary change. This claim is not supported by simulation studies. As mentioned earlier, for the MP method to work well, approximate constancy of evolutionary rate and a small number of nucleotide substitutions per site are necessary.

Comparison of the ML Method with the MP, ME, and NJ Methods

The ML method is generally better than the MP method and is nearly the same as or slightly better than the ME and NJ methods in obtaining the correct tree. However, the performance of this method should be investigated more carefully by considering the cases where the underlying assumptions are violated.

Table 6-8 clearly shows that the ML method is better than the MP method when the evolutionary rate varies extensively. A similar result was also obtained by Hasegawa and Yano (1984), and Saitou (1988) for the case of four sequences. Table 6-8 also shows that the ML method is slightly better than the ME and NJ methods. This is so despite the fact that the pattern of nucleotide substitution simulated by Saitou and Imanishi (1989) was slightly different from the assumptions of the ML method used.

Nevertheless, it should be noted that the actual pattern of nucleotide substitution is much more complicated than the simple model used in the ML method (Gojobori et al., 1982b; Jin and Nei, 1990), so the general applicability of the above conclusion is questionable. Since the ML method is expected to be more sensitive to violation of the assumptions than are some distance methods, it is hoped that this method will be examined more carefully, considering realistic patterns of nucleotide substitution, particularly variation in substitution rate among different nucleotide sites. Previously, Saitou (1988) showed that in the case of $a = c = 1.0$ and $b = d = 0.1$ in model tree G of Figure 6-5, the NJ method ($P_c = 0.74$) is significantly better than the ML method ($P_c = 0.43$). Recently, however, Hasegawa and Saitou (personal communication) improved the performance of the ML method using a slightly different probability model. This indicates that it is very important to use a proper probability model in the ML method.

In the case of distance methods, the problem of substitution pattern becomes important when one estimates the distances between different sequences rather than when a tree is estimated from distance data. As long as the total number of nucleotide substitutions is correctly estimated, many distance methods are capable of producing correct trees. There are several statistical methods of estimating distance that are quite robust unless the distances are very large (see Nei, 1987a: 72). This problem will be discussed later.

Comparison of the EP Method with the MP and NJ Methods

The EP method is inferior to the MP method when nucleotide substitution occurs at random among the four nucleotides (one-parameter model) and

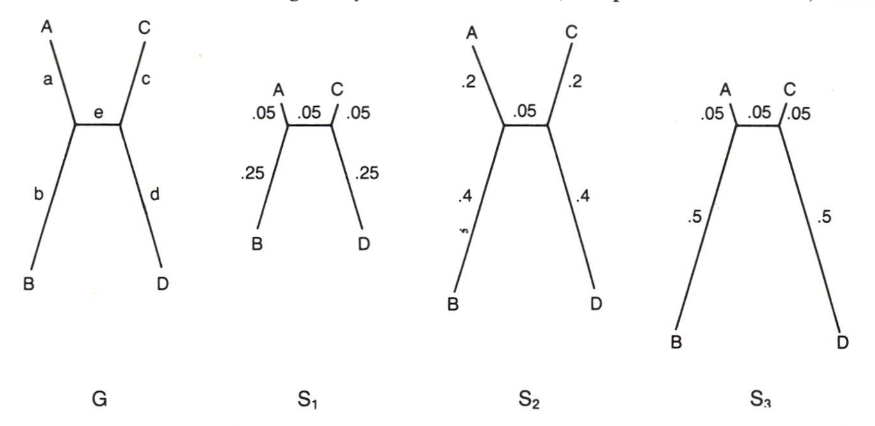

Figure 6-5 Unrooted model trees for four DNA sequences. G; general case. S_1, S_2, S_3; model trees used by Jin and Nei's (1990) computer simulation. The value given to each branch represents the expected number of nucleotide substitutions per site for that branch.

the rate of nucleotide substitution is nearly constant. However, if there is a transition/transversion bias and the rate of substitution varies extensively with evolutionary lineage, the EP method is better than the MP method. The NJ method is almost always superior to the EP method if a proper distance measure is used.

As mentioned earlier, the EP method was developed primarily to be applied to the case of four sequences. Therefore, all computer simulations have been performed for this case. The model tree used is generally of the form given in Figure 6-5. Table 6-9 shows the P_c values for the MP, NJ, and EP methods for the case of Kimura's two-parameter model with model trees S_1, S_2, and S_3 in Figure 6-5. Kimura's model satisfies the assumption of the EP method (Cavender, 1989; Jin and Nei, 1990). Therefore, the EP method is supposed to show a good performance in this case. There are three different sets of P_c's for the NJ method: NJP, NJD, and NJK represent the cases where the p distance, d distance, and Kimura distance [equation (11)], respectively, are used. Since Kimura's two-parameter model was used in generating DNA sequences, NJK is expected to show a better performance than NJP and NJD except when the proportion of transitional changes [$B = \alpha/(\alpha + 2\beta)$] is ⅓. This is indeed the case for almost all parameter sets examined in Table 6-9.

Comparison of the MP and EP methods in Table 6-9 shows that when all branch lengths of the model trees are relatively short (S_1), or when the ratio of branch b to branch a is relatively small (S_2), the former is better than the latter even for the case of a high proportion of transitional changes ($B = 0.9$). Otherwise (S_3), the latter is better than the former. Essentially the same results were obtained for various other sets of parameters (Li et al., 1987a; Jin and Nei, 1990). The performance of the NJ method is expected to depend on the distance measure used, as mentioned earlier. When the appropriate distance (Kimura distance for this case) is used, the NJ method is better than both the MP and EP methods. The same results were obtained for several other sets of parameters as well as for the set of substitution rates estimated from actual data (not the two-parameter model) (Jin and Nei, 1990).

Recently, Sidow and Wilson (1990) modified the EP method, taking into account the unequal frequencies of the four nucleotides (A, T, C, and G) in the sequences. However, since the basic principle of the EP method remains unchanged, our conclusion is expected to apply to this modified version as well.

Effects of Variation in Substitution Rate Among Different Nucleotide Sites

So far we have assumed that every nucleotide site evolves independently and that the pattern of nucleotide substitution is the same for all sites. This assumption is obviously not satisfied with actual data. In the case of protein-coding genes, the third nucleotide positions of codons usually evolve faster

Table 6-9 Probabilities of Obtaining the Correct Unrooted Tree ($P_c \times 100$) for the Two-Parameter Model of Nucleotide Substitution in the Cases of Model Trees S_1, S_2, and S_3 in Figure 6-5*

		Tree-Making Method					EP (χ^2)		
Model Trees	*B*	MP	NJP	NJD	NJK	EP	*X*	*Y*	*Z*
S_1†	0.33	100	98	100	100	99	68	1	1
	0.60	100	97	100	100	91	45	2	2
	0.90	88	85	99	100	61	12	0	0
S_2‡	0.33	91	88	100	100	63	12	6	6
	0.60	86	84	98	97	51	6	4	4
	0.90	67	69	84	86	40	5	1	1
S_3§	0.33	4	0	98	98	72	26	6	4
	0.60	2	0	92	95	57	13	3	3
	0.90	0	0	22	87	42	6	2	0

*Abbreviations used are the same as those of previous Tables. NJP and NJD, neighbor-joining method with *p* and *d* distances, respectively; NJK, neighbor-joining with Kimura distance; EP, evolutionary parsimony; *B*, proportion of transitional changes $[\alpha/(\alpha + 2\beta)]$. EP($\chi^2$) shows the proportion of replications in which only the χ^2 for *X*, *Y*, or *Z* was greater than 3.84. $m = 1{,}000$; $R = 400$.

† $a = c = e = 0.05$ and $b = d = 0.25$.

‡ $a = c = 0.20$, $b = d = 0.40$, and $e = 0.05$.

§ $a = c = e = 0.05$ and $b = d = 0.5$.

(Adapted from Jin and Nei, 1990.)

than the first and second positions, and codons encoding active centers of proteins evolve at a slower rate than other codons except in certain special genes (see Nei, 1987a; Hughes and Nei, 1988). Examining the pattern of amino acid substitution in cytochrome *c*, Uzzell and Corbin (1971) showed that the substitution rate varies from site to site, roughly following a gamma-distribution. This distribution is close to the lognormal distribution for certain parameter values (see Fig. 6-6).

Therefore it is important to examine the effects of variation in substitution rate among different sites on the efficiencies of different tree-making methods. Table 6-10 shows the results of one such study. In this case, substitution rate was assumed to vary according to the lognormal distribution with parameter $\alpha = 8$ (Olsen, 1987). The expected branch lengths used were the same as those for model trees S_1, S_2, and S_3 of Figure 6-5. Table 6-10 indicates that the MP method is better than the EP method for trees S_1 and S_2, but is worse than the latter for tree S_3. Therefore, the conclusion remains the same. The NJ method with the *d* distance and Kimura distance also shows a better performance than either the MP or EP method for S_1 and S_2. For S_3, however, the EP method now shows a better performance.

The poor performance of the NJ method for S_3 is caused by the fact that neither the *d* distance nor the Kimura distance is an additive measure of the number of nucleotide substitutions in this case. However, an approx-

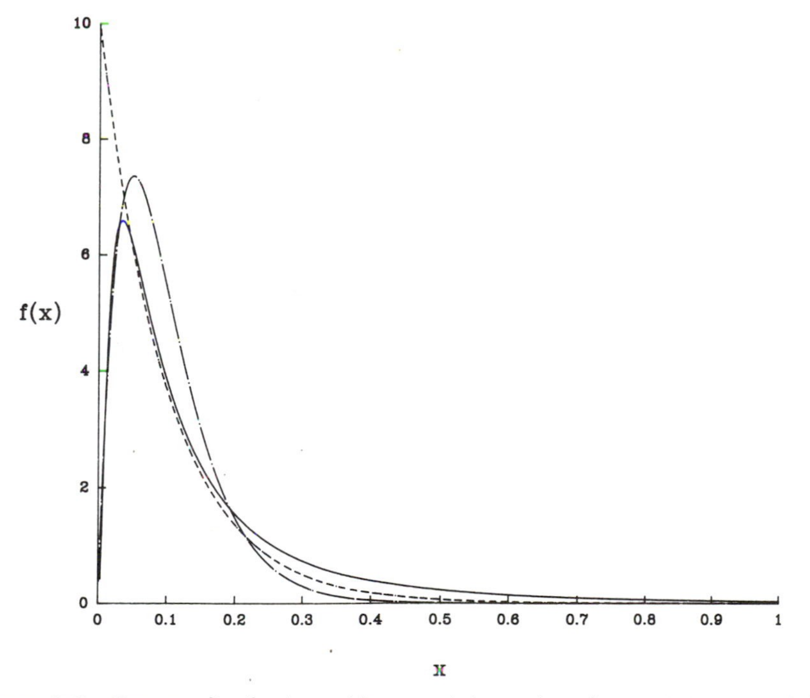

Figure 6-6 Gamma-distribution with $a = 1$ (- - - -) and $a = 2$ (-·-·-) and the lognormal distribution with $\alpha = 8$ (——). X = substitution rate per site.

Table 6-10 Probabilities of Obtaining the Correct Unrooted Tree ($P_c \times 100$) for the Case of Varying Substitution Rate Among Different Nucleotide Sites With the Two-Parameter Model*

		Tree-Making Method						EP (χ^2)		
Model Trees	*B*	MP	NJD	NJK	NJG1	NJG2	EP	*X*	*Y*	*Z*
S_1	0.33	97	98	98	100	99	89	44	3	2
	0.60	94	99	99	100	100	77	22	3	3
	0.90	74	90	92	98	96	53	6	0	0
S_2	0.33	83	90	90	94	93	55	12	4	5
	0.60	83	86	86	90	89	50	11	5	4
	0.90	65	77	75	80	78	41	4	3	3
S_3	0.33	7	30	31	96	75	64	21	5	5
	0.60	5	25	29	94	72	55	9	3	3
	0.90	1	10	27	85	64	41	5	2	0

*Abbreviations used are the same as those in previous tables. NJG1, NJ method with the gamma-distance of $a = 1$. NJG2, NJ method with the gamma-distance of $a = 2$; $m = 1,000$; $R = 400$. Model trees are presented in Figure 6-5. (Adapted from Jin and Nei, 1990.)

imately additive measure can be attained by considering the gamma-distribution (Nei and Gojobori, 1986; Jin and Nei, 1990). Since the log-normal distribution with $\alpha = 8$ is rather close to the gamma-distribution with parameter $a = 1$ or 2 (Fig. 6-6), we developed the following gamma-distance for the two-parameter model (Jin and Nei, 1990).

$$\hat{d} = \frac{a}{2}\left[(1 - 2\hat{P} - \hat{Q})^{-1/a} + \frac{1}{2}(1 - 2\hat{Q})^{-1/a} - \frac{3}{2}\right], \tag{12}$$

where $\hat{P}$ and $\hat{Q}$ are the estimates of P and Q in equations (9) and (10), respectively, and a is the square of the inverse of the coefficient of variation of substitution rate (Nei, 1987a: 234–235). In the case of $a = 1$, $\hat{d}$ becomes

$$\hat{d} = \frac{2\hat{P} + \hat{Q}}{2(1 - 2\hat{P} - \hat{Q})} + \frac{\hat{Q}}{2(1 - 2\hat{Q})}, \tag{13}$$

whereas for $a = 2$ it is

$$\hat{d} = (1 - 2\hat{P} - \hat{Q})^{-1/2} + \frac{1}{2}(1 - 2\hat{Q})^{-1/2} - \frac{3}{2}. \tag{14}$$

Table 6-10 shows that when the gamma-distance with $a = 1$ is used, the NJ method outperforms the EP method, even for S_3. This indicates that as long as a distance measure that gives an approximate additivity of nucleotide substitutions is used, the NJ method performs very well. The same can be said for the ST and ME methods, since these methods are expected to give the correct tree provided that additive distance measures are used.

No study has been done on the effect of intersite variation of substitution rate on the performance of the ML method. However, since it is difficult for this method to take this variation into account, the method will be affected considerably. It is desirable to study this effect quantitatively in the near future.

STATISTICAL TESTS OF THE TREES OBTAINED

Test of Topological Differences

When a tree (topology) is obtained by a tree-making method, one is naturally interested in the accuracy of the tree (i.e., how good the tree is, compared with other alternative trees). A number of authors (e.g., Cavender, 1981; Templeton, 1983; Felsenstein, 1985a,b; Lake, 1987; Prager and Wilson, 1988; Kishino and Hasegawa, 1989) have proposed statistical tests for topological differences. However, all these methods depend on assumptions which in reality do not necessarily hold. Therefore, one has to be cautious in using these methods. The main problem in these tests is that the evaluation of statistical significance of the difference between two topologies depends on the true tree (both topology and branch lengths), which is usually unknown, and the data sets used, which do not necessarily satisfy the assumptions required for a given tree-making method.

For example, Prager and Wilson's (1988) test is for comparing two topologies obtained by parsimony methods. In this test, the significance of the difference between the number of sites in which one topology wins, and the number of sites in which the other topology wins is examined by a binomial test. When the number of substitutions per site is small for all branches, and the trees to be compared are bifurcating, this test seems to be acceptable. In other cases, however, it may identify a wrong tree as the correct one and regard it as statistically established. This is obvious from the computer simulation given for S_3 in Tables 6-9 and 6-10, where a wrong tree was chosen as the correct one in most replications. Templeton's (1983) test for restriction-site data has the same problem as Wilson and Prager's. In this case, even if the rate of nucleotide substitution is constant, it may choose a wrong tree and regard it as statistically established (Nei and Tajima, 1985, 1989; Pamilo, 1990).

Lake (1987) proposed a statistical test in association with his evolutionary parsimony method. As mentioned earlier, this tests examines whether or not quantity X, Y, or Z is significantly different from 0. In practice, all of X, Y, and Z may become nonzero under certain patterns of nucleotide substitution. Thus, this is not a test for comparing different topologies (Jin and Nei, 1990). Indeed, this test may identify a wrong tree as the correct one, as is clear from Tables 6-9 and 6-10. Therefore, this test should not be used.

There is one test that is statistically sound if the rate of nucleotide substitution is constant (Felsenstein, 1985a). It applies to a rooted tree for three DNA sequences (Fig. 6-7). In this case there are three possible trees: (A), (B), and (C) in Figure 6-7. If tree (A) is correct, one would expect that the number of nucleotide differences (n_{12}) between sequences 1 and 2 to be smaller than that (n_{13}) between 1 and 3 or that (n_{23}) between 2 and 3. Therefore, if n_{12} is significantly smaller than n_{13} and n_{23} by a binomial test, one can conclude that it is statistically established. In this case, the tree corresponding to the null hypothesis ($n_{12} = n_{13} = n_{23}$) is the trifurcation tree given in Figure 6-7D. Extension of this method to the case of four or more sequences is complicated, and no studies have been done. Note also that even with three DNA sequences, the above test breaks down if the rate of nucleotide substitution is not constant.

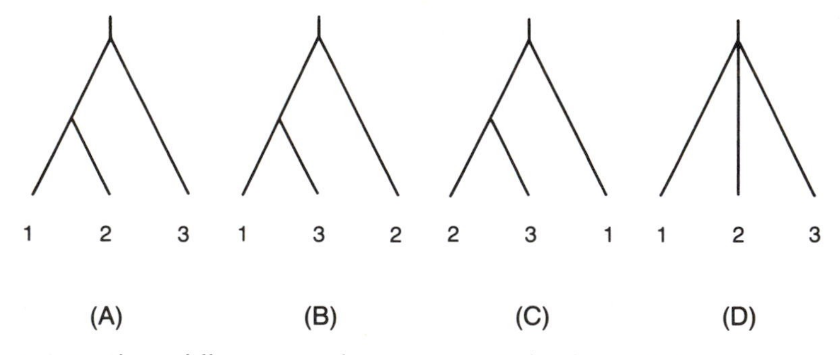

Figure 6-7 Three different rooted trees (A, B, C) for three DNA sequences (1, 2, and 3). (D); null hypothesis tree.

Williams and Goodman (1989) recently extended this approach to the case where the molecular clock does not apply. In this method, however, one must infer the ancestral nucleotide at each "informative site" by parsimony analysis. If this inference is wrong, their test is not reliable. Nevertheless, this test is a promising approach when there are useful outgroup species. This test also applies only to the case of three species of which the branching order is to be determined.

At any rate, it is a very difficult task to assess the significance level for the differences between topologies. Since all statistical methods depend on various assumptions, one should be prepared for the possibility that even a tree statistically established at the 5% or 1% level by a certain method may later turn out to be incorrect. One such example is the phylogeny for humans, chimpanzees, gorillas, orangutans, and gibbons. Templeton (1983) conducted a nonparametric test for restriction-site data of mitochondrial DNAs (mtDNAs) from these organisms and concluded that humans and gorillas are evolutionarily closer than humans and chimpanzees. However, later studies of DNA hybridization data for single-copy genomic DNA have not supported this conclusion (Sibley and Ahlquist, 1984; Caccone and Powell, 1989).

In general, one should not be too confident about a tree obtained from any tree-making method unless the number of nucleotides examined is very large. If there is any doubt about the tree obtained, the first thing to do is to increase the amount of data. Unless the amount of data is large, any sophisticated statistical method may lead to an erroneous conclusion.

Accuracy of the Topology Estimated

Although it is very difficult to test topological differences under realistic conditions, it is easier to test the accuracy of branch lengths estimated for a given topology. This test is primarily for examining the statistical significance of a given branch length, but if a branch length is not significantly different from 0, it casts doubt on the clustering pattern associated with the branch. For example, Figure 6-8 shows a phylogenetic tree for humans, chimpanzees, gorillas, orangutans, and gibbons. If branch length a in this tree is not significantly different from 0, the branching order among humans, chimpanzees, and gorillas becomes questionable.

This type of test was initiated by Nei et al. (1985) for a UPGMA tree. Application of this method to Brown et al.'s (1982) mtDNA data, from which the phylogeny in Figure 6-8 was constructed, indicates that the branch length a is not statistically significant. Therefore, Brown et al.'s data are not sufficient for resolving the branching pattern of humans, chimpanzees, and gorillas. Li (1989) extended this method to the case of other distance methods, though his primary interest was to discriminate among the three possible unrooted trees for four DNA sequences.

Theoretically, the standard error of any branch length can be computed for a UPGMA or any other distance method tree. However, the computation becomes quite complicated when the number of sequences is large.

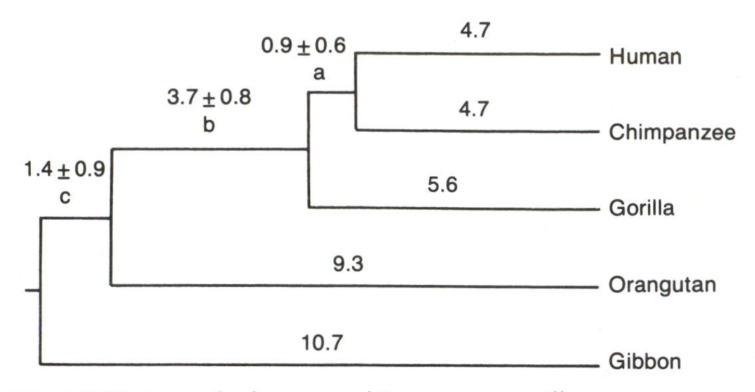

Figure 6-8 UPGMA tree for humans, chimpanzees, gorillas, orangutans, and gibbons. This tree was constructed by using mtDNA sequence data from Brown et al. (1982). Branch length estimates ± SE are given for all internal branches, whereas only branch-length estimates are given for exterior branches. Branch length *b* is significantly larger than 0, whereas branch lengths *a* and *c* are not significant.

In this case, a simpler method would be to use the jackknife or bootstrap method (Efron, 1982). The jackknife method is particularly useful for evaluating the standard error of a given branch length when the number of nucleotides or restriction sites examined is relatively small. Mueller and Ayala (1982) used this method for a tree obtained from gene-frequency data (see also Pamilo, 1990). When this number is large, however, it would be easier to use bootstrapping.

Felsenstein (1985b) used bootstrapping to evaluate the accuracy of a tree obtained by a parsimony method. His method is not to evaluate the standard error of any particular branch length, but to examine how often a particular cluster in a tree appears when nucleotide sites are resampled with replacement many times. This method can be used for any other method. This test can easily be applied, particularly for the NJ method, because the computational time for reconstructing a tree is usually very short.

Nevertheless, one should be cautious about the outcome of this test, because if the data set used does not satisfy the assumption underlying a tree-making method, one may identify a wrong cluster as a correct one. This is particularly so for a maximum-parsimony tree when the rate of substitution varies extensively with evolutionary lineage. It should also be noted that if the original data set is biased for some reason, a cluster may be regarded as statistically significant even if it is a wrong one. This is because the original bias cannot be corrected by the resampling process.

NUMBER OF NUCLEOTIDES TO BE EXAMINED

From the computer simulations mentioned earlier, it is clear that a large number of nucleotides must be examined to obtain the correct phylogenetic tree. The number of nucleotides required to establish a tree statistically depends on the topology and branch lengths of the true tree, the pattern of nucleotide substitution, the tree-making method, and other factors. Saitou and Nei (1986) studied this problem considering mtDNAs from humans, chimpanzees, gorillas, orangutans, and gibbons.

Mammalian mtDNA is known to evolve about ten times faster than nuclear DNA (Brown et al., 1979), and the phylogenetic tree for humans, chimpanzees, gorillas, orangutans, and gibbons (Fig. 6-8) is close to those given in Fig. 6-9. Suppose that tree A or B in Fig. 6-9 is the correct tree. How many nucleotides then must be examined to obtain the correct tree with probability P? The mathematical technique for computing the minimum number of nucleotides required (m^*) for the UPGMA, TD, DW, FM, and MP methods is provided by Saitou and Nei (1986), and it can be shown that m^* for the NJ and ME methods is identical with that of the TD and DW methods for the case of $n \leq 5$ (Saitou and Nei, 1987). One can therefore determine m^* for all these methods.

Some of the results obtained are presented in Table 6-11 for the one- and two-parameter models of nucleotide substitution under the assumption that (*i*) only orangutans are used as an outgroup (four species) and (*ii*) both orangutans and gibbons are used as outgroups. In the case of the two-parameter model, a transition/transversion ratio (α/β) appropriate to mtDNA is assumed to be 20. Table 6-11 shows that m^* for $P = 0.95$ is smallest for the NJ, ME, TD, and DW methods, and largest for the FM method (m^* for the MP method being usually close to that of the former). However, if tree A is correct and $\alpha/\beta = 20$, 2,600 ~ 3,100 nucleotides must be examined even in the NJ method, depending on the number of outgroup species used. This number is much larger than the number of nucleotides examined (895) by Brown et al. (1982). Therefore, Brown et al.'s data are unlikely to resolve the branching pattern of humans, chimpanzees, and gorillas. This conclusion is the same as that obtained from the statistical test of branch length a in Figure 6-8 mentioned earlier.

Saitou and Nei (1986) have also examined the number of nucleotides required when orangutans and gibbons are not available as outgroups. In

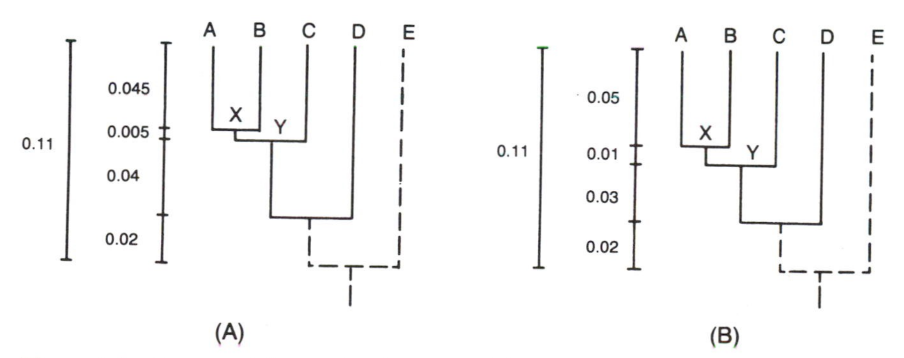

Figure 6-9 Two model trees for the case of four or five species. A, B, C, any of humans, chimpanzees, and gorillas; D, orangutans; E, gibbons. Numbers refer to the expected number of nucleotide substitutions per site per branch.

Table 6-11 Numbers (m^*) of Nucleotides Required for Obtaining the Correct Tree with a Probability of 0.95*

Tree-Making Method	Four Species		Five Species	
	1-p†	2-p	1-p	2-p
		Tree A of Figure 6-9		
NJ, ME, TD, DW	2,100	3,100	1,700	2,600
MP, CP	2,100	3,300	1,700	2,700
UPGMA	4,200	4,700	4,200	4,700
FM	5,000	8,200	3,000	4,900
		Tree B of Figure 6-9		
TD, DW	760	1,200	640	890
MP, CP	790	1,300	680	980
UPGMA	1,400	1,500	1,400	1,500
FM	1,700	2,800	900	1,700

*Abbreviations used are the same as those used in previous tables.
†1-p and 2-p indicate the one- and two-parameter models, respectively. CP, compatibility method. (Adapted from Saitou and Nei, 1986.)

this case, m^* becomes about 4,700 under the assumption of the molecular clock.

Recently, Miyamoto et al. (1987) and Maeda et al. (1988) used about 7,100 and 3,100 nucleotides, respectively, from the nuclear genome, to study the branching pattern of humans, chimpanzees, and gorillas. The total number of nucleotides available is now more than 10,000. However, since the rate of nucleotide substitution in the nuclear genome is about one-tenth that of mtDNA, this number still does not seem to be sufficient for establishing the branching order unquestionably (Maeda et al., 1988).

In this connection, it is interesting that when all cladistically informative sites from these data are used, Williams and Goodman's (1989) test leads to the conclusion that humans and chimpanzees are closer to each other than to gorillas at the 3% level. If we consider the possibility that the ancestral nucleotide inferred is incorrect, however, this conclusion does not seem to be sufficient for establishing the branching order of the three species involved. Using a different statistical method, Li (1989) could not reach the same conclusion from the same set of data. Of course, if we consider all data from DNA hybridization, mtDNA, and nuclear DNA sequences, humans and chimpanzees seem to be genetically closer to each other than to gorillas.

DISCUSSION

As mentioned above, many studies have been conducted on the relative efficiencies of tree-making methods. They indicate that the relative efficiencies depend on various factors such as the shape of the true tree, the numbers of nucleotide substitutions, transition/transversion ratios, and varying rate of nucleotide substitution. However, we can make some general conclusions.

One of the general conclusions is that the maximum parsimony method, which is currently very popular among numerical taxonomists, is as efficient as the ME, NJ, and ML methods only when the number of nucleotide substitutions per site is very small and the number of nucleotides examined is very large. Otherwise, the latter methods are better than the MP method. This seems to be true whether or not the transition/transversion bias is taken into account in the parsimony method (Jin and Nei, 1990). So far, no studies have been conducted on the relative efficiency of such recent parsimony methods as dynamic weighting parsimony (e.g., Williams and Fitch, 1990). Therefore, more simulation work is necessary.

The NJ method is an approximate method of obtaining the minimum evolution tree, but produces the same tree as that obtained by the ME method in most cases; if it does not, the tree obtained is usually very close to the ME tree. The advantage of the NJ method over the ME method is the short computer time required. The ST method depends on a procedure that is similar to that of the NJ method, and these two methods seem to be equally efficient in obtaining the correct tree. The NJ method, however, has one advantage over the ST method; it gives not only the topology but also the branch-length estimates of the tree obtained.

The ML method is also very efficient in obtaining the correct tree when the underlying assumptions are satisfied. When these assumptions are not satisfied, it seems to be less efficient than the NJ or ME method because the former depends on the details of the probability model, whereas the latter are known to be quite insensitive to various aspects of nucleotide substitution (see Table 6-10). Another disadvantage of the ML method is that it requires enormous computer time, even for a small number of DNA sequences examined.

Some investigators (e.g., Czelusniak et al., 1990) seem to be dissatisfied with the fact that the NJ method gives only one topology. They are interested in knowing how good the NJ tree is compared with alternative topologies. Since the NJ method is intended to obtain the minimum evolution tree, comparison of alternative trees can be made by using the total sum of branch lengths, i.e., L in equation (3). In the case of the ME method, L is computed for all topologies, so that one can choose a topology that gives the smallest L value, though it usually takes a large amount of computer time. In practice, however, it is sufficient to examine the L values for several alternative topologies that are close to that of the NJ tree. A simple way to find alternative topologies is to use Robinson and Foulds' (1981) measure (d_T) of computing topological differences. This measure takes a value of 2, 4, 6, 8, etc. I suggest that all topologies which are different from the NJ tree by $d_T = 2$ be examined for this purpose. Since the number of such topologies is not large (Sourdis and Nei, 1988), it would be much simpler to examine alternative topologies in this way. Once the topologies are identified, one can easily compute L for each of the topologies by using Fitch and Margoliash's (1967) method of estimating branch lengths. If this procedure finds a tree which has a smaller L value than that of the NJ tree, then this tree should be used as the final tree. Otherwise, the NJ tree will be used as the final tree. (Note that the ME tree is not necessarily the correct tree).

Although the statistical test of topological differences is besieged with many problems as mentioned earlier, a rough test of the difference in L between two topologies can be made by computing the standard error of the differences by using either the jackknife or bootstrap method. Conceptually, this test is similar to Hasegawa et al.'s (1987) test of maximum-likelihood values, but the computation is much simpler.

When the ME or NJ method is used, one must use an appropriate distance measure. As mentioned earlier, provided that the distance measure gives the correct number of nucleotide substitutions, these methods produce the correct tree. In practice, the pattern of nucleotide substitution is quite complicated, and there is no universally accurate distance measure for this purpose. However, I recommend the following guidelines for measuring nucleotide substitutions (Jin and Nei, 1990).

1. When the Jukes-Cantor estimate of the number of nucleotide substitutions per site (d) between different sequences is about 0.1 or less, use the Jukes-Cantor distance whether there is a transition/transversion bias or the substitution rate (λ) varies with nucleotide site. In this case, the Kimura distance or the gamma-distance gives essentially the same value as the Jukes-Cantor distance. One may also use the p distance (proportion of different nucleotides) for constructing a topology.
2. When d is greater than 0.1 but less than about 0.3, use the Jukes-Cantor distance unless the transition bias is high (e.g., $B > 0.5$). When this bias is high, use the Kimura distance.
3. When $1.0 > d > 0.3$ and there is evidence that λ varies extensively with site, use the gamma-distance. In general, we suggest that the gamma-distance with $a = 1$ be used. However, one may choose a different gamma-distance, estimating a from data. Wilson et al. (1989) recently used a distance with $a = ½$ for restriction-site data of mtDNA in hominoids.
4. When $1.0 > d > 0.3$ and the frequencies of the four nucleotides (A, T, C, and G) deviate substantially from equality, use Tajima and Nei's (1984) distance.
5. When $d > 1.0$ for many pairs of sequences, the phylogenetic tree estimated is not reliable for a number of reasons (e.g., large standard errors of d's, and sequence alignment errors). We therefore suggest that these sets of data should not be used. In this case, one may eliminate the portion of the gene that evolves very fast and use only the remaining region as is often done in studies of the evolution of different kingdoms or phyla using ribosomal RNA genes (e.g., Gouy and Li, 1989). If a coding region of DNA is examined, amino acid sequences rather than DNA sequences should be used. One may also use a different gene which evolves more slowly.
6. When a phylogenetic tree is constructed from the coding regions of a gene, the distinction between synonymous (d_S) and nonsynonymous

(d_N) substitutions (Miyata and Yasunaga, 1980; Li et al., 1985; Nei and Gojobori, 1986) will be helpful, because the rate of synonymous substitution is usually much higher than that of nonsynonymous substitution. When relatively closely-related species with $d_S < 1.0$ are studied for a large number of codons, one may use d_S for constructing a tree. This procedure is expected to reduce the effect of variation in substitution rate among different sites, because synonymous substitutions are apparently largely neutral in higher organisms (Nei, 1987a: 79–86). However, when relatively distantly related species are studied, the use of d_N is recommended.

Finally, it should be mentioned that the simulation study of tree-making methods is far from complete. In most studies, the number of DNA sequences considered is quite small (mostly four to eight) to save computer time. In some tree-making methods (e.g., MP method), the increase in the number of sequences considered may enhance the probability of obtaining the correct tree. Therefore, a more detailed study is necessary to solve this problem. The effects of different patterns of nucleotide substitution, particularly varying rate among different sites, should also be studied more carefully. Another problem that should be studied by computer simulation is the statistical methods for testing topological differences or the accuracy of a tree obtained. Since these methods depend on a number of assumptions about complex evolutionary processes, it would be important to establish the validity of the method by computer simulation.

In this chapter we considered only gene trees. A gene tree is not necessarily the same as the phylogeny of the species or populations from which the gene sequences are sampled (Nei, 1987a; Pamilo and Nei, 1988). The chance that the gene and population trees are different is high when the gene used is polymorphic or when the populations considered split relatively recently. In these cases, one must examine many independently evolving genes from the genome to estimate the correct population tree (Saitou and Nei, 1986; Pamilo and Nei, 1988). In this connection, it is important to note that a phylogenetic tree constructed from mitochondrial or chloroplast DNAs is a gene tree though they include many genes. This is because the genes in the mitochondrial or chloroplast DNA are inherited as a single entity without recombination. Therefore, some caution should be exercised in inferring population phylogenies from mitochondrial or chloorplast DNA. Of course, gene trees are not always studied just for inferring a population phylogeny. In such a case as the study of the evolution of multigene families, a gene tree provides important information.

COMPUTER PROGRAMS

Computer programs for computing various distances for DNA sequences (numbers of nucleotide substitutions per site, synonymous substitutions per synonymous site, nonsynonymous substitutions per nonsynonymous site, etc.) and for constructing (NJDRAW) and testing (NJBOOT) a neighbor-joining tree are available upon request from the Institute of Molecular Evolutionary Genetics, Penn State University, 328 Mueller Laboratory, University Park, PA 16802-5303. Please send IBM compatible 3.5 or 5.25 inch floppy diskettes. Each program requires one 360 Kb diskette.

ACKNOWLEDGMENTS

I thank Li Jin for his comments. This study is supported by research grants from the National Institutes of Health and the National Science Foundation.

REFERENCES

Aquadro, C. F., and B. D. Greenberg. (1983) Human mitochondrial DNA variation and evolution: analysis of nucleotide sequences from seven individuals. *Genetics* **103**:287–312.

Blanken, R. L., L. C. Klotz, and A. G. Hinnebusch. (1982) Computer comparison of new and existing criteria for constructing evolutionary trees from sequence data. *J. Mol. Evol.* **19**:9–19.

Britten, R. J. (1986) Rates of DNA sequence evolution differ between taxonomic groups. *Science* **231**:1393–1398.

Brown, W. M., M. George, Jr., and A. C. Wilson. (1979) Rapid evolution of animal mitochondrial DNA. *Proc. Natl. Acad. Sci. USA* **76**:1967–1971.

Brown, W. M., E. M. Prager, A. Wang, and A. C. Wilson. (1982) Mitochondrial DNA sequences of primates: tempo and mode of evolution. *J. Mol. Evol.* **18**:225–239.

Buneman, P. (1971) The recovery of trees from measurements of dissimilarity. Pp. 387–395 in *Mathematics in the Archeological and Historical Sciences* (F. R. Hodson, D. G. Kendall, and P. Tautu, eds.). Edinburgh University Press, Edinburgh.

Caccone, A., and J. R. Powell. (1989) DNA divergence among hominoids. *Evolution* **43**:925–942.

Cavalli-Sforza, L. L., and A. W. F. Edwards. (1967) Phylogenetic analysis: models and estimation procedures. *Am. J. Hum. Genet.* **19**:233–257.

Cavender, J. A. (1981) Tests of phylogenetic hypotheses under generalized models. *Math. Biosci.* **54**:217–229.

Cavender, J. A. (1989) Mechanized derivation of linear invariants. *Mol. Biol. Evol.* **6**:301–316.

Czelusniak, J., M. Goodman, N. O. Moncrief, and S. M. Kehoe. (1990) Maximum parsimony approach to construction of evolutionary trees from aligned homologous sequences. Pp. 601–615 in *Molecular Evolution: Computer Analysis of Protein and Nucleic Acid Sequences* (R. F. Doolittle, ed.). Methods in Enzymology, vol. 183. Academic Press, New York.

Eck, R. V., and M. O. Dayhoff. (1966) *Atlas of Protein Sequence and Structure*. National Biomedical Research Foundation, Silver Spring. Md.

Efron, B. (1982) *The Jackknife, the Bootstrap and Other Resampling Plans*. Society for Industrial and Applied Mathematics, Philadelphia.

Faith, D. P. (1985) Distance methods and the approximation of most-parsimonious trees. *Syst. Zool.* **34**:312–325.

Farris, J. S. (1970) Methods for computing Wagner trees. *Syst. Zool.* **19**:83–92.

Farris, J. S. (1972) Estimating phylogenetic trees from distance matrices. *Am. Natur.* **106**:645–668.

Farris, J. S. (1977) On the phenetic approach to vertebrate classification. Pp. 823–850 in *Major Patterns in Vertebrate Evolution* (M. D. Hecht, P. C. Goody, and B. M. Hecht, eds.). Plenum Press, New York.

Farris, J. S. (1981) Distance data in phylogenetic analysis. Pp. 3–23 in *Advances in Cladistics. Proceedings of the First Meeting of the Willi Hennig Society* (V. A. Funk and D. R. Brooks, eds.). New York Botanical Garden, Bronx.

Felsenstein, J. (1978) Cases in which parsimony or compatibility methods will be positively misleading. *Syst. Zool.* **27**:401–410.

Felsenstein, J. (1981a) Evolutionary trees from DNA sequences: a maximum likelihood approach. *J. Mol. Evol.* **17**:368–376.

Felsenstein, J. (1981b) Evolutionary trees from gene frequencies and quantitative characters: finding maximum likelihood estimates. *Evolution* **35:** 1229–1242.

Felsenstein, J. (1985a) Confidence limits on phylogenies with a molecular clock. *Syst. Zool.* **34**:152–161.

Felsenstein, J. (1985b) Confidence limits on phylogenies: an approach using the bootstrap. *Evolution* **39**:783–791.

Felsenstein, J. (1986) Distance methods: a reply to Farris. *Cladistics* **2**:130–143.

Felsenstein, J. (1988) Phylogenies from molecular sequences: inference and reliability. *Ann. Rev. Genet.* **22**:521–565.

Fitch, W. M. (1971) Toward defining the course of evolution: minimum change for a specific tree topology. *Syst. Zool.* **20**:406–416.

Fitch, W. M. (1981) A non-sequential method for constructing trees and hierarchical classifications. *J. Mol. Evol.* **18**:30–37.

Fitch, W. M., and E. Margoliash. (1967) Construction of phylogenetic trees. *Science* **155**:279–284.

Gojobori, T., K. Ishii, and M. Nei. (1982a) Estimation of average number of nucleotide substitutions when the rate of substitution varies with nucleotide. *J. Mol. Evol.* **18**:414–423.

Gojobori, T., W.-H. Li, and D. Graur. (1982b) Patterns of nucleotide substitution in pseudogenes and functional genes. *J. Mol. Evol.* **18**:360–369.

Gouy, M., and W.-H. Li. (1989) Molecular phylogeny of the kingdoms Animalia, Plantae, and Fungi. *Mol. Biol. Evol.* **6**:109–122.

Graur, D. (1985) Pattern of nucleotide substitution and the extent of purifying selection in retroviruses. *J. Mol. Evol.* **21**:221–231.

Hasegawa, M., H. Kishino, and T. Yano. (1987) Man's place in Hominoidea as inferred by molecular clocks of DNA. *J. Mol. Evol.* **26**:132–147.

Hasegawa, M., and T. Yano. (1984) Maximum likelihood method of phylogenetic inference from DNA sequence data. *Bull. Biometric Soc. Japan* **5**:1–7.

Hughes, A., and M. Nei. (1988) Pattern of nucleotide substitution at major histocompatibility complex class I loci reveals overdominant selection. *Nature* **335**:167–170.

Jin, L., and M. Nei. (1990) Limitations of the evolutionary parsimony method of phylogenetic analysis. *Mol. Biol. Evol.* **7**:82–102.

Jukes, T. H., and C. R. Cantor. (1969) Evolution of protein molecules. Pp. 21–132 in *Mammalian Protein Metabolism* (H. N. Munro, ed.). Academic Press, New York.

Kim, J., and M. A. Burgman. (1988) Accuracy of phylogenetic-estimation methods under unequal evolutionary rates. *Evolution* **42**:596–602.

Kimura, M. (1980) A simple method for estimating evolutionary rate of base substitutions through comparative studies of nucleotide sequences. *J. Mol. Evol.* **16**:111–120.

Kimura, M. (1983) *The Neutral Theory of Molecular Evolution*. Cambridge University Press, Cambridge.

Kishino, H., and M. Hasegawa. (1989) Evaluation of the maximum likelihood estimate of the evolutionary tree topologies from DNA sequence data, and the branching order in Hominoidea. *J. Mol. Evol.* **29**:170–179.

Klotz, L. C., and R. L. Blanken. (1981) A practical method for calculating evolutionary trees from sequence data. *J. Theor. Biol.* **91**:261–272.

Klotz, L. C., N. Komar, R. L. Blanken, and R. M. Mitchell. (1979) Calculation of evolutionary trees from sequence data. *Proc. Natl. Acad. Sci. USA* **76**:4516–4520.

Lake, J. A. (1987) A rate-independent technique for analysis of nucleic acid sequences: evolutionary parsimony. *Mol. Biol. Evol.* **4**:167–191.

Lake, J. A. (1988) Origin of the eukaryotic nucleus determined by rate-invariant analysis of rRNA sequences. *Nature* **331**:184–186.

Li, W.-H. (1981) A simple method for constructing phylogenetic trees from distance matrices. *Proc. Natl. Acad. Sci. USA* **78**:1085–1089.

Li, W.-H. (1989) A statistical test of phylogenies estimated from sequence data. *Mol. Biol. Evol.* **6**:424–435.

Li, W.-H., M. Tanimura, and P. M. Sharp. (1987b) An evaluation of the molecular clock hypothesis using mammalian DNA sequences. *J. Mol. Evol.* **25**:330–342.

Li, W.-H., K. H. Wolfe, J. Sourdis, and P. M. Sharp. (1987a) Reconstruction of phylogenetic trees and estimation of divergence times under nonconstant rates of evolution. *Cold Spring Harbor Symp. Quant. Biol.* **52**:847–856.

Li, W.-H., C.-I. Wu, and C.-C. Luo. (1985) A new method for estimating synonymous and nonsynonymous rates of nucleotide substitution considering the relative likelihood of nucleotide and codon changes. *Mol. Biol. Evol.* **2**:150–174.

Maeda, N., C.-I. Wu, J. Bliska, and J. Reneke. (1988) Molecular evolution of intergenic DNA in higher primates: pattern of DNA changes, molecular clock and evolution of repetitive sequences. *Mol. Biol. Evol.* **5**:1–20.

Miyamoto, M. M., J. L. Slightom, and M. Goodman. (1987) Phylogenetic relationships of human and African apes from DNA sequences of the ψη-globin region. *Science* **238**:369–373.

Miyata, T., and T. Yasunaga. (1980) Molecular evolution of mRNA: A method for estimating evolutionary rates of synonymous and amino acid substitutions from homologous nucleotide sequences and its application. *J. Mol. Evol.* **16**:23–36.

Mueller, L. D., and F. J. Ayala. (1982) Estimation and interpretation of genetic distance on empirical studies. *Genet. Res.* **40**:127–137.

Nei, M. (1987a) *Molecular Evolutionary Genetics*. Columbia University Press, New York.

Nei, M. (1987b) Genetic distance and molecular phylogeny. Pp. 193–223 in *Population Genetics and Fishery Management* (N. Ryman and F. Utter, eds.). University of Washington Press, Seattle.

Nei, M., and T. Gojobori. (1986) Simple methods for estimating the numbers of synonymous and nonsynonymous nucleotide substitutions. *Mol. Biol. Evol.* **3**:418–426.

Nei, M., and A. L. Hughes. (1991) Polymorphism and evolution of the major histocompatibility complex loci. Pp. 222–247 in *Evolution at the Molecular Level* (R. K. Selander, A. G. Clark, and T. S. Whittam, eds.). Sinauer Associates, Sunderland.

Nei, M., J. C. Stephens, and N. Saitou. (1985) Methods for computing the standard errors of branching points in an evolutionary tree and their application to molecular data from humans and apes. *Mol. Biol. Evol.* **2**:66–85.

Nei, M., and F. Tajima. (1985) Evolutionary change of restriction cleavage sites and phylogenetic inference for man and apes. *Mol. Biol. Evol.* **2**:189–205.

Nei, M., F. Tajima, and Y. Tateno. (1983) Accuracy of estimated phylogenetic trees from molecular data. II. Gene frequency data. *J. Mol. Evol.* **19**:153–170.

Olsen, G. J. (1987) Earliest phylogenetic branchings: comparing rRNA-based evolutionary trees inferred with various techniques. *Cold Spring Harbor Symp. Quant. Biol.* **52**:825–837.

Pamilo, P. (1990) Statistical tests of phenograms based on genetic distances. *Evolution* **44**:689–697.

Pamilo, P., and M. Nei. (1988) Relationships between gene trees and species trees. *Mol. Biol. Evol.* **5**:568–583.

Peacock, D., and D. Boulter. (1975) Use of amino acid sequence data in phylogeny and evaluation of methods using computer simulation. *J. Mol. Biol.* **95**:513–527.

Penny, D. (1982) Towards a basis for classification: the incompleteness of distance measures, incompatibility analysis and phenetic classification. *J. Theor. Biol.* **96:**129–142.

Prager, E. M., and A. C. Wilson. (1988) Ancient origin of lactalbumin from lysozyme: analysis of DNA and amino acid sequences. *J. Mol. Evol.* **27**:326–335.

Robinson, D. F., and L. R. Foulds. (1981) Comparison of phylogenetic trees. *Math. Biosci.* **53**:131–147.

Rohlf, F. J., and M. C. Wooten. (1988) Evaluation of the restricted maximum-likelihood method for estimating phylogenetic trees using simulated allele frequency data. *Evolution* **42:**581–595.

Saitou, N. (1987) Patterns of nucleotide substitutions in influenza A virus genes. *Jpn. J. Genet.* **62**:439–444.

Saitou, N. (1988) Property and efficiency of the maximum likelihood method for molecular phylogeny. *J. Mol. Evol.* **27**:261–273.

Saitou, N., and T. Imanishi. (1989) Relative efficiencies of the Fitch-Margoliash, maximum-parismony, maximum-likelihood, minimum-evolution, and neighbor-joining methods of phylogenetic tree construction in obtaining the correct tree. *Mol. Biol. Evol.* **6**:514–525.

Saitou, N., and M. Nei. (1986) The number of nucleotides required to determine the branching order of three species with special reference to the human-chimpanzee-gorilla divergence. *J. Mol. Evol.* **24**:189–204.

Saitou, N., and M. Nei. (1987) The neighbor-joining method: a new method for reconstructing phylogenetic trees. *Mol. Biol. Evol.* **4**:406–425.

Sattath, S., and A. Tversky. (1977) Additive similarity trees. *Psychometrika* **42**:319–345.

Shoemaker, J. S., and W. M. Fitch. (1989) Evidence from nuclear sequences that invariable sites should be considered when sequence divergence is calculated. *Mol. Biol. Evol.* **6**:270–289.

Sibley, C. G., and J. E. Ahlquist. (1984) The phylogeny of the hominoid primates, as indicated by DNA-DNA hybridation. *J. Mol. Evol.* **20**:2–15.

Sidow, A., and A. C. Wilson. (1990) Compositional statistics: an improvement of evolutionary parismony and its application to deep branches in the tree of life. *J. Mol. Evol.* **31**:51–68.

Sneath, P. H. A., and R. R. Sokal. (1973) *Numerical Taxonomy*. Freeman, San Francisco.

Sober, E. (1985) A likelihood justification of parsimony. *Cladistics* **1**:209–233.

Sokal, R. R., and C. D. Michener. (1958) A statistical method for evaluating systematic relationships. *Univ. Kansas Sci. Bull.* **28**:1409–1438.

Sourdis, J., and C. Krimbas. (1987) Accuracy of phylogenetic trees estimated from DNA sequence data. *Mol. Biol. Evol.* **4**:159–166.

Sourdis, J., and M. Nei. (1988) Relative efficiencies of the maximum parsimony and distance-matrix methods in obtaining the correct phylogenetic tree. *Mol. Biol. Evol.* **5**:298–311.

Swofford, D. L. (1981) On the utility of the distance Wagner procedure. Pp. 25–43 in *Advances in Cladistics. Proceedings of the First Meeting of the Willi Hennig Society* (V. A. Funk and D. R. Brooks, eds.). New York Botanical Gardens, Bronx.

Swofford, D. L., and G. J. Olsen. (1990) Phylogeny reconstruction. Pp. 411–501 in *Molecular Systematics* (D. M. Hillis and C. Moritz, eds.). Sinauer Associates, Sunderland.

Tajima, F. (1990) A simple graphic method for reconstructing phylogenetic trees from molecular data. *Mol. Biol. Evol.* **7**:578–588.

Tajima, F., and M. Nei. (1984) Estimation of evolutionary distance between nucleotide sequences. *Mol. Biol. Evol.* **1**:269–285.

Tateno, Y., M. Nei, and F. Tajima. (1982) Accuracy of estimated phylogenetic trees from molecular data. I. Distantly related species. *J. Mol. Evol.* **18**:387–404.

Templeton, A. R. (1983) Phylogenetic inference from restriction endonuclease cleavage site maps with particular reference to the evolution of humans and the apes. *Evolution* **37**:221–244.

Uzzell, T., and K. W. Corbin. (1971) Fitting discrete probability distributions to evolutionary events. *Science* **172**:1089–1096.

Williams, P. L., and W. M. Fitch. (1990) Phylogeny determination using dynamically weighted parsimony methods. Pp. 615–626 in *Molecular Evolution: Computer Analysis of Protein and Nucleic Acid Sequences* (R. F. Doolittle, ed.). Methods in Enzymology, vol. 183. Academic Press, New York.

Williams, S. A., and M. Goodman. (1989) A statistical test that supports a human/chimpanzee clade based on noncoding DNA sequence data. *Mol. Biol. Evol.* **6**:325–330.

Wilson, A. C., E. A. Zimmer, E. M. Prager, and T. D. Kocher. (1989) Restriction mapping in the molecular systematics of mammals: a retrospective salute. Pp. 407–419 in *The Hierarchy of Life. Molecules and Morphology in Phylogenetic Analysis* (B. Fernholm, K. Bremer, and H. Jornval, eds.). Elsevier Science Publishers, B. V., Amsterdam.

7

Compositional Statistics Evaluated by Computer Simulations

AREND SIDOW AND ALLAN C. WILSON

Phylogenetic hypotheses testable with DNA sequences span the entire range of evolutionary relationships, from trees relating individuals within a species (Vigilant et al., 1989), to the tree of life (Fox et al., 1980; Lake, 1988; Gouy and Li, 1989; Sidow and Wilson, 1990). A number of methods are available for inferring evolutionary relationships among homologous and aligned DNA sequences. This chapter is about a new method called compositional statistics, which is most suitable for elucidating relationships among highly diverged sequences. In contrast to most other methods, it takes into account the sequences' base compositions.

Our discussion emphasizes the idea that biases in the base composition of the compared sequences may affect a phylogenetic analysis and produce systematic errors if not properly corrected for. Such biases are especially strong in animal mitochondrial DNA (mtDNA) and in the heavy compartments of homoiotherm genomes (Perrin and Bernardi, 1987). Biases may arise as a consequence of selection (Bernardi et al., 1985) or mutation pressure (Sueoka, 1988), but there is some uncertainty as to which of these two processes is actually responsible for producing them (Li and Graur, 1991; M. Bulmer, personal communication). Compositional statistics was developed to take into account biases that are due to global selection on the DNA's primary structure, but it may also function in cases where mutation pressure produces the bias.

In order to point out the most useful applications of compositional statistics, we first discuss some important strengths and weaknesses of the most commonly used methods of phylogenetic inference.

METHODS OF PHYLOGENETIC INFERENCE

Parsimony

Parsimony, the most widely used method of phylogenetic inference, lacks a stochastic model of sequence evolution; it simply chooses an evolutionary tree among all trees as the correct one if it requires the smallest number of substitutions. Parsimony would be the natural method for inferring phylogenetic relationships among organisms if parallelisms and reversals were impossible in evolution. The correct phylogeny could immediately be inferred from the existence of shared characters between related organisms. It is now clear that DNA sequences generally do not meet this requirement (Felsenstein, 1978; Shoemaker and Fitch, 1989; Saccone et al., 1989). Parsimony is most likely to give positively misleading results when the amount of evolution is very unequal in the branches of the phylogenetic tree (Felsenstein, 1978; Lake, 1987). Nevertheless, it is useful and reliable for closely related sequences in which parallelisms and reversals are rare.

Distance Methods

Distance methods are also widely used for inferring evolutionary relationships. Sophisticated algorithms are available that, with a high probability, find the best fit of evolutionary relationships to a distance matrix (Saitou and Nei, 1987; Saitou and Imanishi, 1989; Jin and Nei, 1990).

For sequences that are not closely related, a major complication is that *observed* pairwise distances are insufficient in an analysis. Multiple-hit corrections are necessary to approximate *actual* pairwise distances between the sequences (Gojobori et al., 1990). A simple two-parameter model, distinguishing transitions from transversions (Kimura, 1980), can give sufficient correction in many cases. For distantly related sequences, proper corrections require extensive knowledge about the substitution process and evolutionary constraints (Fitch, 1986; Olsen, 1987; Shoemaker and Fitch, 1989; Jin and Nei, 1990). In practice, it is usually impossible to take into account differences in the rate of evolution among the sites, or estimate the number of variable sites in a sequence—independent from the data set that is used for the analysis—to make such corrections.

Maximum Likelihood Methods

These methods (Felsenstein, 1981) compute the likelihoods of all possible trees, based on stochastic models of sequence evolution whose parameters can either be specified by the user or estimated by the algorithm. It is then tested whether the tree with the greatest likelihood is significantly better than others. Potentially, maximum likelihood is the most powerful and statistically most reliable method of phylogenetic inference. When likelihood calculations are based on multiple-parameter models of sequence evolution (e.g., Kishino and Hasegawa, 1990), the base composition of the sequences is considered. Present limitations are partly computational because of the complexity of likelihood calculations. Drawbacks of available maximum likelihood methods include the assumption that every site in the sequence undergoes evolution at the same rate or that rate differences among sites have to be specified by the user (Felsenstein, 1989).

Evolutionary Parsimony

Since highly divergent sequences are less likely to obey simple models of evolution, solving distant evolutionary relationships requires the most general methods available. A major step in this direction was the invention of evolutionary parsimony (EP) (Lake, 1987). Evolutionary parsimony's property of **rate invariance** renders it particularly invulnerable to any variation in the number of substitution events among branches of the evolutionary tree, as well as variations in the evolutionary rate among different sites in the sequences. For a thorough explanation of how EP achieves rate invariance, see Sidow and Wilson (1990).

Evolutionary parsimony, like parsimony, relies on the concept of shared states of homologous characters (Fig. 7-1). Briefly, the crucial differences are the following: (1) In EP, informative nucleotide positions are all those in which two sequences differ from the other two by a transversion. In parsimony, transitional differences are informative as well. (2) In parsimony, all informative patterns are at the same time supportive. In EP, informative patterns need not be supportive. Evolutionary parsimony tests the parsimony assignments relative to other informative patterns (referred to as "background patterns") present in the aligned sequences. Parsimony allows no test independent of the parsimony positions (Fig. 7-2).

Evolutionary parsimony statistically distinguishes misleading nucleotide patterns that arose on peripheral branches, from those patterns that arose on the central branch of the phylogenetic tree, as explained in Figure 7-2 (Holmquist et al., 1988; Sidow and Wilson, 1990). Parsimony (as well as distance methods), because it lacks an internal comparison, may support an incorrect phylogenetic tree when the substitution process in peripheral branches creates parallel substitutions (Felsenstein, 1978).

Compositional Statistics

Compositional statistics (CS) is a new method of phylogenetic inference that is related to EP but also incorporates base compositional components

```
                         Informative Sites
Sequence
number          Parsimony                 EP and CS

    1           ...G...G............G..G...G..G...G..
    2           ...G...G............G..G...G..G...A..
    3           ...T...C............T..C...T..C...T..
    4           ...T...C............T..C...C..T...C..
```

Figure 7-1 Phylogenetically informative positions in four aligned, hypothetical sequences. Those nucleotide positions are shown whose patterns are informative for the grouping of sequences 1 with 2, and of 3 with 4. On the left, are those patterns which are informative in parsimony, on the right, are those informative for EP and CS.

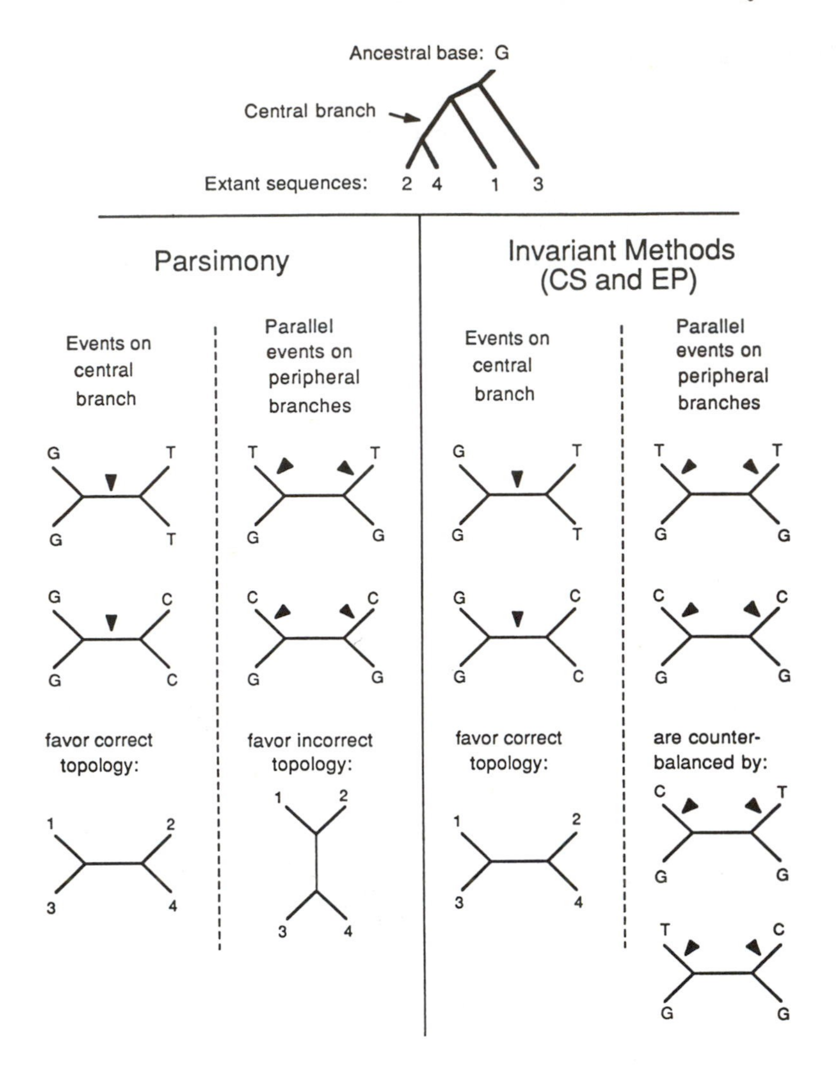

Figure 7-2 Comparison of parsimony and invariant methods of tree analysis. (Top) The correct, rooted evolutionary tree for taxa 1, 2, 3, and 4. (Bottom) For both parsimony and invariant methods (CS and EP), events on the central branch lead to correct support of the network relating sequences 1 and 3 to the exclusion of 2 and 4 (as shown in the left half of each panel). Parallel events on the peripheral branches support the incorrect topology when parsimony is used but not when the invariant methods (EP and CS) are used (as shown in the right half of each panel).

of the stochastic models that underlie maximum likelihood calculations. Since describing this method and applying it to two sets of DNA sequences (Sidow and Wilson, 1990), we have conducted computer simulations aimed at comparing its performance with that of EP. Besides briefly reviewing CS and its relation to EP, this chapter gives the results of some of these simulations. Compositional statistics is shown to be less likely than EP to make errors when base compositions are biased.

In comparison with other methods, a drawback of CS is that it can be applied to only four sequences at a time. In practice, the solution to this problem is to test all four-taxon trees that pertain to the phylogenetic hypothesis to be solved. An example showing how CS was applied to ten taxa appears in Sidow and Wilson (1990).

Limitations of Evolutionary Parsimony

Evolutionary parsimony assumes the two transversions from one base (e.g., $G_i \rightarrow T_i$ and $G_i \rightarrow C_i$) to be roughly equally probable. That is,

$$P(G_i \rightarrow T_i) = P(G_i \rightarrow C_i), \tag{1}$$

where subscript i designates the position in the sequence. An exception to this assumption is if T and C are maintained at different frequencies because of selective constraints on the base composition. For example, if

$$F(T) > F(C), \tag{2}$$

then

$$P(G_i \rightarrow T_i) > P(G_i \rightarrow C_i) \tag{3}$$

in order to maintain proper frequencies of T and C. Compositional constraints, like those in (2), are found to exist in a wide range of sequences. Examples include animal mtDNA (where the coding strand displays a significant bias against G) or the heavy compartments of homoiotherm genomes, which show a significant bias toward a high G + C content (Anderson et al., 1981; Bernardi et al., 1985; Saccone et al., 1989). Thus, certain kinds of sequences may render EP inapplicable to their analysis. The same limitation holds true for distance and parsimony methods.

CS's Model of Sequence Evolution and its Phylogenetic Test

The central difference to EP is CS's model of sequence evolution. It is identical to the stochastic model described in Hasegawa et al. (1985) and Kishino and Hasegawa (1990). It distinguishes explicitly between mutation and selection on the base composition of the evolving sequence. It also distinguishes between transitions and transversions. In base compositional equilibrium, transitions are assumed to obey the compositional balance within the pool of purines, or pyrimidines:

$$P(\mathbf{G} \rightarrow \mathbf{A}) = P(\mathbf{A} \rightarrow \mathbf{G}) \tag{4}$$

and

$$P(\mathbf{T} \rightarrow \mathbf{C}) = P(\mathbf{C} \rightarrow \mathbf{T}). \tag{5}$$

Transversions also obey compositional balance, and they are assumed to proceed as follows:

1. Random choice of a position in the sequence to undergo a transversion. The probability of a given base to be chosen for a transversion mutation (tv) thus depends linearly on its frequency in the sequence:

$$P(\mathrm{tv_N}) = F(\mathrm{N}). \tag{6}$$

2. Choice of the particular substitution. The conditional probabilities of the two competing transversions are directly proportional to the frequencies of the two possible bases. If G, for example, undergoes a transversion, selection will probabilistically balance the two possible outcomes.

$$P(\mathbf{G} \rightarrow \mathbf{T}|\mathrm{tv_G}) \approx F(\mathrm{T}) \tag{7}$$

and

$$P(\mathbf{G} \rightarrow \mathbf{C}|\mathrm{tv_G}) \approx F(\mathrm{C}). \tag{8}$$

Thus,

$$P(\mathbf{G} \rightarrow \mathbf{T}) = \delta\ F(\mathrm{G})\ F(\mathrm{T}) \tag{9}$$

and

$$P(\mathbf{G} \rightarrow \mathbf{C}) = \delta\ F(\mathrm{G})\ F(\mathrm{C}) \tag{10}$$

where δ is a proportionality coefficient for the overall probability of any transversion. Remarkably, the model leads to "reversibility." For example, a substitution from G to T becomes as likely as a substitution from T to G because

$$P(\mathbf{G} \rightarrow \mathbf{T}) = \delta\ F(\mathrm{G})\ F(\mathrm{T}) = P(\mathbf{T} \rightarrow \mathbf{G}) = \delta\ F(\mathrm{T})\ F(\mathrm{G}). \tag{11}$$

The two steps of the substitution process are modeled after two processes that exist in sequence evolution: mutation and selection.[1]

1. Mutation as the initial step in sequence evolution. The frequency of a given mutation is dependent on the frequency of the original base and on the error rate of the DNA replication and repair machinery. Assuming that the latter distinguishes only transitions from transversions, our model accurately describes the process of mutation.

[1]This model may not apply in cases where the compositional bias is due to mutation pressure instead of selection. Under those circumstances, the condition necessary for CS to work is that one type of base (e.g., G) does not prefer to change to one specific base (e.g., C) while its counterpart (in this example, A) prefers to change to the other one (here, T). The condition may be expressed as

$$P(\mathbf{G} \rightarrow \mathbf{C}) \,/\, P(\mathbf{G} \rightarrow \mathbf{T}) = P(\mathbf{A} \rightarrow \mathbf{C}) \,/\, P(\mathbf{A} \rightarrow \mathbf{T}).$$

As long as this condition is met, CS should also apply to sequences whose base compositional biases are caused by directional mutation pressure (Sueoka, 1988; M. Bulmer, personal communication), rather than selection.

2. Global selection on the base composition. Given that the base composition has to be maintained in equilibrium, modeling selective pressure according to the base frequencies is the simplest way of describing this step of sequence evolution.

Based on this model, CS calculates the expected frequencies of phylogenetically informative patterns from the additional information present in the sequences. First, a base composition matrix is constructed. The frequencies of all bases at those positions in which one and only one sequence differs by a transversion from the other three sequences are entered in this matrix. The expected frequencies of occurrence of all informative nucleotide patterns are then calculated. We recognize that in this step, a multiple-hit correction should be incorporated. Such a correction is not available at present, however.

We stress that these frequencies are expected under evolution on the peripheral branches only. Thus, the null hypothesis of the test can be paraphrased as "no central branch." If there is no central branch, the distribution of informative patterns of the topology which is being tested will not exhibit significant deviations from the expected values. However, if one of the topologies is correct, there will have been transversions on its central branch. These contribute a number of parsimony patterns that shift the observed distribution of informative patterns away from the expected one. Given a long enough central branch, the null hypothesis ("no central branch") will be rejected, leading to significant support for that topology.

The statistical test of CS is the following chi-square (χ^2) test:

$$\chi^2 = (\text{pars} - np_{\text{pars}})^2/np_{\text{pars}} + (\text{back} - np_{\text{back}})^2/np_{\text{back}} \tag{12}$$

where "pars" and "back" are observed numbers of parsimony and background patterns, respectively; p_{pars} and p_{back} are the expected frequencies of parsimony and background patterns (as calculated by CS), with $p_{\text{pars}} + p_{\text{back}} = 1$; and n is the sample size with $n = \text{pars} + \text{back}$.

In CS and EP, the difference between observed and expected background patterns will follow a distribution whose mean should be 0 for an incorrect topology. If t equivalent tests are conducted (for example, with t different genes from the same four organisms), for one half of all differences d_i,

$$d_i = \text{back} - np_{\text{back}} \tag{13}$$

where $0 < i \leq t$, the inequality

$$d_i \geq 0 \tag{14}$$

should hold. We call this an "overrepresentation of background patterns." For the other half, $d_i \leq 0$. By contrast, none of the tests of the correct topology should show an overrepresentation of background patterns because the substitutions on the central branch moved the distribution of d_i

away from its mean at 0. Thus, for all i,

$$d_i \leq 0 \tag{15}$$

should hold, unless multiple transitions on the informative patterns randomized the distribution of parsimony and background patterns. It is therefore possible to identify the correct topology by comparing the number of times equivalent topologies display an overrepresentation of background patterns. An example is given in Sidow and Wilson (1990) where two incorrect topologies of the RNA polymerase tree were identified using this approach. This approach thus adds power to the statistical tests of EP and CS.

RESULTS

Comparison of CS and EP

In order to determine whether CS's modification of EP is valuable under a range of conditions of sequence evolution, we conducted a series of computer simulations. Table 7-1 and Figure 7-3 show the parameters used in the simulations. The probability of a transition occurring was twice that of a transversion.[2] The branch lengths were chosen such that the total number of topology-relevant patterns (i.e., the sample size in the statistical test) was, on average, the same for both the correct tree and the (incorrect) tree that was favored by parallel events on paraphyletic branches. After each simulation, the generated sequences were analyzed with both EP and CS.

The performance of both methods was evaluated by three criteria: (1) rates of failure (i.e., the percentage of simulations in which the incorrect topology received a χ^2 value of greater than 3.84, which is the boundary for 95% confidence with one degree of freedom); (2) rates of success (i.e., the percentage of simulations in which the evolutionarily correct topology received a χ^2 value of greater than 3.84); and (3) overrepresentation of background patterns (i.e., the percentage of simulations in which the incorrect topology shows an overrepresentation of background patterns).

We are showing only the results for the simulations in which all 10,000

[2]We also conducted simulations using a fivefold transition bias. Under the conditions used, the error rates of CS and EP are very similar to those that we found for the twofold transition bias. However, there is a substantial loss of information for the correct topology because a high transition bias randomizes the distribution of true parsimony patterns that were created during evolution on the central branch. Real sequences with a high transition bias, such as third positions of animal mtDNA codons, thus pose a serious dilemma to the phylogeneticist. In such cases it is probably advisable to convert the sequences to two-character-state data, distinguishing only between purines and pyrimidines. A distance analysis may then give sufficient resolution, but it has to be borne in mind that "the proper distance measure" (*sensu* Jin and Nei, 1990) may be impossible to find in real data. Compositional statistics and EP should be immune to the problem if applied to nuclear genes because their mode of evolution does not show a high transition bias.

Table 7-1 Simulation Parameters in the Comparison of Evolutionary Parsimony and Compositional Statistics

Parameter	Value
Constant	
Number of replications	200
Transition bias	2*
Number of bases	10,000
Fraction of purines	0.5†
Variable	
Base composition (G or T)	25% to 45%‡
Variability of positions	1 or 1/100 to 1§
Transition probabilities	Calculated from‖ base composition

*For every transversion, two transitions were simulated. On average, therefore, each transition was four times as likely as each transversion.
†The sequences always contained 50% purines and 50% pyrimidines; setting the frequency of one base within each class therefore constrains the frequency of the other one.

‡The proportions of G and T were changed in increments of 5%. Therefore, there are 25 possible base compositions, from [25%G ; 25%T] to [45%G ; 45%T]. Ten of them (e.g., [30%G ; 25%T] and [25%G ; 30%T]) are twofold redundant, and simulations were only performed once for each pair.

§In the case of equal variability, all 10,000 positions had a probability of one, of accepting a substitution. When the variability varied, an equal number of positions in the sequence accepted substitutions with a probability of 100/100, 99/100, 98/100, . . ., 1/100. If a substitution was rejected, the search for a position was repeated until a position accepted a substitution.

‖The conditional probabilities of transitions were calculated by the formula $P(\mathrm{G} \rightarrow \mathrm{A} | \mathrm{ts}) = P(\mathrm{A} \rightarrow \mathrm{G} | \mathrm{ts}) = F(\mathrm{G})\,F(\mathrm{A}) / (F(\mathrm{G})\,F(\mathrm{A}) + F(\mathrm{T})\,F(\mathrm{C}))$ and analogously for T and C. Therefore, they are also dependent on the base composition. If, for example, G is much more abundant than A, there will be significant mutation pressure toward equilibration. In our model, this process will be counterbalanced by selection.

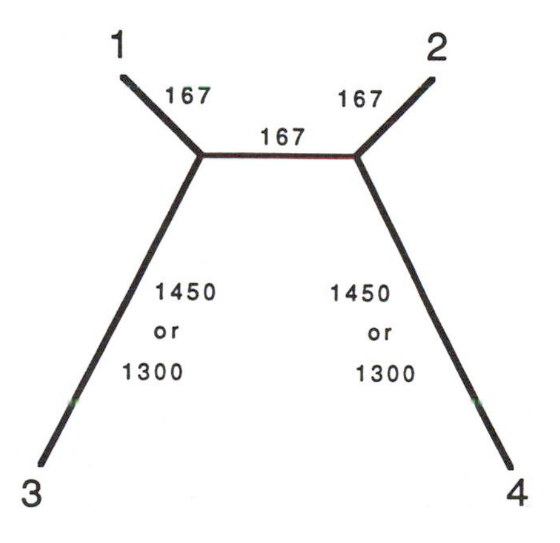

Figure 7-3 The branch lengths used in all analyses. Shown are the number of transversions on each branch. In the simulations in which the probability of nucleotide positions to change varied 100-fold, branches leading to 3 and 4 were slightly shorter than for the other analyses to generate a roughly equal number of informative patterns for the correct and incorrect topologies.

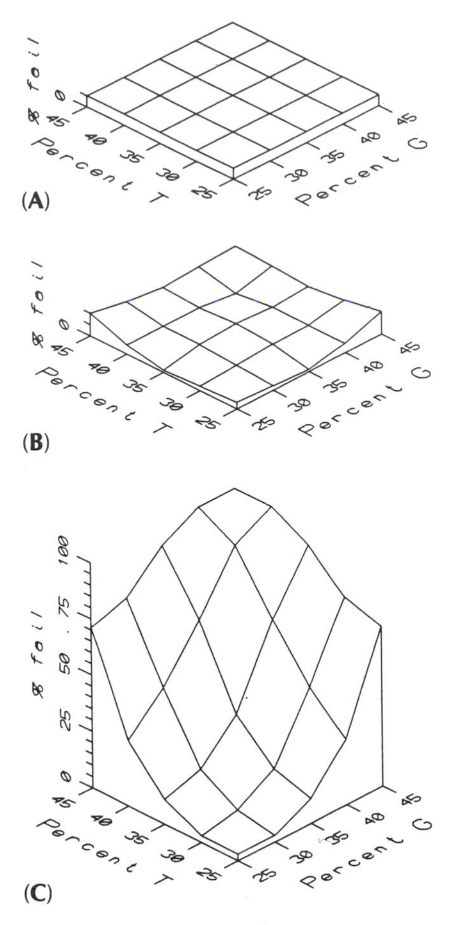

Figure 7-4 Three-dimensional plots showing the low rate of failure of CS compared to EP. The percentage of times the incorrect topology was chosen as shown on the vertical (Z) axis is plotted against a range of different base compositions (X and Y axes). (A) Expected distribution of failure when the confidence limit is set to 95%. (B) Failure by CS. (C) Failure by EP.

nucleotide sites were equally variable. The results for those simulations in which the variability differed 100-fold among positions were very similar to those shown (see below).

Rates of Failure

In Figure 7-4, the number of times a method chose the incorrect topology at statistical significance is plotted against the base compositional bias. Five percent failure is expected because the confidence interval is set to 95% (Fig. 7-4A). The error rate of CS exceeds the 5% ceiling only under extreme compositional inequalities (Fig. 7-4B). This is probably due to the lack of a multiple-hit correction during construction of the base composition matrix. By contrast, EP fails at a higher frequency than 5% (Fig. 7-4C), even under relatively mild compositional inequalities.

Rates of Success

The results (with respect to the correct topology) are shown with the same kind of plot as for the incorrect topology (Fig. 7-5). The performance of EP (Fig. 7-5A) has to be considered in context with the rates of failure in

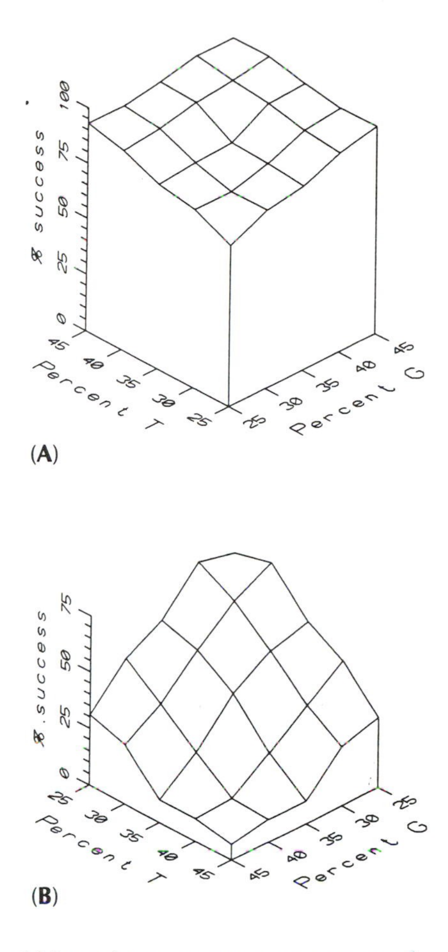

Figure 7-5 Success plots depicting the percentage of times EP and CS calculate a χ^2 value of greater than 3.84 for the correct topology. (A) EP. (B) CS. These plots are analogous to those in Figure 7-4, but note the reversal of the values on the X and Y axes (B).

Figure 7-4C. As the base compositional biases become more pronounced, EP chooses both the correct and the incorrect topology at statistical significance. This reflects the systematic error due to violation of EP's assumptions. Generally, the higher χ^2 value still identifies the correct topology, but with less consistency than does CS (not shown). Compositional statistics chooses the correct topology at statistical significance, but the performance decreases as base compositional inequalities become extreme (Fig. 7-5B). The latter phenomenon can be attributed to the loss of power of the method when base compositional inequalities, like the ones we simulated, cause convergent evolution in conjunction with large rate differences among lineages. Naturally, no method of phylogenetic inference is robust against convergent evolution. Therefore, data sets with extreme base compositional biases should be avoided.

Rejection of the Incorrect Topology by an Overrepresentation of Background Patterns

The simulations are in agreement with the theoretical considerations above. The expected 50% value (Fig. 7-6A) is not approximated by EP's plot

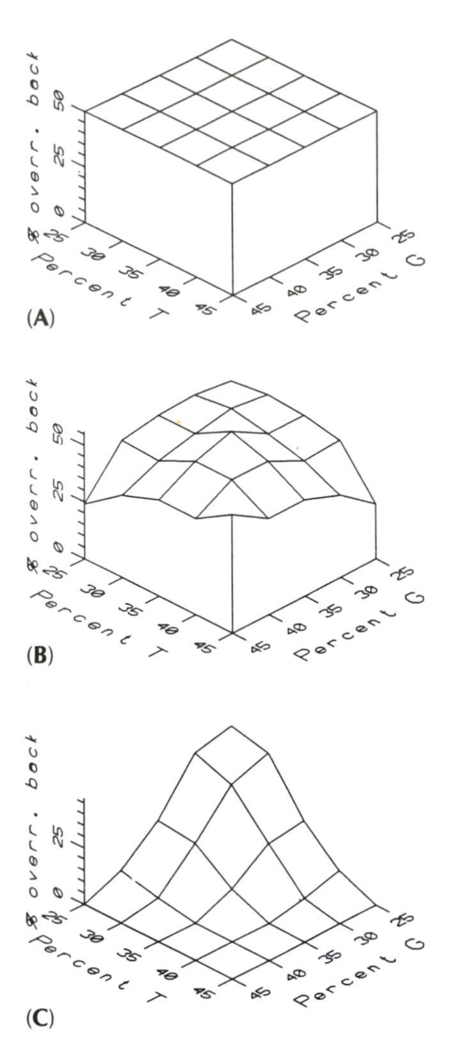

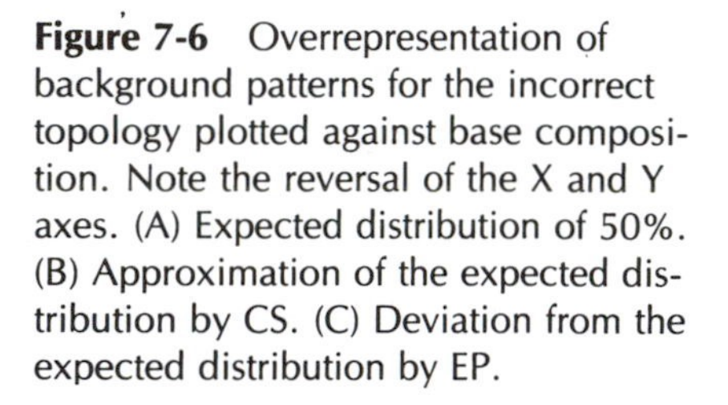
Figure 7-6 Overrepresentation of background patterns for the incorrect topology plotted against base composition. Note the reversal of the X and Y axes. (A) Expected distribution of 50%. (B) Approximation of the expected distribution by CS. (C) Deviation from the expected distribution by EP.

(Fig. 7-6C). Compositional statistics, however, shows a distribution closer to the expected one (Fig. 7-6B) and identifies an overrepresentation of background patterns in about 50% of all incorrect topologies. When the correct topology was tested with CS, the values rose above 10% only under the most extreme base compositional inequalities (not shown). The distribution remained generally low despite the significant probability of multiple hits under the branch lengths used. Thus, CS can still identify incorrect topologies even when the correct one is not chosen at statistical significance any more.

Comparison of CS With a Distance Method

Using the same simulation parameters as those for comparing EP and CS (see Table 7-1; Fig. 7-3), we repeated the simulations for analysis by a

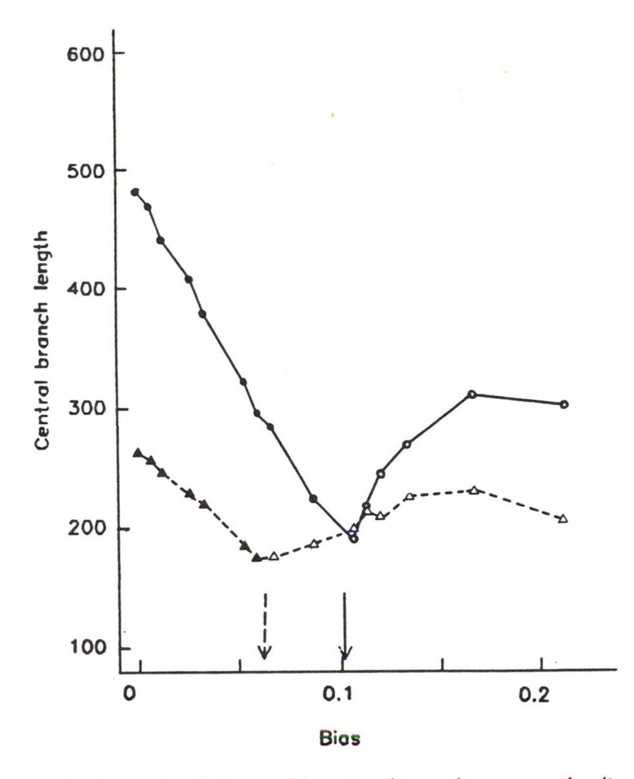

Figure 7-7 Results from simulations addressing the performance of a distance method (neighbor-joining) as a function of base compositional bias. X axis, base compositional bias as calculated by formula (16). Y axis, length of the central branch as calculated by the distance method. (Actual length = 500 substitutions.) Solid symbols denote central branch lengths for the correct topology, open symbols denote central branch lengths for the incorrect topology. The continuous line connects data points of simulations in which all positions in the sequences were equally likely to change. The dashed line connects data points of those simulations in which the positions varied in their likelihood to undergo change (Table 7-1). Arrows show critical values of the base compositional bias. For values to the right of the arrows, the method begins to assign a central branch to the incorrect topology.

distance method. Because of the computational time required, we decided to use a shortcut in assessing the performance of the method. From 50 simulations, the average of the observed distances was corrected for multiple hits, distinguishing transitions from transversions (Kimura, 1980). This is a valid approach because the length of the sequences and the number of substitutions are great enough for the variance in observed distances to become small, when individual simulations are compared. The resulting distance matrix was analyzed by the neighbor-joining program (Saitou and Nei, 1987). The length of the central branch as calculated by the neighbor-joining algorithm was plotted against the base compositional bias (Fig. 7-

7) as calculated by the formula

$$B = 4/3\ \Sigma(0.25 - b_i)^2 \tag{16}$$

where subscript i denotes the bases (G, A, T, or C), and b_i the frequency of base i in the sequence. The analysis was then repeated for the same 15 base compositions as used for CS and EP and both types of positional variabilities (see Table 7-1). The following observations are of importance.

1. When all positions are equally variable (Fig. 7-7, solid line), the method underestimates the length of the central branch of the correct topology as soon as the base composition begins to be skewed. When the base compositional bias becomes greater than 0.1, it starts choosing the incorrect tree (solid arrow).
2. When the positions in the sequence are assumed to vary with respect to probability of change (Fig. 7-7, dashed line), the distance method significantly underestimates the length of the central branch in all cases, even when the base composition is not skewed at all. This is in agreement with Lake's simulation study, which specifically addressed positional variability (J. Lake, unpublished). When base compositional biases become greater than 0.06, the distance method chooses the incorrect tree[3] (dashed arrow). Thus, under conditions of compositional bias, CS appears to be significantly more robust than a distance method.

DISCUSSION

Comparing CS and EP

Our computer simulations show that CS is a robust and general method of phylogenetic inference. Just like EP, it functions under a variety of different conditions. Most importantly, it is unaffected by a wide range in variability among different positions of the sequence. This is a major strength of both CS and EP. Distance measures, by contrast, require corrections based on significant assumptions about the substitution process to achieve additivity (Jin and Nei, 1990).

Another characteristic of both CS and EP is that three separate tests are required to identify the correct topology. This is due to the nature of the informative nucleotide patterns. The sets of informative positions do not overlap, and are independent of one another. This may seem to be a drawback, but it has an advantage: systematic errors are usually identified when there are significant results for more than one topology. Other methods choose a topology by a process that cannot determine whether other hypotheses would be equally consistent with the data.

[3]Any reduction in the length of the central branch of the simulated tree would increase the likelihood of the distance method choosing the other (incorrect) tree. Evolutionary parsimony or CS would not choose an incorrect tree but would simply indicate that no decision can be made.

The main difference from EP is that CS takes into account base compositional biases in its model of sequence evolution. We have shown that CS performs significantly better than EP under biased conditions. Compositional statistics also appears to be robust in cases in which the substitution process satisfies reversibility *independent* of the base composition. That is, when $P(\mathbf{A} \rightarrow \mathbf{T}) = P(\mathbf{T} \rightarrow \mathbf{A})$, $P(\mathbf{G} \rightarrow \mathbf{T}) = P(\mathbf{T} \rightarrow \mathbf{G})$, etc., for all transversions, CS's performance was very similar to that reported here (not shown; see footnote 1).

When there are no base compositional biases, and/or transversions are balanced, CS becomes identical to EP. In this case, CS gives $p_{\text{pars}} = p_{\text{back}} = 0.5$, the value assumed by EP. Thus, CS is unlikely to do worse than EP. In addition, since EP tends to fail more often than in 5% of all tests even in cases when the base composition is not highly biased (see Fig. 7-4), we recommend that CS be used, if applicable. In practice, however, it may be easier to use EP, if a preliminary analysis using CS's algorithm shows that there is no deviation from the equality $p_{\text{pars}} = p_{\text{back}}$.

Compositional statistics' performance also appeared to be more reliable than that of a distance method, when base compositional bias was present. It is especially noteworthy that the bias and different positional variability affect the performance of the distance method synergistically, as can be seen in Figure 7-7.

Under conditions of base compositional disequilibrium, CS's assumptions (as well as those of most other methods of phylogenetic inference) are violated. Since disequilibrium is common between distantly related sequences, this is a potential problem to be aware of. Compositional statistics may identify certain kinds of disequilibrium (Sidow and Wilson, 1990: appendix II), but the specific circumstances under which disequilibrium affects a phylogenetic analysis have not been studied.

Applications for CS

The robustness of CS renders it particularly useful for:

1. orthologous genes with a wide range of variability among nucleotide positions within the sequences;
2. paralogous genes in which unequal rate effects, base compositional inequalities, or other phenomena may render other methods inapplicable; and
3. analyses involving highly divergent sequences with strongly varying branch lengths.

The recent advancement of phylogenetic inference into the realm of statistical analysis, together with an explosion of nucleotide sequences, is generating high standards for phylogenetic analysis. Statistically significant results are now important to lend support to one's conclusions. A drawback of both CS and EP is that they are conservative methods; they need a lot of sequence data to attain sufficient sample sizes for their statistical tests to be reliable. However, the conditions under which we tested them are rather extreme and the length of the central branch was small compared to the peripheral branches (see Fig. 7-3). The number of homologous nucleotides needed to obtain a significant result with real sequences depends on a number of factors; for example, the length of the central branch compared to the peripheral ones, the base compositional biases, and the number of available taxa.

As more sequences become available from a range of organisms, the relatively low sensitivity of CS and EP will be less of a limitation. An example for this is our phylogeny of RNA polymerase sequences representing the three deepest branches in the tree of life (Sidow and Wilson, 1990). By two criteria, CS was successful in that analysis: (1) only one topology was chosen more often than the expected 5% and was therefore inferred to be the correct one; and (2) for both other topologies, background patterns were overrepresented in approximately 50% of all tests, in contrast to the selected one which did not show this pattern. Of course, not nearly all problems in phylogenetic analysis will require application of CS. Maximum likelihood methods are better if the sequences' mode of evolution is closely approximated by the parameters used in the analysis—a requirement that is usually hard to meet. For closely related sequences, parsimony or distance methods are clearly superior (Saitou and Imanishi, 1989; Jin and Nei, 1990). Yet when distant evolutionary relationships are addressed, several phenomena may cause such methods to produce serious systematic errors. Such errors can be avoided by using a more general method such as CS, the assumptions of which are less likely to be violated by the substitution process.

SUMMARY

The performances of EP and CS were directly compared using computer-generated DNA sequences. Both methods are robust under two adverse (but frequently occurring) conditions: (1) a wide range in variability among nucleotide positions; and (2) large differences in lengths of the terminal branches. Therefore both methods are generally rate-invariant. A weakness is their requirement for large data sets.

The major difference between CS and EP is their performance under conditions that require maintenance of base compositional inequalities (caused in our model by selective constraints on the DNA's primary structure). As the bias becomes more pronounced, two effects become apparent: (1) EP begins to choose the incorrect tree at higher frequencies than allowed by the confidence interval of its statistical test (CS, in contrast, remains robust in such cases); and (2) identification of the incorrect topology by an overrepresentation of background patterns becomes impossible with EP, whereas CS retains this ability. A distance method also showed decreased performance under the same simulation conditions.

ACKNOWLEDGMENTS

We thank David Irwin, James Lake, Trang Nguyen, Cecilia Saccone, Naruya Saitou, and Terry Speed for helpful discussion, and Thomas Kocher for the formula for base compositional bias. Supported by grants from the National Science Foundation and National Institutes of Health to ACW.

REFERENCES

Anderson, S., A.T. Bankier, B.G. Barrell, M.H.L. de Bruijn, A.R. Coulson, J. Drouin, I.C. Eperon, D.P. Nierlich, B.A. Roe, F. Sanger, P.H. Schreier, A.J.H. Smith, R. Staden, and I.G. Young. (1981) Sequence and organization of the human mitochondrial genome. *Nature* **290**:457–465.

Bernardi, G., B. Olofsson, J. Filipski, M. Zerial, J. Salinas, G. Cuny, M. Meunier-Rotival, and F. Rodier. (1985) The mosaic genome of warm-blooded vertebrates. *Science* **228**:953–958.

Felsenstein, J. (1978) Cases in which parsimony or compatibility methods will be positively misleading. *Syst. Zool.* **27**:401–410.

Felsenstein, J. (1981) Evolutionary trees from DNA sequences: a maximum likelihood approach. *J. Mol. Evol.* **17**:368–376.

Felsenstein, J. (1989) *Phylogenetic Inference Programs (PHYLIP)*, Manual 3.2. University of Washington, Seattle, and University Herbarium of the University of California, Berkeley.

Fitch, W. M. (1986) The estimate of total nucleotide substitutions from pairwise differences is biased. *Philos. Trans. R. Soc. Lond.* [*Biol. Sci.*] **316**:317–324.

Fox, G.E., E. Stackebrandt, R.B. Hespell, J. Gibson, J. Maniloff, T.A. Dyer, R.S. Wolfe, W.E. Balch, R.S. Tanner, L.J. Magrum, L.B. Zablen, R. Blakemore, R. Gupta, L. Bonen, B.J. Lewis, D.A. Stahl, K.R. Luehrsen, K.N. Chen, and C.R. Woese. (1980) The phylogeny of prokaryotes. *Science* **209**:457–463.

Gojobori, T., E.N. Moriyama, and M. Kimura. (1990) Statistical methods for estimating sequence divergence. Pp. 531–550 in *Molecular Evolution: Computer Analysis of Protein and Nucleic Acid Sequences* (R. F. Doolittle, ed.). Methods in Enzymology, vol. 183. Academic Press, New York.

Gouy, M., and W.-H. Li. (1989) Phylogenetic analysis based on rRNA sequences supports the archaebacterial rather than the eocyte tree. *Nature* **339**:145–147.

Hasegawa, M., H. Kishino, and T. Yano. (1985) Dating of the human-ape splitting by a molecular clock of mitochondrial DNA. *J. Mol. Evol.* **22**:160–174.

Holmquist, R., M.M. Miyamoto, and M. Goodman. (1988) Analysis of higher-primate phylogeny from transversion differences in nuclear and mitochondrial DNA by Lake's methods of evolutionary parsimony and operator metrics. *Mol. Biol. Evol.* **5**:217–236.

Jin, L., and M. Nei. (1990) Limitations of the evolutionary parsimony method of phylogenetic analysis. *Mol. Biol. Evol.* **7**:82–102.

Kimura, M. (1980) A simple method for estimating evolutionary rates of base substitutions through comparative studies of nucleotide sequences. *J. Mol. Evol.* **16**:111–120.

Kishino, H., and M. Hasegawa. (1990) Converting distance to time: application to human evolution. Pp. 550–570 in *Molecular Evolution: Computer Analysis of Protein and Nucleic Acid Sequences* (R. F. Doolittle, ed.). Methods in Enzymology, vol. 183. Academic Press, New York.

Lake, J.A. (1987) A rate-independent technique for analysis of nucleic acid sequences: evolutionary parsimony. *Mol. Biol. Evol.* **4**:167–191.

Lake, J.A. (1988) Origin of the eukaryotic nucleus determined by rate-invariant analysis of rRNA sequences. *Nature* **331**:184–186.

Li, W.-H., and D. Graur. (1991) *Fundamentals of Molecular Evolution*. Sinauer Associates, Sunderland.

Olsen, G. (1987) Earliest phylogenetic branchings: comparing rRNA-based evolutionary trees inferred with various techniques. *Cold Spring Harbor Symp. Quant. Biol.* **52**:825–837.

Perrin, P., and G. Bernardi. (1987) Directional fixation of mutations in vertebrate evolution. *J. Mol. Evol.* **26**:301–310.

Saccone, C., G. Pesole, and G. Preparata. (1989) DNA microenvironments and the molecular clock. *J. Mol. Evol.* **29**:407–411.

Saitou, N., and T. Imanishi. (1989) Relative efficiencies of the Fitch-Margoliash, maximum-parsimony, maximum-likelihood, minimum-evolution, and neighbor-joining methods of phylogenetic tree construction in obtaining the correct tree. *Mol. Biol. Evol.* **6**:514–525.

Saitou, N., and M. Nei. (1987) The neighbor-joining method: a new method for reconstructing phylogenetic trees. *Mol. Biol. Evol.* **4**:406–425.

Shoemaker, J.S., and W.M. Fitch. (1989) Evidence from nuclear sequences that invariable sites should be considered when sequence divergence is calculated. *Mol. Biol. Evol.* **6**:270–289.

Sidow, A., and A.C. Wilson. (1990) Compositional statistics: an improvement of evolutionary parsimony and its application to deep branches in the tree of life. *J. Mol. Evol.* **31**:51–68.

Sueoka, N. (1988) Directional mutation pressure and neutral molecular evolution. *Proc. Natl. Acad. Sci. USA* **85**:2653–2657.

Vigiliant, L., R. Pennington, H. Harpending, T.D. Kocher, and A.C. Wilson. (1989) Mitochondrial DNA sequences in single hairs from a southern African population. *Proc. Natl. Acad. Sci. USA* **86**:9350–9354.

8

Weighted Parsimony: Does it Work?

WALTER M. FITCH AND JIA YE

In 1969, Farris suggested a relatively unbiased way of weighting the value of a character for systematic purposes based upon the proposition that characters that frequently change their state are unreliable guides to relationships, whereas those that change only once are perfect guides to relationships.

In nucleotide sequences, one can apply the same philosophy not only to the various characters (homologous positions in the sequences), but to the character changes as well. Thus, if transition changes (e.g., C ↔ U) are more common than transversion changes (e.g., C ↔ A), then one might give less weight to the former events than to the latter. In its extreme form this leads to transversion parsimony where transition changes are not counted at all.

Williams and Fitch (1989) devised a scheme to generalize such weighting of character-state changes which proved to be identical to that of Sankoff and Cedergen (1983).

A computer program has been developed that permits one to perform either kind of weighting, or both simultaneously, plus some other options which are explained in detail in Williams and Fitch (1990). (Some further details of the program are given in the section, Weighted Tree Reconstruction.) The program is freely available by writing to the senior author.

In principle, weighted parsimony should give superior results to parsimony with uniform weighting of characters and of state changes, because weighting reduces the impact of noisy positions by giving emphasis to the rarer events. In this work, we use simulation to test whether the principle works in practice. More tests will be required before the value of weighting can be assessed reliably, although the tests presented here suggest the procedure is useful.

METHODS

The Simulation Program

Descent with change of an ancestral nucleotide sequence was accomplished by computer. An ancestral sequence of 300 nucleotides is obtained by random sampling of a pool of nucleotides of whatever composition the experimenter chooses. In these tests, all nucleotides were equiprobable.

The descent was according to the tree shown in Figure 8-1 where the numbers on the branches are the number of nucleotide substitutions that are accepted between branchings. The experimenter may choose other values.

The choice of the nucleotide substitution was random with the probability of the transition being 0.8, and the probability of the two transversions being equal to 0.1 each. Again the experimenter may choose other values.

The choice of the nucleotide to be substituted was random, but only from among the variable positions. The experimenter may decide on any fraction, v, of the nucleotides to be variable and thus $1 - v$ of them are invariable. A set of nucleotide sites from the sequence is randomly assigned to the variable set.

To model the property of the variable sites changing as nucleotide substitutions are fixed [concomitantly variable codons or "covarions," (Fitch and Markowitz, 1970); concomitantly variable nucleotides or "covariotides," (Fitch, 1986; Shoemaker and Fitch, 1989)], a persistence (of variability) parameter, p, is set by the experimenter. Thus, $1 - p$ is the probability that, after each substitution of a variable nucleotide, one randomly chosen member of the currently variable set will be exchanged with a randomly chosen member of the currently invariable set of nucleotides. There is no restriction on the number of times a position may change its variability status. This program is also available from the senior author.

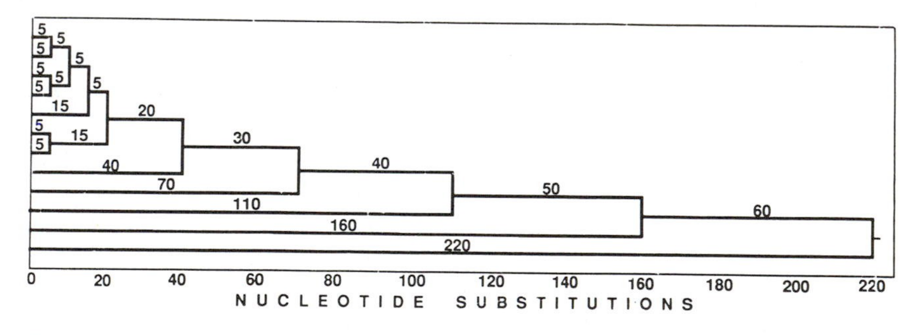

Figure 8-1 The simulated tree. An ancestral sequence of 300 nucleotides was allowed to evolve as depicted, the number of nucleotide substitutions between nodes being shown on the figure. The resulting sequences at the 12 tips were used to test weighted and uniform parsimony procedures that attempt to recover the correct tree. Branch lengths are proportional to the number of substitutions thereon so that for this model the molecular clock is perfect. The length of the tree is 880 nucleotide substitutions.

Weighted Tree Reconstruction

As a result of the preceding process, one obtains a set of sequences (in this case 12) corresponding to the tips of whatever tree one gives the program (Fig. 8-1). It is necessary then to try to recover, from the sequences alone, the phylogeny depicted in Figure 8-1. The process normally involves providing a starting tree, either by the program or the investigator, and determining that tree's length, that is, determine the minimum number of nucleotide substitutions required to account for the descent of these sequences given the topology of the tree. This is a simple and well-known procedure (Fitch, 1971). This tree topology is then modified by neighbor-branch swapping, and the process repeated to see if a more-parsimonious tree (= shorter length = fewer substitutions) can be found. The process of topology modification continues until a tree is found for which none of the allowable swaps produces a more-parsimonious tree. That minimal tree is then considered to be the best estimate of the true topology.

Establishing Parameters

In the initial phases, we wished to determine the value of v (fraction of sequence that are covariotides), p (persistence), and the lengths of branches that would produce errors when uniform parsimony was applied, so that we could see if weighted parismony did better. Thus, we always gave the program the correct tree to start from so that we would be guaranteed that its shortest tree was no longer than the correct tree. Once we had such values, a similar procedure was used for the test of the method, but now we examined all ten sets of sequences produced whether or not uniform parsimony obtained the correct tree. This is because we need to know not only how often weighted parismony improves upon the errors of uniform parsimony, but also how often it does worse than uniform parsimony.

RESULTS

The effect of the number of covariotides and their persistence seemed to have little effect in terms of causing uniform parsimony to get the wrong tree compared to the effect of branch length. Analysis showed that until the number of positions changing more than once in two contiguous branches of the tree was sizeable, uniform parsimony kept getting the correct tree. Hence, we ended up with a tree containing nearly three substitutions per site in order to get incorrect trees with any frequency.

Nature of the Test

The test actually involves only ten simulations. These represent our first efforts, and the small sample size means that all inferences are weak and are only suggestive. To understand the results, a few details should be understood about the weighting procedure. Subject to the initial conditions, the procedure looks for the best tree it can find. On the basis of that tree, it establishes new weights for the positions and for the nucleotide substitutions. Using those new weights, it takes a starting tree (in this work the starting tree of successive passes was always the best tree of the previous pass) and again looks for a better tree. The succession of passes continues until the same (best) tree topology is obtained on two passes in a row. (There is provision for detecting the rare occurrence of cycling which did not occur in this study.)

Our weighting procedure has 20 variant forms in addition to uniform parsimony. In uniform parsimony, which is what most investigators use, only one pass is performed and all positions have the same weight, as do all nucleotide changes. The other 20 variants are a function of three properties: (1) initial conditions; (2) substitutional change symmetry; and (3) type of position weighting.

Initial (first pass) conditions for nucleotide change weighting are of three forms. The first is all changes equal (uniform). The second is based upon the frequency of positions that have different pairs of nucleotides (existential). The third is like the second except the count is over the number of ways of selecting a nucleotide pair in each position, summed over all positions (combinational) (see Williams and Fitch, 1989, 1990, for further explication).

Substitutional change symmetry is whether or not one requires that the weight of a change in one direction must be equal to the weight in the reverse direction, (e.g., $A \rightarrow C = C \rightarrow A$?).

Type of position weighting can be uniform, but after the first pass can be related inversely to the number of changes in that position observed in the best tree of the previous pass. For the latter, positions can be weighted inverse linearly or inverse quadratically. In these tests, we performed the procedure for all 21 parameter combinations and asked how many of the 20 non–uniformly-weighted ways gave the correct tree. Unlike uniform weighting, two different topologies seem never to give the same weighted score, although we could synthesize data that would.

Outcome of the Tests

The ten tests gave results summarized in Table 8-1. Although there are four combinations of p and V, we held $V(1 - p) = \text{constant} = 25$. Tests 1, 2, 5, and 6 represent a standoff in that both uniform parsimony and all 20 versions of weighted parsimony were all right or all wrong. Test 3 might seem to be a bad case for weighting in that three of the 20 weighted trees were incorrect, but they differed only by the single swap of sequences 10 and 11.

Test 4 is in fact a worse case for weighting. Four of the five most-parsimonious uniform trees are, of course, wrong, but none of the nine, wrong, weighted, best trees are among the four most-parsimonious trees

Table 8-1 Uniform versus Weighted Parsimony*

Test	V	p	Uniform		Weighted		E	Length
			+	−	+	−		
1	100	0.75	1		20		220	340
2	100	0.75	1		20		220	370
3	50	0.5	1		17	3	440	315
4	25	0.0	1	4	11	9	880	353
5	100	0.75		1		20	220	352
6	50	0.5		3		20	440	313
7	100	0.75		1	14	6	220	331
8	50	0.5		1	14	6	440	313
9	250	0.9		1	14	6	88	502
10	100	0.75		1	18	2	220	341

*Test is a simple numbering of the ten cases; V is the number of covariotides among the 300 nucleotides; p is the persistence of variability; under "uniform" and "weighted" are given the number of times the correct (+) and incorrect (−) tree was obtained; E is the expected number of times a covariotide exchanged places with an invariable nucleotide; length is the number of substitutions in the most-parsimonious tree, using uniform weighting irrespective of whether it was the correct tree. Tests 4 and 6 had five and three equally most-parsimonious trees under uniform weighting; one of the five was the correct tree. The correct tree was never worse than the best tree by more than four substitutions, nor by more than three neighbor branch swaps.

that are incorrect. In fact, they are one extra branch swap farther away topologically from the correct tree than are the four incorrect most-parsimonious trees of uniform parsimony. Tests 7 to 10 get increasingly larger numbers of correct trees (where uniform parsimony gets the wrong tree) and thus support weighting.

DISCUSSION

The first observation is that parsimony, even unweighted, is clearly robust for the tree presented. We had to simulate a tree with many substitutions before parsimony gave the wrong answer, and even then the tree was nearly right, usually being only one branch swap from the correct tree. Of course the trees would have gone wrong sooner if we had had fewer substitutions on the internodal branches.

On the other hand, there was poor recovery of the actual substitutions, generally only about 40% of the total, and not all of those were the correct changes in the correct positions. Of course, recovery would have been better had there been fewer substitutions along the branches.

We had thought that reducing the number of covariotides would, by increasing the frequency of second substitutions in the same site in neighboring branches, cause the parsimony procedure to go wrong with fewer substitutions. This appears to be true, but with less strength than anticipated. We suspect that this arises, at least partly, because the occasional removal of a changed covariotide from the variable group makes it impossible to destroy the evidence it bears by additional homoplasious changes.

The second observation is that weighting indeed improves the performance. Excluding the standoff cases (tests 1, 2, 5, and 6), weighting more commonly got a better tree. Even in test 6, where none of the weighted trees was correct, in all 20 cases the tree was the same and was only one branch swap removed from the correct tree, whereas two of the three incorrect best uniformly weighted trees were two branch swaps away from the correct tree. Only in test 4 were the weighted trees two branch steps removed from the correct tree, while the incorrect uniform trees were but one swap away.

What would be of greater interest is to discover if the benefits of weighting are more pronounced for some of the 20 options than others. This turns out to be the case, although, not surprisingly, no single option is invariably superior to all other options.

Table 8-2 shows how well each of the 21 choices fared for the six nontrivial cases. The pair of values in the upper left corner (two right, four wrong) are the results of uniform weighting (usually misleadingly called unweighted) of both positions and transformations. The column, as it continues below that cell, shows the effects of nonuniform-transformation weighting while maintaining uniform positional (character) weighting. The top row to the right of that cell shows the effect of nonuniform-character weighting while maintaining uniform-transformation weighting. The remainder of the table shows the effect of combinations of nonuniform positional and transformation weighting.

From this, it appears that: (1) if there is uniform transformational weighting, nonuniform positional weighting does not help. We believe this may be because the simulation did not have a sufficiently great enough variance

Table 8-2 Frequencies of Getting the Correct Tree

	Positions*						
	U		*L*		*Q*		
Matrix†	+	−	+	−	+	−	Initial‡
U	2	4	2	4	2	4	*u*
	4	2	3	3	3	3	*u*
Ls	4	2	6	0	5	1	*e*
	4	2	5	1	5	1	*c*
	5	1	3	3	3	3	*u*
La	5	1	6	0	6	0	*e*
	5	1	5	1	5	1	*c*

*Positions may be weighted uniformly (*U*), inversely proportional to the number of substitutions they have (linear, *L*), or inversely proportional to the square of the number of substitutions they have (quadratic, *Q*).

†The transformation matrix may be weighted uniformly (*U*) or inversely proportional to the number of times a particular kind of nucleotide changed to any particular other nucleotide (linear, *L*) except that it can be either symmetric, *Ls*, or asymmetric, *La*, depending upon whether or not one counts, e.g., A → C with C → A. For *Ls*, one adds them together and divides by two.

‡For the initial pass, when there is no count of changes from a best tree, the starting transformation matrix may be uniform (*u*), based on the number of positions containing a pair (existential, *e*), or based on the number of ways a particular pair of nucleotides might be drawn at random (combinatorial, *c*).

in the number of substitutions per nucleotide. Nevertheless, such weighting often improves the result when there is also transformational weighting; (2) transformational weighting was effective even in the absence of positional weighting; (3) having the initial transformation matrix set to something other than all one's (rows with *e* or *c* on the right) was beneficial; (4) both positional and substitutional weighting should be used; (5) quadratic weighting does not seem more likely to lead to the correct tree than linear weighting; and (6) an asymmetric substitutional weighting does slightly better than symmetric weighting. The last choice may, however, be a consequence of using the correct root. In cases where there is not a clear outgroup, asymmetric weighting might be counterproductive as the directionality of each change will remain questionable.

The third observation, to be detailed elsewhere, is that application of the Fitch and Markowitz (1970) procedure to estimate the number of covariotides in these trees almost always obtained the correct value within 10%. Thus, the parameter that complicates the model, the number of covariotides, is in fact estimable from the sequence data.

Our conclusion from these tests, although subject to the performing of many more tests, is that weighted parsimony is more likely to get the correct or a better tree. There is added force to this conclusion to the extent that the model is more realistic because of its inclusion of covariotides, a feature we regard as an important and necessary reflection of biological reality and a feature not heretofore an important aspect of simulations. The model is unrealistic in other ways, particularly in its clocklike regularity, a condition imposed to cause all trees in the test to have the same underlying structure.

ACKNOWLEDGMENTS

This work was supported by National Science Foundation grant BSR-9096152.

REFERENCES

Farris, J.S. (1969) A successive approximations approach to character weighting. *Syst. Zool.* **18**:374–385.

Fitch, W.M. (1971) Toward defining the course of evolution: minimum change for a specific tree topology. *Syst. Zool.* **20**:406–416.

Fitch, W.M. (1986) The estimate of total nucleotide substitutions from pairwise differences is biased. *Philos. Trans. R. Soc. Lond.* [*Biol. Sci.*] **316**:317–324.

Fitch, W.M., and E. Markowitz. (1970) An improved method for determining codon variability in a gene and its application to the rate of fixations of mutations in evolution. *Biochem. Genet.* **4**:579–593.

Sankoff, D. and R.J. Cedergren. (1983) Simultaneous comparison of three or more sequences related by a tree. Pp. 253–263 in *Time Warps, String Edits, and*

Macromolecules: The Theory and Practice of Sequence Comparison (D. Sankoff and J.B. Kruskal, eds.). Addison-Wesley, Reading.

Shoemaker, J.S., and W.M. Fitch. (1989) Evidence from nuclear sequences that invariable sites should be considered when sequence divergence is calculated. *Mol. Biol. Evol.* **6**:270–289.

Williams, P.L., and W.M. Fitch. (1989) Finding the minimal change in a given tree. Pp. 453–470 in *The Hierarchy of Life. Molecules and Morphology in Phylogenetic Analysis* (B. Fernholm, K. Bremer, and H. Jornvall, eds.). Elsevier Science Publishers B.V., Amsterdam.

Williams, P.L., and W.M. Fitch. (1990) Phylogeny determination using a dynamically-weighted parsimony method. Pp. 615–625 in *Molecular Evolution: Computer Analysis of Protein and Nucleic Acid Sequences* (R. F. Doolittle, ed.). Methods in Enzymology, vol. 183. Academic Press, New York.

9

Testing the Theory of Descent

DAVID PENNY, MICHAEL D. HENDY, AND MICHAEL A. STEEL

> If it could be demonstrated that any complex organ existed which could not possibly have been formed by numerous, successive, slight modifications, my theory would absolutely break down (Darwin, 1859: 189).

The comment of Popper (1976:168) that "Darwinism is not a testable scientific theory, but a *metaphysical research program*—a possible framework for testable scientific theories" is sometimes used to question the scientific status of evolutionary theory. The original comment was not in any way "antievolution," and indeed, Popper noted "the strange similarity between my theory of the growth of knowledge and Darwinism" (Popper, 1976:169). Later, Popper (1978, 1984) limited these criticisms, but this has not always satisfied critics. One of our interests has been to determine, without ambiguity, if evolutionary theory could meet Popper's criteria for the demarcation of science.

The introductory quote given above is one of many examples of how Charles Darwin considered his theory of evolution to be falsifiable. Although Darwin is remembered today for the general aspects of theory of evolution, in his day-to-day research, he made many predictions from his theory—then sought (or made observations) to test the predictions. Some of the best known involve his experiments on plants, including work on orchid flowers (both in structure and function), pin and thrum flowers of primrose (and other dimorphic and trimorphic forms of flowers), and on the power of movement of plants (Ghiselin, 1969; Allan, 1977; Penny, 1985). This last example of the power of movement in plants is particularly interesting in that he had left a record of his reasoning. The first extract is from his autobiography.

> For in accordance with the principles of evolution, it was impossible to account for climbing plants having been developed in so many widely different groups, unless all kinds of plants possess some slight power of movement of an analogous kind.

And again, the delightful comment in a notebook from 1839,

> Is there any very sleepy mimosa, nearly allied to the Sensitive Plant?

(Details of sources are in Penny, 1985). The prediction that all plants should have some "slight power of movement" led to two books on his research, *The Movements and Habits of Climbing Plants* and *The Power of Movement in Plants*. Many other examples of how Darwin used predictions to direct research could be given, including features of human evolution.

It was not just a coincidence that Charles Darwin was looking for predictions from theories. During the early stages of the development of his theory (after the voyage of the Beagle), Darwin read widely in many areas of science. This was partly in order to understand what was required of a good scientific theory. His reading included works on the philosophy of science by Herschel, Whewell and Comte (Ruse, 1975; Schweber, 1977). Each of these authors emphasized the importance of prediction in science and, judging from letters to Charles Lyell (Schweber, 1977), Darwin was well aware of the importance of a good scientific theory leading to predictions.

In this century, Karl Popper (1963, 1972) has developed this theme further and emphasized the potential falsifiability of *scientific* theories. A theory is more useful (and therefore better) if it prohibits (excludes) a larger proportion of possible observations. The approach claims to be both descriptive and prescriptive. It is descriptive in that it claims to describe how the most effective scientists have worked, and it is prescriptive in that it advocates how scientists should aim to work. Hypotheses (or conjectures) are considered tools that are judged on their effectiveness in helping scientists devise new and more powerful tests whether the tests be observational, experimental, or analytical. None of this denies that sociological and cultural factors play a role in science—as long as such factors are considered descriptive of scientific procedures and not prescriptive of how scientists should select theories. During the past two decades there has been a strong anti-intellectual movement suggesting that hypotheses are largely determined by cultural factors, and consequently, to this extent are arbitrary. Such a movement has failed to develop falsifiable predictions and is not scientific by Popper's criterion for the delimitation of science.

In this chapter we review our approach to the study of evolutionary trees. This has been developed within a strong Popperian framework (Ridiford and Penny, 1984) of aiming to develop falsifiable hypotheses. After discussing some of the general issues involved, we then discuss the question of how good methods are for inferring trees, particularly from molecular data.

IS EVOLUTION A SCIENTIFIC THEORY?

Similar Trees From Different Sequences

The simple prediction from the "theory of descent" is that, because the sequences share the same tree pattern of ancestry, the optimal trees from different sets of data should be similar. (From the proposed stochastic nature of the mechanism of mutation and selection it would be surprising if the trees were identical. Indeed, it would be more devastating to Darwinism if different sets of short sequences always gave identical trees). Comparing results from different data sets had been used previously (e.g., Mickevich, 1978) and our interest was in getting quantitative results.

Our first major project was to compare trees derived for the same 11 mammalian taxa but from different sets of sequence data. We saw three requirements for being able to test the prediction of similar trees from different sequences. These were the ability to:

1. find the optimal tree(s) for 10 or more taxa;
2. find a tree comparison measure to compare trees objectively; and
3. derive the distribution of this tree comparison measure.

Finding the Optimal Tree

Sufficient taxa were required so that it would be most unlikely to get the same tree by chance. Fortunately, sequence data for 11 taxa were available for five proteins or peptides.

For 11 taxa there are 34,459,425 (17!!) binary trees. The number of trees increases exponentially with the number of taxa; therefore, any method which considers all trees cannot be efficient. It was shown quite early in the project that searching trees for the optimal tree(s) is an example of a set of problems known to be NP-complete (Graham and Foulds, 1982). This result implies that it is most unlikely that an efficient method will ever be found for a complete search on a large number of taxa. William Day has extended this analysis to other related problems with trees (see Day et al., 1986).

Finding optimal trees requires a program which is unbiased toward any subclass of trees. For example, it could happen that a program tended to group together adjacent taxa in the data matrix. An error of this nature could lead to the trees from different sequences being more similar then expected, not through common descent, but through program limitations. A quite different problem is that an optimality criterion, for example parsimony, could tend to favor some trees by bringing together isolated taxa. This will be discussed later.

In 1980, a search through all trees for 11 taxa was estimated to require 55 days with the computers we had available. The solution to this problem was the development of a branch and bound algorithm (Hendy and Penny, 1982), which is now widely used. It reduced the computing time for these data to just under five minutes, while still guaranteeing to have found all optimal trees. This met our first objective of finding optimal trees for the five sets of data.

Finding a Tree Comparison Measure

When the parismony branch and bound program was applied to each of the five different proteins, the minimal trees were not identical, but did

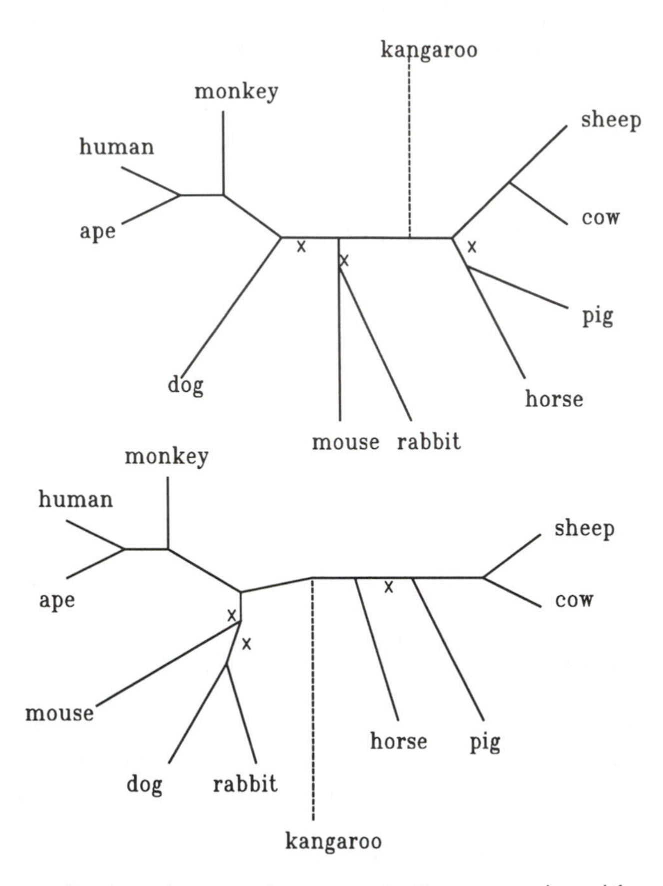

Figure 9-1 Application of symmetric tree metric. Two trees selected from the minimal trees from β-hemoglobin (top) and α-hemoglobin (bottom) sequences. On the symmetric difference metric there are six differences. These are indicated by a small cross on the three nonequivalent edges of each tree.

indeed look "similar" (Penny et al., 1982). The second objective was to find a useful tree comparison metric that would allow an objective comparison of these trees. The symmetric difference metric had been developed (Robinson and Foulds, 1981) from a parallel interest in trees and had an efficient method for its calculation (Penny and Hendy, 1985b). A more efficient method was developed by Day (1985).

The symmetric difference metric on two trees counts the number of edges that occur in one, but not both, trees. Edges are equivalent if they partition the taxa into the same two subsets (see Edge Bipartitions). This is repeated for each edge of the tree in turn. Figure 9-1 shows two minimal trees from different sequences which are distance six differences apart. The differences between the two trees are indicated by a small cross on edges which have no equivalent edge on the other tree.

Deriving the Distribution of the Tree Comparison Measure

At this point, it is unclear whether finding six differences is significant. What is the probability that two randomly selected trees would have six differences? In order to answer that question, the problem of deriving the distribution of the tree comparison metric must be solved. The expected distribution has been calculated for up to 16 taxa (Hendy et al., 1984) and is shown graphically in Penny and Hendy (1986) and Hendy et al., (1988). The distribution is highly asymmetric, which is useful for many biological applications because it is particularly sensitive for closely-related trees. We find the probability of randomly selecting trees on 11 taxa with six or fewer differences (as in Fig. 9-1) is 4×10^{-5}.

Similar results were found from comparisons of minimal trees from other sequences (Penny et al., 1982), where each pair of minimal trees from different sets of sequences was more similar than expected by chance. Thus, it is fair to claim that the original prediction that minimal-length trees from different data sets would be similar, is supported. The method of analysis allowed the possibility for the theory of descent to fail. We have more confidence in the theory if it passes quantitative tests.

Our conclusion is that the theory of descent can meet the same quantitative standards as expected in other areas of science. There is no need for "special pleading" that evolution is hard to quantify. The project led to improved techniques in several areas for studying trees. The problem of finding optimal trees was shown to be NP-complete (Graham and Foulds, 1982). Branch and bound methods were developed (Hendy and Penny, 1982) so that optimal trees could be found in reasonable time for up to at least 16 taxa. A tree comparison metric (Robinson and Foulds, 1981) was implemented by showing it could be calculated efficiently (Penny and Hendy, 1985b). The expected distribution of this metric was derived (Hendy et al., 1984) so that quantitative tests can be made. The analysis of a complex scientific problem in order to get falsifiable predictions can be a productive approach to science.

Additional Predictions

Are Shorter Trees Better?

The parsimony criterion for an optimal tree assumes that shorter trees (those requiring fewer changes) are better estimates of evolution than longer trees. As a corollary of this we would expect that trees shorter for one data set should also be shorter on other data sets. This is different from the previous section which compared minimal trees from independent data sets.

A test of this prediction is shown in Figure 9-2 where trees requiring 124 to 133 changes on the β-hemoglobin data were selected. The lengths of these trees were then determined on the combined sequences from cytochrome *c*, fibrinopeptides A and B, and α-hemoglobin.

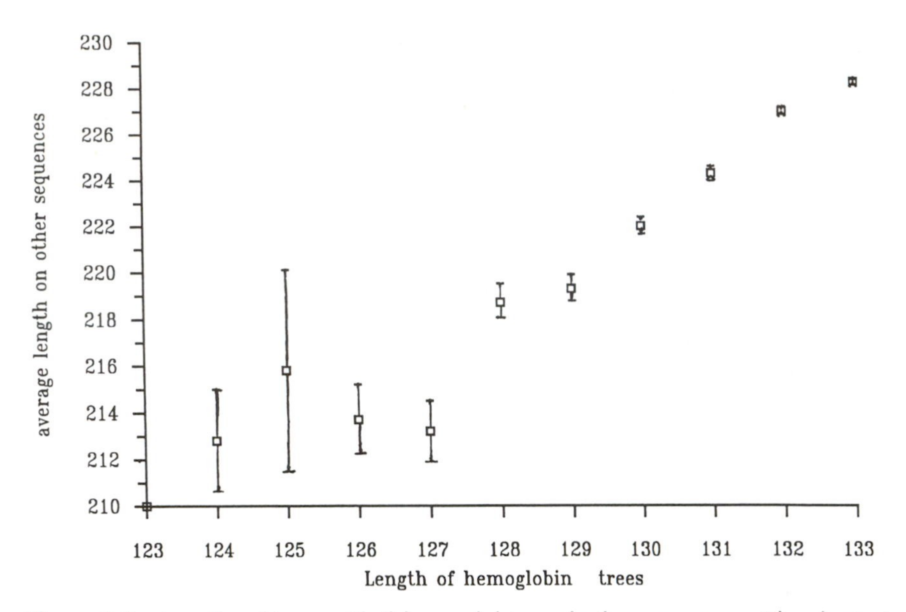

Figure 9-2 Lengths of trees with β-hemoglobin and other sequences. The shortest trees for β-hemoglobin sequences were selected. The lengths of these trees were then recalculated on the combined sequences from cytochrome c, fibrinopeptides A and B, and α-hemoglobin. *On average*, trees that require fewer mutations with β-hemoglobin, require fewer mutations with the new sequences. The bars are twice the standard error of the mean. (Adapted from Penny and Hendy, 1985a.)

The results in Figure 9-2 show that, *on average*, trees requiring fewer mutations with β-hemoglobin require fewer mutations with other sequence data. The test was repeated for each of the other four sequences (Penny and Hendy, 1985a), giving even better results. Evolution is a stochastic process, and the shortest tree on an individual sequence cannot be guaranteed correct.

Sampling Error and Convergence

It is generally recognized that the sequences used in a particular study may be too short to give an accurate prediction. If the only problem is that the sequences are too short, then it is expected that the optimal tree should become a better estimate as the sequences become longer. We cannot, of course, measure this difference directly. However, what can be measured is whether optimal trees from different data sets become more similar as sequences become longer.

Perhaps the best method for selecting subsets of data is by the random resampling of columns. This can be done by either bootstrapping or jackknifing. In bootstrapping (Felsenstein, 1985; Penny and Hendy, 1985a), subsets of columns from the data matrix are randomly selected *with* replacement. These subsets can have the same number of columns as the original data. A column may be omitted from a particular subset, or selected more than once. With jackknifing (Penny and Hendy, 1985a, 1986), the subsets are randomly selected *without* replacement. Consequently, the subsets must be shorter than the original sequence. Jackknifing methods can also be divided into those where subsets may overlap and those with disjoint subsets (where no column occurs in both subsets). In this latter group, which we call hobbits or halflings, each subset contains no more than half the columns.

Trees can be formed from these subsets by standard methods and the results analyzed using a tree comparison metric. There have been two ways of comparing results. Felsenstein (1985) determined the internal edges that occur in at least 95% of the trees. Penny and Hendy (1985a, 1986) studied the rate of convergence as longer sequences (subsets) were used.

With either form of jackknifing we have measured the average distance between optimal trees (using the symmetric difference tree comparison metric) from each subset of columns. What we would expect is that trees from different subsets would become more similar as the subsets contained more columns. This indeed is the case as is shown in Figure 9-3. Instead of showing the value from the symmetric difference metric directly, we have converted it, using the calculated distribution of the metric (Hendy et al., 1984, 1988), to the equivalent number of trees. These results show that even the five sequences for each mammal is insufficient to allow

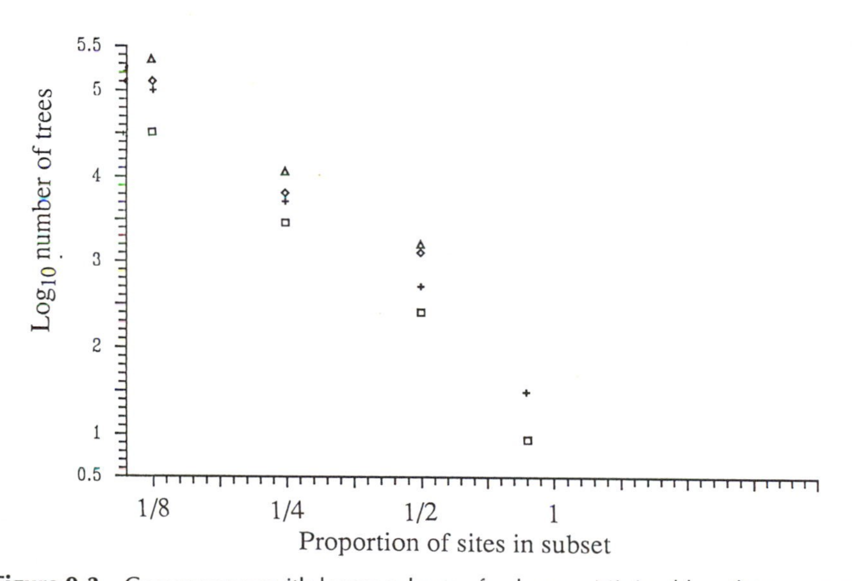

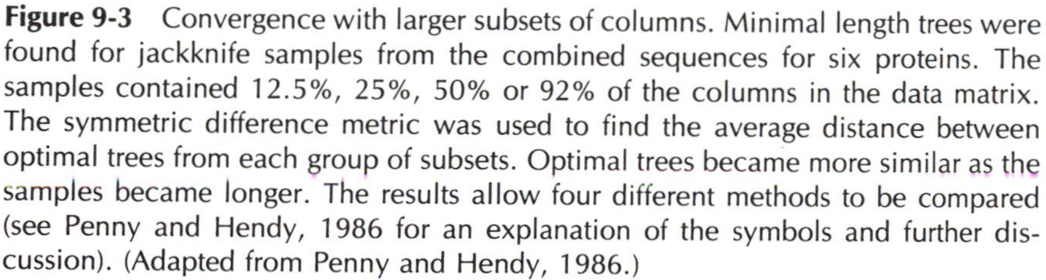
Figure 9-3 Convergence with larger subsets of columns. Minimal length trees were found for jackknife samples from the combined sequences for six proteins. The samples contained 12.5%, 25%, 50% or 92% of the columns in the data matrix. The symmetric difference metric was used to find the average distance between optimal trees from each group of subsets. Optimal trees became more similar as the samples became longer. The results allow four different methods to be compared (see Penny and Hendy, 1986 for an explanation of the symbols and further discussion). (Adapted from Penny and Hendy, 1986.)

convergence to a single tree. In the cases we have studied (Penny and Hendy, 1985a, 1986), the trees, as expected, become more similar as the subsets become larger. This is independent evidence for evolutionary information in the sequences.

These resampling approaches allow tentative answers to several interesting questions. Is it likely that a different optimal tree will be found if more information is gathered for each taxon? If so, which trees? How large a subset of trees is necessary to be confident of including the correct tree? Do some methods for inferring trees converge faster than others as longer sequences are used?

Is bootstrapping better than jackknifing with larger-and-larger samples? The answer may depend on the application. In a taxonomic study, an author may not wish to propose a new taxonomic category unless confident that future work will support it. This is using stability of a single edge of the tree as a criterion. In such a case, bootstrapping may be the preferred test since it does not accept an edge if there is reasonable doubt. If, however, the intent is to find the best estimate of the phylogeny (the tree which gives the best prediction as more data become available), then convergence may be suitable. Under these circumstances the aim may be to find how large a subset of trees is required to be confident that the subset includes the correct tree. It must be noted that although the test allows a decision as to whether convergence has occurred, it is still possible for methods to converge to an incorrect tree. This is discussed later (see Hadamard Transformations).

Testing Other Models

From a Popperian viewpoint, an important feature of the scientific approach is that scientists should be able to give rational explanations of why they use a particular theory. There is an asymmetry here in that it is not claimed that arriving at the original theory was necessarily rational—the creative process is much more complex than that. The aim is for decisions between competing ideas to be rational, and that we should be able to give conditions under which we would reject a currently favored hypothesis.

In the first section we discussed testing the prediction that trees from different sequences led to similar trees. Could we test nonevolutionary models? One well-known astronomer, Sir Fred Hoyle, had ideas of the earth being bombarded with influenza viruses from passing comets (Hoyle, 1984). We called this the unhealthy falling object (UFO) model. From the UFO theory it is not expected that viral sequences would arrive in a particular order consistent with an evolutionary tree. The details of the theories examined are described in Henderson et al. (1989).

One step of the analysis required the comparison of how well a column of data fitted a binary tree, compared to the null model of the star (or big bang) tree (Fig. 9-4). On this null method we could imagine a single ancestor from which all existing species had been independently derived. That is, no pair of species is more closely related than any other pair. Even with

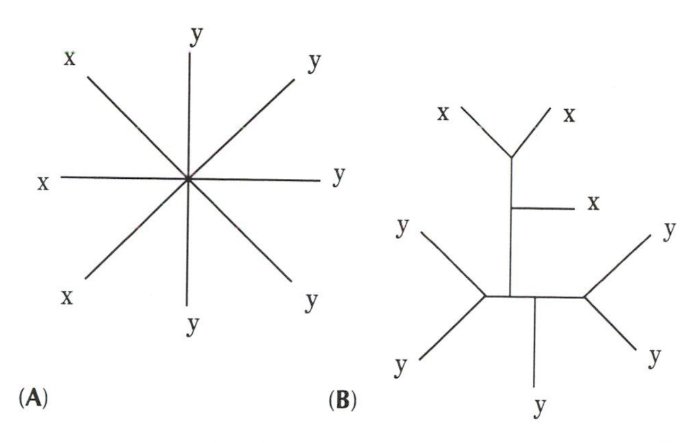

Figure 9-4 (A) The star (or big bang) tree which is a common null hypothesis (Thompson, 1975). (B) is used to illustrate that even if sequences were generated by a star tree process, it is still possible to fit the data with fewer changes to a binary tree.

data generated independently from a common origin (the star tree model) it would still fit more closely to a binary tree. For example, Figure 9-4 has a 3,5 character [three of one color (code or character state), five of the second] and this would require at least three changes on a star tree (Fig. 9-4A). But even if the star tree was correct, many binary trees could be drawn that required only one change on the binary tree. The number can be calculated from Figure 9-4B by forming a tree with eight taxa from the two subtrees with a new edge. It thus became important to know the expected number of changes required to fit a column of data to a random binary tree. If a binary tree model is a good representation of the data, then the number of changes required to best fit the data to a tree should be significantly less than the number required for data generated from the star tree model.

For 2-state colors (codes or character states) with frequencies a and b, on a single column of n taxa ($n = a + b$), we found $f_m(a,b)$ binary trees whose minimal coloring required m changes (Carter et al., 1990).

$$f_m(a,b) = (m-1)!(2n-3m)(2n-5)!!N(a,m)N(b,m)/(2n-2m-1)!! \quad (1)$$

where, a is the number of pendant vertices with the first color (y in Fig. 9-4),

b is the number of pendant vertices with the second color (x in Fig. 9-4),

$a \geq b$, $a + b = n$, the number of taxa,

m (≥ 1) is the minimal number of changes on the tree,

!! is the double factorial [$(2n-5)!! = 1 \times 3 \times 5 \ldots \times (2n-5)$; $0!! = -1!! = 1$], and

$$N(n,m) = \begin{cases} (2n-m-1)!/(n-m)!(m-1)!2^{n-m} & \text{if } n \geq m, \text{ and} \\ 0 & \text{if } n < m. \end{cases}$$

Table 9-1 Values of $F_m(a,b)$ for Four to Eight Taxa*

Number of Taxa			Single Column Distribution				Weighted Average Distribution				
			# Trees for m = 1,2,3,4				t	#Trees for m = 1,2,3,4			
	a	b	1	2	3	4		1	2	3	4
4	2	2	1	2			z	1	2		
5	3	2	3	12			z	3	12		
6	4	2	15	90			z	12	78		
							y	3	12		
	3	3	9	54	42		z	9	45	36	
							y	0	9	6	
7	5	2	105	840			z	60	570		
							y	45	270		
	4	3	45	360	540		z	36	234	360	
							y	9	126	180	
8	6	2	945	9,450			z	360	4,680		
							y	270	2,250		
							x	270	2,250		
							w	45	270		
	5	3	315	3,150	6,930		z	180	1,440	3,420	
							y	45	855	1,620	
							x	90	720	1,710	
							w	0	135	180	
	4	4	225	2,250	5,544	2,376	z	144	1,152	2,592	1,152
							y	72	432	1,440	576
							x	0	648	1,296	576
							w	9	18	216	72

*$F_m(a,b)$ is the number of trees with m changes with a and b occurrences of two character-states. The values for the single column distribution are calculated from equation (1). For the weighted average distribution, t indicates the topology (unrooted binary tree) using the convention in Henderson et al. (1989); z is the caterpillar (unbranched) tree (Fig. 9-6); y the topology with a single branch at the end (Fig. 9-6); x has a single branch in the middle; and topology w has two branches. The numbers for the weighted average distribution are normalized to the number of times each topology occurs (see Fig. 9-6). For any pair of values a and b (e.g., 5 and 3), the columns of the weighted average method will sum to the values for the single column distribution.

In the example in Figure 9-4 [$n = 8, a = 5, b = 3$ (eight taxa, one color occurring five times and the other three times)] the probability of observing $m = 1, 2$, and 3 changes on the tree is 0.030, 0.303, and 0.667, respectively (Table 9-1). Le Quesne (1989) has derived similar results by direct counting.

Little progress has been made for $r > 2$ colors (codes). If the third and subsequent codes occur only once, then the problem is trivial, for example, $f_m(a,b,1) = (2n-3)f_{m-1}(a,b)$. The single third color can be added to any of the $2n-3$ edges on the trees containing the first two codes. With the third or subsequent colors occurring more than once, the only cases derived are when $m = r-1$ (i.e., with only the least possible number of changes on the tree) and where $m = r$.

$$f_{r-1}(a_1,a_2 \ldots a_r) = (2n-5)!!N(a_1,1) \ldots . N(a_r,1)/(2n-2r-1)!! \quad (2)$$

(Carter et al., 1990), and

$$f_r(a_1,a_2 \ldots a_r) = (2n-5)!!N(a_1,1) \ldots . N(a_r,1)$$
$$(r-1)(4(n-r)^2-2n+r)/(2n-2r+1)!! \quad (3)$$

(Steel, 1992).

Distributions for all cases up to 16 taxa and four colors have been estimated by simulation.

Equations (1), (2), and (3) are restricted to a single column (character) which limits their usefulness. The reason for this limitation is that the probabilities for each column are not independent. For example, consider a simple two-color (code) case with six taxa where each color occurs three times (*aaabbb*). Figure 9-5 shows the two unrooted topologies for six taxa. (We define topologies as unlabeled trees, rooted or unrooted.) It is not possible to fit such a column to the second topology (Fig. 9-5B) with only a single $a \leftrightarrow b$ change on the tree. If such a column (three a's and three b's) does fit a tree with only one $a \leftrightarrow b$ change, then the tree must be the topology shown in Figure 9-5A. This knowledge will affect the probabilities for subsequent columns of data. Consequently, the above formulae only

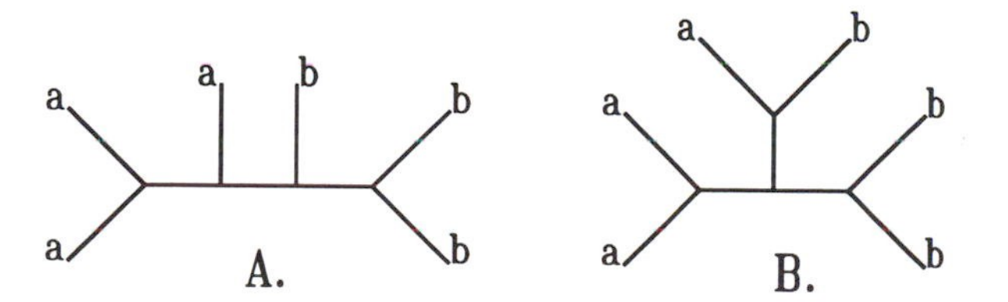

Figure 9-5 Nonindependence of characters. The figure gives the two unrooted topologies (unlabeled trees) for six taxa. A 2-state character with three *a*'s and three *b*'s cannot be fitted to the second topology (B) with only one $a \leftrightarrow b$ change on the tree. If such a column (three *a*'s and three *b*'s) does fit a tree with only one $a \leftrightarrow b$ change, then the tree must be topology (A) whereas four *a*'s and two *b*'s can be fitted to both topologies with a single $a \leftrightarrow b$ change. This indicates how the distribution of changes on different columns of data cannot be independent. The formulae described in the text can only be applied to single columns of data.

apply to single columns of data, although they give good approximations when the data set is large (Steel et al., 1991). We refer to this as the single column distribution.

An alternative approach has been developed by Steel et al. (1991), which takes into account the lack of independence between columns. This evaluates the probabilities for each topology separately. A weighted average is made using the number of trees that can be derived from each topology. The formula for the number of trees from a topology is,

$$n!/2^x6^y \qquad (3)$$

(Hendy et al., 1984),
where x is the number of twofold centers of symmetry in the topology and y (≤ 1) is the number of threefold centers of symmetry (which occurs only with $n = 3,6,9, \ldots$). Figure 9-6 gives the six topologies for $n = 9$ taxa, identifies the centers of twofold and threefold symmetry, and applies the above formula to each topology. As a check, it is shown that the total number of trees over the six topologies sums to $(2n\text{-}5)!!$ or 135,135. This "weighted average" approach is more difficult to calculate as it requires a separate calculation for each topology, but the results can be combined for many columns of data. It was the method referred to earlier (Henderson et al., 1989), and used for testing nonevolutionary models.

Results of both approaches for up to eight taxa are given in Table 9-1. Weighted average values for $n = 9$ are given in Henderson et al. (1989). Archie (1989) has recently used simulation to estimate these values.

These methods that calculate the probabilities of finding columns of given lengths on the tree allow some models of evolution that do not assume an evolutionary tree to be tested. In the influenza virus case referred to above, several nonevolutionary models could be eliminated. In the present context, the important point is that a tree is a falsifiable model.

One additional point needs to be considered when calculating the expected number of changes on a tree. Formula (3) assumes that all trees have the same chance of occurring, the "all trees equiprobable" model. (In this context, trees are binary unrooted trees with end-points labeled.) An alternative "Markov model" assumes the trees are derived from a process that includes random speciation and extinction (Simberloff, 1987). Under these circumstances, it is not valid to use formula (3). This second model has been used in biogeographic studies where there are a small number of areas being analyzed. However, the assumptions of the Markov model are not met in many (most?) phylogenetic studies where only a small proportion of possible taxa are used. The subset of taxa in such studies are *not* chosen at random.

In the set of 11 mammals referred to earlier, the taxa used were not randomly selected from all mammalian species. For example, there is only one rodent and no bat sequences. In a random sample of mammalian species, most would be drawn from these two orders. Rather, the taxa have been "selected" to cover a wider range of taxa and so the assumptions

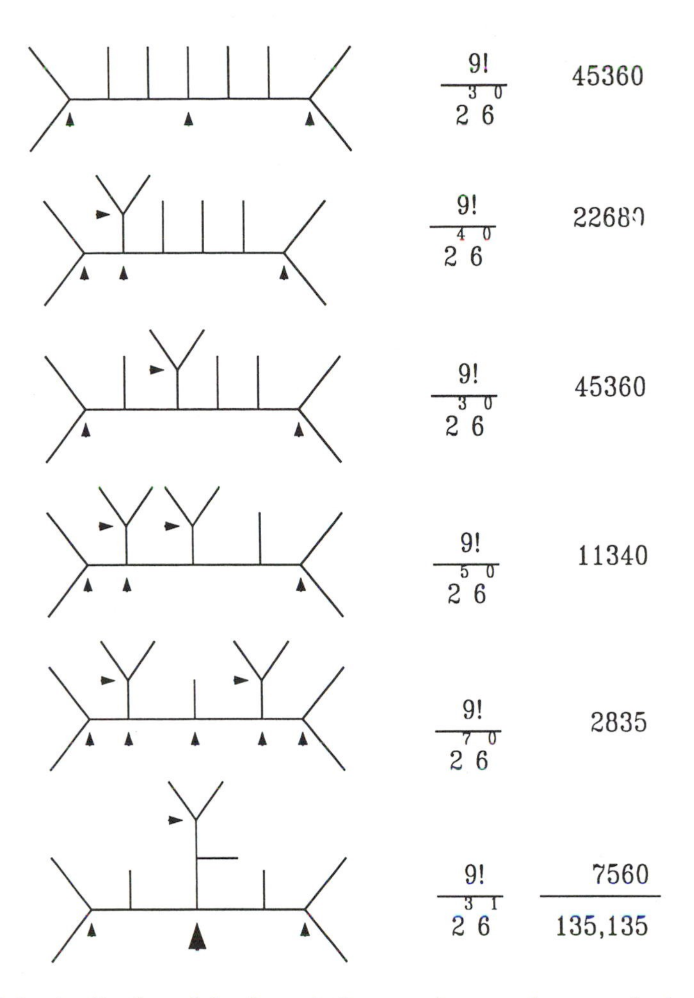

Figure 9-6 Application of the formula for counting trees from topologies. The six unrooted topologies for $n = 9$ taxa are given. The x centers of twofold symmetry are marked with a small arrow, and the one center of threefold symmetry ($y = 1$) is marked by a large arrow. The numbers of phylogenetic trees that can be derived from each topology ($n!/2^x6^y$) are shown, together with the observation that these sum to the expected number of trees, $(2n - 5)!!$.

of the Markov model are not met. Under these conditions the neutral "all trees equiprobable" assumption is more appropriate.

As an additional precaution, it should be noted that the probabilities found for the symmetric difference metric vary with the topology. The published values (Hendy et al., 1984, 1988) are a weighted average over all topologies. This is the reason the distribution was initially calculated only for up to 16 taxa where there are about 500 topologies, although more recently some properties have been determined for larger numbers of taxa (Steel, 1988). It is not valid to find one tree and then use the distribution to estimate the probability that a second tree has x differences. The distribution refers to a pair of trees, selected at random.

VALIDITY OF METHODS FOR INFERRING TREES

Hadamard Transformations

One major conclusion from Popper's approach to science is that more progress is made by trying to find the limits of a hypothesis or conjecture, rather than "testing" the hypothesis in areas where it is expected to apply. In the present context, the hypothesis or conjecture is that a particular tree-building method is expected, given sufficient data, to reconstruct the correct tree. In retrospect, the previous sections are examples of the limitations of making and testing simple predictions. In each case we would have been very surprised if, for example, the trees had not become more similar as longer sequences were used. In our own defense we would say that it is useful to be able to estimate the number of trees that should still be considered possible. Nevertheless, a far more powerful test would have been to try to find conditions under which a method of tree reconstruction would fail.

The work of Felsenstein (1978) and Cavender (1978) has introduced an improved approach which does this. It allows a search for models of evolution where tree building methods would **not** be expected to find the correct tree. That is, it allows a search for conditions where a tree reconstruction method will fail. A method is said to be inconsistent if, under some conditions of the model, it can converge to an incorrect solution as longer sequences are taken. The paradox is that a method may by chance find the correct tree with short sequences, but as longer sequences are used, the probability that the method finds the correct tree goes to 0.

A model of evolution has three parts,

1. a tree (or more generally a graph),
2. an assumed "mechanism" of change to the sequences, and
3. "edge lengths" (probabilities of change along the edges of the tree).

A frequently assumed mechanism (Farris, 1973; Cavender, 1978) is that changes occurring in the sequence are "independent and identically distributed" (i.i.d.). Changes at any position along the sequence, and anywhere on the tree, are independent. All positions (nucleotide or amino acid) have the same chance of changing state (identically distributed). In addition, some mechanisms assume the same rate of change along each edge of the tree—the molecular clock. We will use the term "standard model" for a mechanism of independent and identically distributed changes on a tree but which does not assume the molecular clock. Felsenstein (1978) showed that under some conditions, a model with four taxa and uneven rates of evolution could give data for which parsimony would be expected to reconstruct the wrong tree. Thus, parsimony is, in general, an inconsistent method even though there are many specific models where it will work correctly. An important question is whether the conditions that lead to inconsistent performance are common (and therefore, parsimony should seldom be used) or unusual (in which case the inconsistency may not be a problem in practice).

To extend Felsenstein's analysis to $n > 4$ taxa we developed the Hadamard transformation (Hendy and Penny, 1989; Hendy, 1991) for 2-state characters, a and b (Table 9-2). We will describe this in three parts, each of which is now straightforward (this was probably not true of the original

Table 9-2 Bipartitions and Vectors for the Hadamard Transformations, Illustrated for the Five Taxa in the Model in Figure 9-8B*

					Probabilities	
Index	Subsets	**p**	**q**	**δ**	**r**	**s**
1	{1}	0.100	0.1116	0.1116	0.0000*	0.0583+
2	{1,2}	0.005	0.0050	0.0050	0.2232	0.0113
3	{1,3}			0.0000	0.2332	0.0081
4	{1,2,3}			0.0000	0.2332	0.0155
5	{1,4}			0.0000	0.2332	0.0081
6	{1,2,4}			0.0000	0.2332	0.0155
7	{1,3,4}			0.0000	0.2332	0.0155
8	{1,2,3,4}	0.200	0.2554	0.2554	0.4463*	0.1301+
9	{1,5}			0.0000	0.3720	0.0155
10	{1,2,5}	0.005	0.0050	0.0050	0.3720	0.0113
11	{1,3,5}			0.0000	0.3720	0.0081
12	{1,2,3,5}	0.100	0.1116	0.1116	0.5952*	0.0583+
13	{1,4,5}			0.0000	0.3720	0.0081
14	{1,2,4,5}	0.100	0.1116	0.1116	0.5952*	0.0583+
15	{1,3,4,5}	0.100	0.1116	0.1116	0.5952*	0.0583+
16	{1,2,3,4,5}			−0.7118	0.5952*	0.5197+
				0.0000		1.0000

*The 16 (2^{n-1}) subsets containing 1, of {1,2,3,4,5} for $n=5$ taxa. In each case the subset with its complement (the remaining taxa) forms a bipartition. Each edge of a tree splits the set of taxa into complementary subsets. Similarly, any 2-state character also splits the set of taxa. We refer to these as "edge bipartitions" and "character bipartitions." The edge bipartitions of the tree of Figure 9-8B have indices 1, 2, 8, 10, 12, 14, and 15. For each edge e_i we list the probability p_i that a character will have different states at its endpoints. The q_i value is the number of changes expected under a Poisson model on that edge. **δ** is obtained from **q** by inserting 0 where there is no corresponding q_i value and one negative value for δ_{2n-1} so that $\Sigma\delta_i = 0$. **r** is an intermediate vector which contains the actual distance, per nucleotide, of the minimal path between even-sized subsets of taxa. The Hadamard transformation (Hendy and Penny, 1989) allows us to compute **s** from **δ** where s_j is the probability of obtaining the character bipartition with index j. Thus, for c characters, cs_j will be the expected frequency of this character bipartition. Note in this case $s_{16} = 0.5197$, so we would expect approximately 52% of characters to be constant, while bipartition 3 should occur for less than 1% (0.81%) of characters. The inverse Hadamard transformation produces a vector **δ**, with $q_i = \delta_i > 0$ identifying the edges e_i of the tree.

Parsimony uses only "informative" characters, those which group taxa together (that is, ignoring singleton and constant characters). Their corresponding s_i values are marked "+." In choosing the parsimony tree(s) for these data we find four minimal length trees (expected length 0.0902c), none of which is the tree used to generate the data. The expected length of the correct tree is 0.0934c, the third longest of the 15 possible binary trees. This is an example of inconsistency in parsimony, even with equal rates of evolution.

description). The three parts are the model, the expected form of the data, and a method for interconverting between these two (that is, calculating the data from parameters of the model, and *vice versa*.

We note that each edge e_i of a tree partitions the taxa into two subsets, the set of taxa to the left of the edge and the set of taxa to the right. This pair of sets is called an edge bipartition (split). There are 2^{n-1} possible bipartitions. Also, each column of the sequences induces a bipartition of the taxa, the two subsets being those taxa with state *a* and those with *b*. These we call sequence bipartitions.

We express the bipartitions by the taxa (represented by numbers) they contain. For example, if the character-states for four taxa are *a*, *b*, *b* and *b* (summarized as *abbb*), this is the bipartition {1} and {2, 3, 4}. (The character-states *baaa* also give this bipartition.) There is no need to list both subsets of the bipartition. We normally list only the subset containing taxon 1.

There are four (2^{n-2}) ways in which the sequence bipartition {1} {2, 3, 4} can be generated on a tree and these are shown in Figure 9-7. They correspond to the four ways of labeling two ($n-2$) internal points with the two codes. The probabilities for each of the eight (2^{n-1}) bipartitions can be calculated in this way. In all there are 2^{2n-3} sets of calculations (2^{n-2} ways for 2^{n-1} partitions). However, the whole procedure can be carried out using Hadamard transformations. Assume a tree T where for each edge e_i there is a probability p_i that the character-states at each end of the edge are different. With these values given we use the Hadamard transformation to calculate the probability of obtaining any particular sequence bipartition.

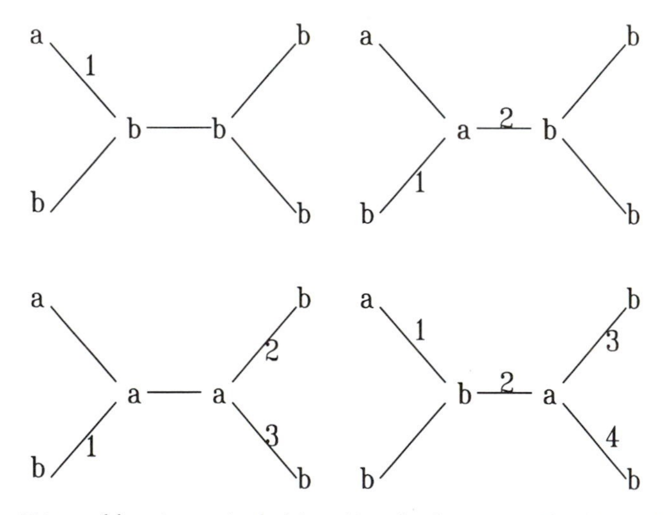

Figure 9-7 Ways of forming a single bipartition for four taxa. The figure shows four combinations of changes along edges of a tree that all lead to the bipartition {1} {2,3,4}.

The transformation is

$$\mathbf{s} = (\mathbf{H}^t \ln(\mathbf{H}\ \boldsymbol{\delta}))/2^{n-1} \qquad (4)$$

where **H** is a Hadamard matrix of 2^{n-1} rows and columns, and **s** and **δ** are defined in Table 9-2. Equation (4) is easily inverted as $\mathbf{H}^{-1} = \mathbf{H}^t/2^{n-1}$ giving

$$\boldsymbol{\delta} = \mathbf{H} \exp((\mathbf{H}^t\ \mathbf{s})/2^{n-1}). \qquad (5)$$

The calculations are illustrated for a case with $n = 5$ in Table 9-2 with an example derived from Figure 9-8B. In practice it is not necessary to construct the Hadamard matrix, as there is now a simple algorithm to carry it out. From equation (4) we can find values of edge lengths on a tree where parsimony will be inconsistent; that is, it will be guaranteed to select the wrong tree as sequences become longer.

δ and **s** are equivalent for a specified mechanism of change in that each can be obtained from the other, and are not dependent on a particular tree. Calculating the expected data in this way quickly allows the tree to be identified to which a tree-building method will converge. Then the rate

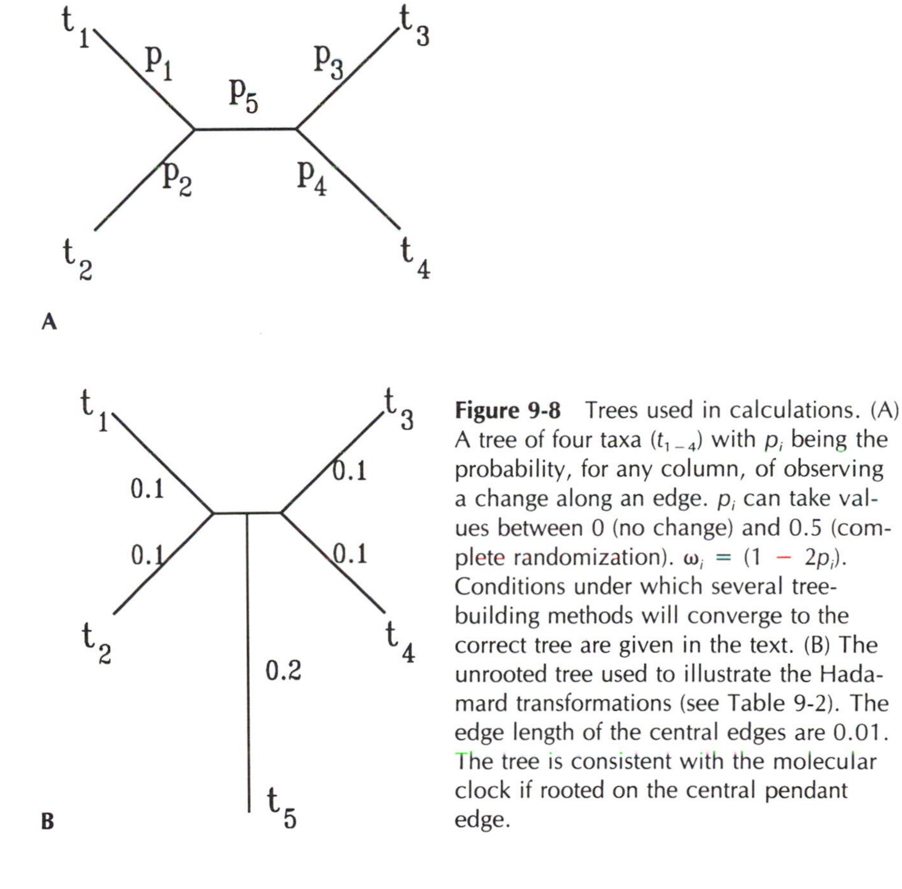

Figure 9-8 Trees used in calculations. (A) A tree of four taxa (t_{1-4}) with p_i being the probability, for any column, of observing a change along an edge. p_i can take values between 0 (no change) and 0.5 (complete randomization). $\omega_i = (1 - 2p_i)$. Conditions under which several tree-building methods will converge to the correct tree are given in the text. (B) The unrooted tree used to illustrate the Hadamard transformations (see Table 9-2). The edge length of the central edges are 0.01. The tree is consistent with the molecular clock if rooted on the central pendant edge.

of convergence to this tree can be studied separately by randomly selecting larger data sets from s.

The computational advantage of the Hadamard transformations over existing maximum likelihood methods is that they allow the data (sequences or bipartitions) to be used directly for estimation of parameters of the model. This is indicated in Figure 9-9a to show that it is possible to go from the model to give predicted sequence bipartitions, or from sequence bipartitions to estimate lengths of edges on trees.

By contrast, existing maximum likelihood methods start with a single tree and some initial guesses for edge lengths. It then repeatedly (maybe thousands of times) makes slight adjustments to edge lengths and repeats the calculation to see if the observed data are now more likely. The process is indicated in a general way in Figure 9-9b to indicate the repeated calculations. Figure 9-9b has been modified from the usual calculation for maximum likelihood to allow an easier comparison to the Hadamard transformations. One interesting development would be to combine the Hadamard transformations with maximum likelihood as the optimality criterion. Such a process would set a good estimate of the optimal branch lengths with a single computation.

The optimality criterion we have favored is finding the closest tree (Hendy, 1991). Its advantages come from its being characterized mathematically.

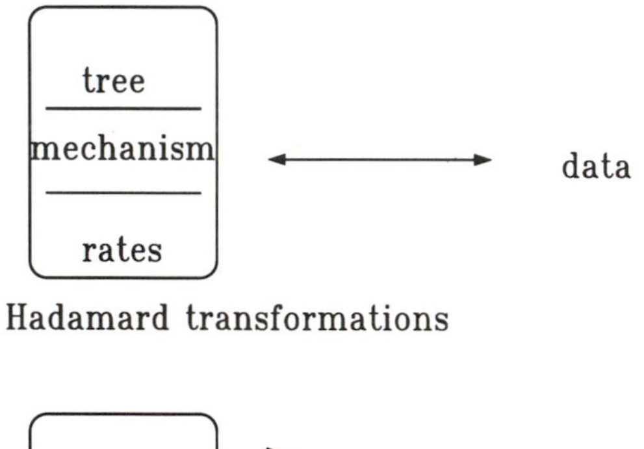

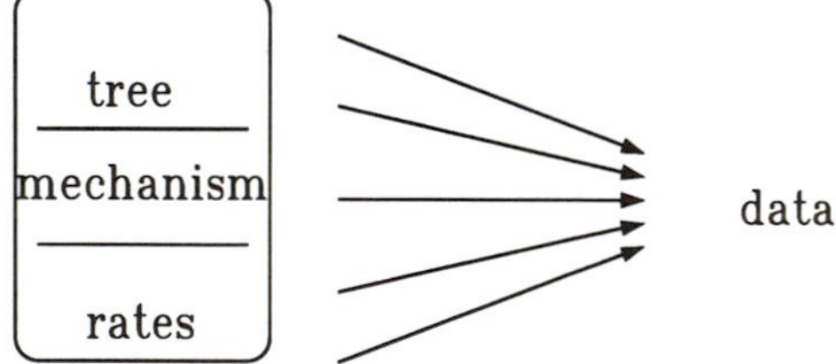

Figure 9-9 The advantage in being able to invert the calculation. With the Hadamard transformations each calculation need only be calculated once, not repeatedly as for maximum likelihood.

This allows the global optimal tree to be found exactly (by branch and bound methods). Thus far, the closest tree method is the only one based on the idea of general invariants for $n > 4$. We have used the closest tree criterion to find the optimal tree for 20 taxa (Penny et al., 1990). For 20 taxa there are $> 10^{20}$ trees! By contrast, current maximum likelihood methods would not search more than 100 (10^2) trees—an advantage of 10^{18} times for methods based on the Hadamard transformations.

Conditions for Consistency—Four Taxa

Some progress has been made with four-taxa, 2-state codes (colors) on the general conditions under which a method will, or will not, converge to an incorrect tree on the standard model. Using the terminology of Figure 9-8A and defining $\omega_i = (1 - 2p_i)$ we find some common tree-building methods *will* converge to the correct tree (be consistent) if and only if, the following conditions are met (Steel, 1989).

Cluster Analysis. For a simple clustering procedure which joins the pair of taxa having the smallest observed distance, the condition for consistency is:

$$\omega_5 < \frac{\max\{\omega_1\omega_2, \omega_3\omega_4\}}{\max\{\omega_1\omega_3, \omega_1\omega_4, \omega_2\omega_3, \omega_2\omega_4\}}.$$

It can be shown that this is a special case of the next condition shown below. Thus, for example, if parsimony fails to converge to the correct tree with four taxa, so too will a simple clustering procedure. The converse does not hold. The special case, with $p_1 = p_3$ and $p_2 = p_4 = p_5$ was derived by Felsenstein (1978).

Parsimony or Compatibility Methods and Methods Using the Four-Point Distance Criterion. This last method selects $\min\{d_{1,2} + d_{3,4}, d_{1,3} + d_{2,4}, d_{1,4} + d_{2,3}\}$. These are consistent, if and only if,

$$\omega_5 < \min\left\{\frac{(\omega_1\omega_2 + \omega_3\omega_4)}{(\omega_1\omega_3 + \omega_2\omega_4)}, \frac{(\omega_1\omega_2 + \omega_3\omega_4)}{(\omega_1\omega_4 + \omega_2\omega_3)}\right\}.$$

With equal rates of evolution (the molecular clock) this condition is always met and parsimony will be consistent with four taxa.

Maximum Likelihood, Closest Tree (from Hadamard Transformations), and the Corrected Four-Point Distances Metric. These methods will always be consistent with four taxa on the standard model. The corrected four-point condition is to select,

$$\min\,(d_{ij} + d_{kl} - 2\,d_{ij}\,d_{kl}/c) : \{i, j, k, l\} = \{1, 2, 3, 4\}$$

where c is the number of columns.

It should be noted that, at least with parsimony, the conditions for consistency are more restrictive as the number of taxa increases. Parsimony with:

four taxa, can only be inconsistent with unequal rates of evolution;
five taxa, with equal rates of evolution (the molecular clock) can be inconsistent if the root is on the central pendant edge which is adjacent to a short and long edges;
six taxa, with arbitrarily low rates of change can be inconsistent;
n large, with all branch lengths (internal and pendant) are both small and equal can be inconsistent.

The first three cases (four to six taxa) are from Hendy and Penny (1989) and the last is from Steel (1989). The conclusion is that the range of cases where parismony is inconsistent increases as more taxa are added.

It is desirable that this type of study be extended to other optimality criteria. With four taxa, methods using the four-point distance criterion have identical performance (with respect to consistency) as parismony (see above). Again with four taxa, simple clustering methods are worse than parsimony in that they will fail under a wider range of conditions. We know less about the performance of cluster analysis with additional taxa except that it is not identical to parsimony. With five or more taxa we can find examples where parsimony fails and simple clustering is consistent, and *vice versa*.

Loss of Information

We have commented elsewhere (Penny et al., 1990) that all methods ignore some of the information in sequences. With four taxa, distances use nearly all the information, but the proportion of information in distances declines very rapidly (Steel et al., 1988) as the number of taxa increases.

Table 9-2 also indicates information that is omitted by parsimony and standard distance methods. Singletons and constant partitions (marked "+" in Table 9-2) are omitted by parsimony methods. Entries in **r** which do *not* correspond to a path between a pair of taxa (marked "*") are omitted by distance methods. There is no general correspondence between these omissions in **r** and **s**. The losses of information in parsimony and distances are not equivalent. Parsimony methods ignore any singletons and constant columns, corresponding to $n + 1$ elements in **s**. Distance methods use all the **s** values to construct a distance matrix which corresponds to only $n(n - 1)/2$ of the **r** values. The Hadamard transformations use all of these 2^{n-1} values. Table 9-3 contrasts the increase in the numbers of bipartitions used by the Hadamard transformation with those used by parsimony and distance methods for small values of n. The example in Table 9-2 and Figure 9-8B is a case where the loss of information is sufficient for parsimony to converge to the incorrect tree (Hendy and Penny, 1989). The correct tree found by selecting partitions {1, 2} and {1, 2, 5} is not the shortest, but the third longest tree (out of 15).

Table 9-3 Comparative Use of Information by Parsimony and Distance Methods*

	Bipartitions		
Taxa	Total	Used in Parsimony	Entries in Distance Matrix
4	8	3	6
5	16	10	10
6	32	25	15
7	64	56	21
8	128	119	28
9	256	246	36
10	512	501	45
11	1024	1012	55
12	2048	2035	66
13	4096	4082	78
14	8192	8177	91
15	16384	16368	105
16	32768	32751	120

*For $n = 4\text{–}16$ we list the number of bipartitions, together with the numbers considered by parsimony and distance matrix methods. Parsimony methods omit $n + 1$ bipartitions from **s** which rapidly become a minute proportion of the total. The omitted bipartitions include the n singletons and the number of constant columns. These omitted bipartitions are important under any model that includes estimates of rates of change. Because they occur so often, the bipartitions omitted by parsimony are those with the most accurate estimates of their frequency. Distance methods include some information from all bipartitions, but the proportion of entries of **r** that are used, rapidly becomes very small (Steel et al., 1988).

Twelve Mammals, Seven Sequences

In earlier papers, we used a data set from 11 mammals and five (later extended to six) sequences. The data are amino acid sequences converted to "best-guess" nucleotide sequences. These data have been particularly useful for developing and testing new methods of analysis. However, even six sets of sequences were insufficient for the optimal tree to be stable (that it would not change as longer sequences became available). When the minimal and near-minimal trees were analyzed (Penny and Hendy, 1985b) it was found that the position of the carnivore (dog) was the least stable. The carnivore was attached by a long "unbranched edge" which can be (Hendy and Penny, 1989) comparatively unstable on the tree. Would adding a second carnivore improve the stability? Would we then get a tree that would be expected to remain constant as longer sequences became available? In other words, would we have convergence to a single tree?

We have now added a second carnivore to the tree and an additional sequence, α-crystallin, for all 12 taxa. Thus the data set now has seven sets of sequences from 12 mammals (Table 9-4). The sequences from the earlier studies were α- and β-hemoglobins, fibrinopeptides A and B, cytochrome c, and myoglobin. In some cases a "taxon" was made up of sequences from two species. One example is using either mouse or rat to represent the rodent taxon. The additional carnivore taxon is made up

Table 9-4 Combined Data from Seven Sequences for 12 Mammals*

	Columns 1–100
	1234567890123456789012345678901234567890123456789012345678901234567890123456789012345678901234567890
	GACGGGGGGAUUACGCAAAAAACCCGGGGCCUCUGCAGAGCGGCCAGGCACGCCCCGAUCACGCGGGCGAGACCAAACAAAAGGCUUAUAACGAGCAGAC
1. Monkey	CU.............G...CU.GGAAA....G..............CAA..A....C...U.C......G......C...C.A...G.....C.......
2. Sheep	..G.....C......GGC....G....A.A..G.AACAGA..AA..........U.......G...A...U..A....G.G.A..G.G.C...UCAU.C.
3. Horse	............G......CU.G......A.GU.AAC....U..GU...UU....A.C...AC.....C.UGAGC.C...GC...G.G..C...C..C..
4. Kangaroo	.GG.A.....CA..A...GC..AA...AAA.G...A..GC.......AA..AG.AAACCA.A.A..C..U.G.A...A....AAAA........A.GC..
5. Rodent	.U.A..A....C...G..G..C...........A..C.......AU...U..G.......U...A..GAC......C...UC.AAA..C.C..C..U...
6. Rabbit	C..A.AA....GAA........G.....A..AA.U....A....G.AU.......A...U..A.AA..C..A..GC.........A......C.....C.
7. Dog	C..AAAA..CCA........C.........U.GAA...GCU......U.U..G...AC....C..A.GA.U......A....A...G.....C....AU.
8. Pig	.G....C...GA..G....G.A............AACA.A....U...UU.G........A........U.A............AG.G.CCA.UC...GU
9. Cat	.U..U.A...CG....C..C..A.AC...U.GA.....GCUU.U.....G.A.G.......U...AA.GA.U.....A.G......G..CCAA...U.GU
10. Human	C.....C........G.....C.GAAA....G..............CU..............C.......G....GCC....C....G....CC....U.
11. Cow	.GG.....C......GGC....A......A....AGC..A..AA.........GUA...CA....AA...U..AC...GGGC...G.GCC...UCAU.C.
12. Ape	C.....C.C......G.....C.GAAA....G..............CU..............C.......G....GCC....CA...G....CC....U.

	Columns 101–191
	1234567890123456789012345678901234567890123456789012345678901234567890123456789012345678901
	GGGAGGAACCGAGCCGGAGAGGAUUGAGCAGCUUACACGAGCGUGAGCAGGUACCAAACCCGAGCAAUGGACCGGAAACAAGAAAAUCGCC
1. Monkey	.......C......AA....A.CC.AGAAUU.A.C.CUA.......C........GU......U..........U....G.A.C..A.UA.
2. Sheep	.U.G.............CA.CUCC.................A.C.G..C..AC....GAGA.C..CCGA.G....C.U.555555555555
3. Horse	.A.......A......ACA..U.C..........A...C..A......C..AC...CC..AC..CC...C..A.......G....C.A..G
4. Kangaroo	AA...AC.A........UA.A..GG........A.UC.....A.AG.G........UC....C......C.A..........UC...AA.A
5. Rodent	..ACA...AAC...GCA..U.A...AG....G....C........C.G.AA...G......A.........G..A.....G..U.G.....
6. Rabbit	....AAC.AAC...GC.....UU.C......G....C.A......C...AC..............CC..CG..A.C...G.A..G......
7. Dog	AA......A....A..AU.UA..CG..............CA.A.AC...AAA.GG..C....C......C.A.A......G.U.GC.G...
8. Pig	CAA.............CCA..UCAG................A...G...........C............G.........G...GC.G...
9. Cat	......GCA....A.....U..G.G..............CA...AC...A...GG..C...A...CCG.C.A.A....A.G.U.GC.G...
10. Human	.......C...UA.AA......U..AGAAUU.A.CUCUA.......C........GC......UG.......A.U.C..G.A.C..A.UAA
11. Cow	CCAG...C........CCA..UGG..G..............A.C.G..C..AC....GAGAAC..CCGA.G..U...UA.GA...C.G...
12. Ape	.......C...UA.AA......U..AGAAUU.A.CUCUA.......C........GC......UG.......A.U.C..666666666666

*Columns 1–42 and 43–97 are derived from α- and β-hemoglobins; 98–110 and 111–127 from fibrinopeptides A and B; 128–142 from cytochrome *c*; 143–179 from myoglobins; and 180–191 from α-crystallin. Columns 1–142 are (with the addition of cat) from Penny et al. (1982), columns 143–179 from Penny and Hendy (1986), and columns 180–191 are derived from deJong et al. (1977). α-crystallin sequences were not available for sheep or apes and so additional singleton character-states ('5 and 6') were assigned to these taxa. This does not distinguish between any tree, and so does not affect the results obtained. The amino acid sequence for lion cytochrome *c* was supplied by Professor R.P. Ambler, Edinburgh. The data, with some additional information, are available by e-mail from D.Penny @ massey.ac.nz.

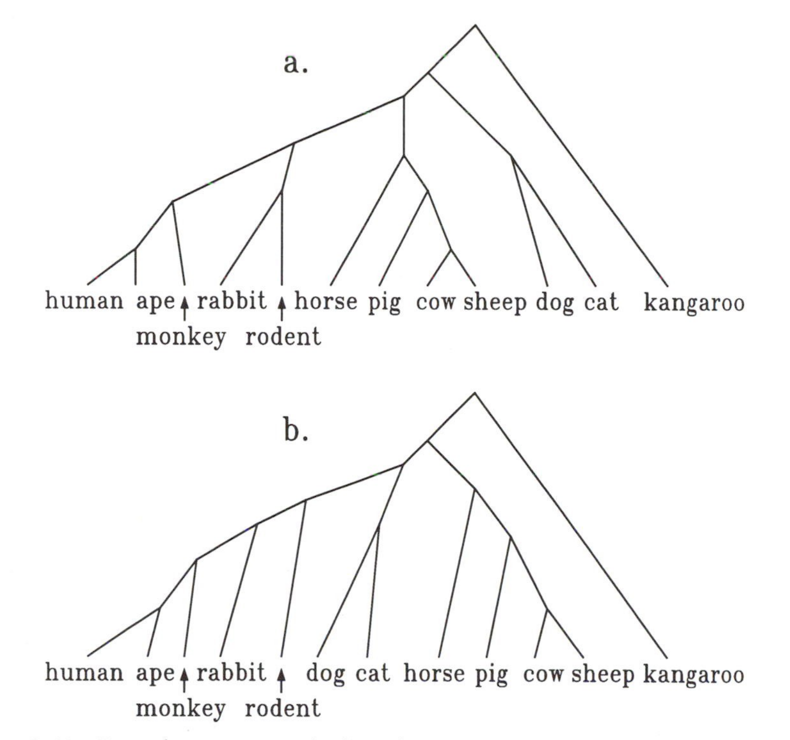

Figure 9-10 Trees for 12 mammals, based on seven sequences. (a) is the minimum length tree (472 mutations) based on the 191 columns in Table 9-4. It is also the consensus or median tree (Penny et al., 1982). (b) is three changes longer (475 mutations) but may be correct if the "long edges attract" problem of parsimony is leading to incorrect convergence. One additional change results from transferring carnivores and two changes from separating the rabbit-rodent neighboring pair.

from six feline sequences (five from cat and lion cytochrome *c*) plus one from a seal. The procedure of using composite taxa is legitimate in this study as it cannot result in trees from different sequences being more similar than expected. It can only introduce more dissimilarity between trees if the composite taxon was phylogenetically incorrect. For the reasons discussed below we suggest the best tree is that shown in Figure 9-10b.

Another approach is possible by analyzing the frequency of the edges in minimal trees from bootstrap samples. This is illustrated in Figure 9-11 with results from 132 minimal trees (in this case using parsimony) from bootstrap samples. Only the relationships between three groups; primates, lagomorphs (rabbit), and rodents are shown. The most common subtree is (rabbit, rodent), primate (Fig. 9-11b) which occurs 74 times. The next most frequent is (primate, rabbit), rodent (Fig. 9-11a), 44 times; and (primate, rodent), rabbit subtree (Fig. 9-11c), which occurs seven times. Figure 9-11b conforms to the Glires model (Novacek, 1985) which links rodents and lagomorphs (rabbits).

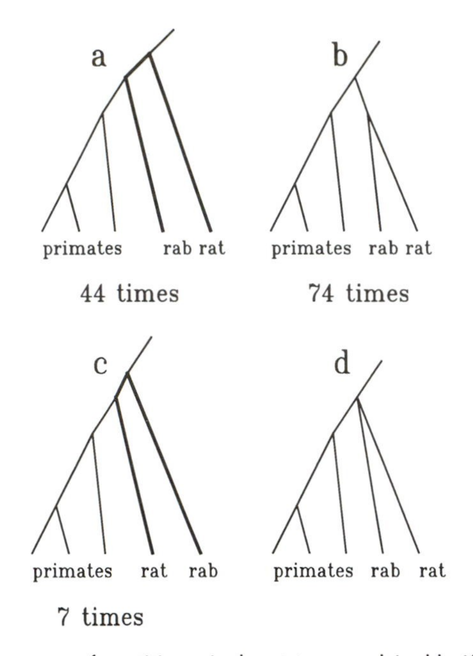

Figure 9-11 Frequency of partitions in bootstrap and jackknife samples. (a), (b), and (c) are the three subtrees for primates, rabbit (rab), and rodents (rat), and the frequency of occurrence of these subtrees in the minimal length trees from 132 bootstrap samples. (d) is the unresolved trichotomy. An anomaly is the low frequency of just one of the three subtrees, (c) compared to (a). This can be explained as an example of the error with parsimony involving a path of long-short-long edges, as in the heavy lines in (a). With parsimony these long edges would tend to "attract" (Hendy and Penny, 1989) leading to tree (b).

The question is to interpret these frequencies (74, 44, 7). If there was a short edge linking these three groups (that is, close to a trichotomy, Fig. 9-11d) then all three subtrees (Figs. 9-11a–c) would be expected to occur with about equal frequency. If Figure 9-11b was correct, then we would expect Figures 9-11a and 9-11c with equal frequency. The observed frequencies are significantly different from those expected for either the trichotomy (Fig. 9-11d) or Glires (Fig. 9-11b) models.

Qualitatively we can account for the observed distribution of Figure 9-11a–c if Figure 9-11a ((primate, rabbit), rodent) is the correct relationship. This tree has a long-short-long series of edges (rabbit, internal edge, rodent) which is shown as dark lines in Figure 9-11a. Parsimony would lead to the long edges being drawn together to give Figure 9-11b. This would account for two of the three subtrees being found more frequently than the third. A prediction from this hypothesis is that bootstrap samples from data sets which include sequences from a quite different lagomorph (or rodent) should resolve the issue. Note that convergence from the other two trees (Fig. 9-11b and 9-11c) could not lead to the observed frequencies in that neither should lead to the observed frequencies. If Figure 9-11b were correct, then any tendency for incorrect convergence should have equal affects on both rodents and lagomorphs, leading equally to Figures 9-11a and 9-11c. We suggest the tree in Figure 9-10b shows the correct relationship of the three groups (rodents, lagomorphs, and primates).

A similar problem occurs with the position of the two carnivores, relative to the ungulates and the metatherians (kangaroo). The bipartition with the two carnivores is almost always found, but their position on the tree varies. It is on the first division of the eutherians (83 times), after the separation of ungulates (40 times), or on the first division of the ungulate lineage (eight times). Again we do not have the pattern expected from a near-trichotomy (all three equal, or one more common and the next two equal). The difference is statistically significant in a χ^2 test.

Is this another case of incorrect convergence? Consider what would happen if the correct tree had the carnivores separating after the ungulate line has diverged (that is, in the position shown in Fig. 9-10b). We again have a case of a long lineage (kangaroo), a short internal edge, and a medium lineage leading to the two carnivores. This may be sufficient to indicate incorrect convergence. Again, if either of the other two possibilities for kangaroo, ungulates, and carnivores was the correct tree, they would not have the long-short-long sequence of edges that can give incorrect convergence.

These conclusions need to be made quantitative. A possible correction for parsimony is available (Penny and Hendy, in preparation). However it requires a reasonable estimate of rates of change along the edges of a tree. This depends on the numbers of singletons and constant columns.

We note that the "long edges attract" problem that leads parsimony to converge to a wrong tree may be more frequent than anticipated. The initial work by Felsenstein (1978) with four taxa suggested the problem may occur with very unequal rates. With five taxa it can occur with equal rates (Hendy and Penny, 1989) but required a short edge(s) between long edges. Even this knowledge is inadequate to indicate whether or not incorrect convergence occurs in a specific case.

Even with these data we were unable to reach a firm conclusion. The time of separation of these 12 mammals is probably 60 to 80 million years ago. If seven sequences are insufficient to resolve such a recent divergence, would we expect a single sequence (even quite a long sequence) to be sufficient for a divergence that occurred much earlier—say 1 billion, or even 3 billion years ago? Such a rhetorical question is expected to be answered in the negative. It is of course possible that we are very fortunate and a single sequence [e.g., small-subunit ribosomal RNA (rRNA)] is sufficient. But we have no rational reason to *assume* it without testing. Work on the small-subunit rRNA has vastly improved our knowledge of the phylogeny of organisms but there is no reason to assume it is sufficient.

What concerns us is the scientific myth (and at present, it can only be considered a myth) that a single sequence is sufficient to reconstruct the whole history of life. We call this the "Myth of a Universal Tree from One

Gene" (MUTOG). The myth is strong enough to get evolutionary trees into textbooks which lack qualifying statements. The main problem is that the myth inhibits testing of ideas by not taking additional sequences and testing for convergence. If the tree is "believed" to be correct, there is little motivation to undertake new work to test the tree. Such a state of affairs is disturbing to anyone using a Popperian framework for science. An idea (trees in this case) should be a stimulus to new measurements and better, more rigorous tests.

DISCUSSION

By discussing our work from a Popperian viewpoint, we do not wish to imply a concept of a single monolithic framework in which scientists work. Evolutionists in particular should be well aware of the problem of essentialism. They are aware of the importance of diversity in evolution (Mayr, 1982), and consequently should be skeptical of any attempt to define "one true scientific method." Medawar, who supported a Popperian view of science (Medawar, 1974), has commented that

> Among scientists are collectors, classifiers and compulsive tidiers-up. Many are detectives by temperament and many are explorers; some are artists and others artisans. There are poet-scientists and philosopher-scientists and even a few mystics (Medawar, 1967:132).

Attempts by philosophers to describe a single mechanism of research that all successful scientists follow, are doomed to failure. A diversity of approaches to science is still accommodated on the Popperian model in which no scientific hypothesis is ever absolutely proven, where no hypothesis should be "believed." A similar comment on the diversity of approaches, using the analogy of diversity within a species appears in Hull (1988). The important point is the attitude toward hypotheses: they should never be accepted as beyond questioning. They are tools to aid in the design of harder and more rigorous tests.

We have recently discussed tree-building methods as requiring at least five criteria. Methods should be consistent, robust, efficient, fast, and falsifiable (CREFF) (Penny et al., 1990). These criteria appear incompatible in that efficient methods (which increase in a polynomial manner with the number of taxa) for real data are not known. Little is known on the robustness of methods relative to deviations from the assumptions about the mechanism of change. Improvements are still needed to increase the power of tree-building methods. Perhaps the most urgent is an analog of the Hadamard transformations for four-state characters. So far, an initial approach has been developed (Penny et al., 1990) which still needs further development. Another area that is important to test is the effect of deviations from the "standard model." This would allow a test of the robustness of tree building.

Whatever the approach that is taken, it appears to us that it is important to find the limits of any tree-building method and to find under what conditions it will break down. Such a Popperian approach should help identify problem areas and assist the search for better methods. Improved methods will almost certainly depend on a better understanding of their mathematical basis. This makes the study of evolutionary trees an exciting part of modern biology.

REFERENCES

Allan, M. (1977) *Darwin and His Flowers: The Key to Natural Selection*. Faber and Faber, London.

Archie, J.W. (1989) A randomization test for phylogenetic information in systematic data. *Syst. Zool.* **38**:239–252.

Carter, M., M.D. Hendy, D. Penny, L.A. Székely, and N.C. Wormald. (1990) On the distribution of lengths of evolutionary trees. *SIAM J. Disc. Math.* **3**:38–47.

Cavender, J. (1978) Taxonomy with confidence. *Math. Biosci.* **40**:271–280.

Darwin, C. (1859) *On the Origin of Species by Means of Natural Selection, or the Preservation of Favoured Races in the Struggle for Life*. John Murray, London.

Day, W.H.E. (1985) Optimal algorithms for comparing trees with labelled leaves. *J. Classif.* **2**:7–28.

Day, W.H.E., D.S. Johnson, and D. Sankoff. (1986) The computational complexity of inferring rooted phylogenies by parsimony. *Math. Biosci.* **81**:33–42.

de Jong, W.W., J.T. Gleaves, and D. Boulter. (1977) Evolutionary changes of α-crystallin and the phylogeny of mammalian orders. *J. Mol. Evol.* **10**:123–135.

Farris, J.S. (1973). A probability model for inferring evolutionary trees. *Syst. Zool.* **22**:250–256.

Felsenstein, J. (1978) Cases in which parsimony or compatibility methods will be positively misleading. *Syst. Zool.* **27**:401–410.

Felsenstein, J. (1985) Confidence limits on phylogenies: an approach using the bootstrap. *Evolution* **39**:783–791.

Ghiselin, M.T. (1969) *The Triumph of the Darwinian Method*. University of Chicago Press, Chicago.

Graham, R.L., and L.R. Foulds. (1982) Unlikelihood that minimal phylogenies for a realistic biological study can be constructed in reasonable computational time. *Math. Biosci.* **60**:133–142.

Henderson, I.M., M.D. Hendy, and D. Penny. (1989) Influenza viruses, comets, and the science of evolutionary trees. *J. Theor. Biol.* **140**:289–303.

Hendy, M.D. (1991). A combinatorial description of the closest tree algorithm for finding evolutionary trees. *Disc. Math.* (in press).

Hendy, M.D., C.H.C. Little, and D. Penny. (1984) Comparing trees with pendant vertices labelled. *SIAM J. Appl. Math.* **44**:1054–1067.

Hendy, M.D., and D. Penny. (1982) Branch and bound algorithms to determine minimal evolutionary trees. *Math. Biosci.* **59**:277–290.

Hendy, M.D., and D. Penny. (1989) A framework for the quantitative study of evolutionary trees. *Syst. Zool.* **38**:297–309.

Hendy, M.D., M.A. Steel, D. Penny, and I.M. Henderson. (1988) Families of trees and consensus. Pp. 355–362 in *Classification and Related Methods of*

Data Analysis (H.H. Bock, ed.). Elsevier Science Publishers. B.V., Amsterdam.

Hoyle, F. (1984) *Living Comets*. Cardiff University Press, Cardiff.

Hull, D. (1988) *Science as a Process: An Evolutionary Account of the Social and Conceptual Development of Science*. University of Chicago Press, Chicago.

Le Quesne, W.J. (1989) Frequency distribution of lengths of possible networks from a data matrix. *Cladistics* **5**:395–407.

Mayr, E. (1982) *The Growth of Biological Thought: Diversity, Evolution and Inheritance*. Harvard University Press, Cambridge.

Medawar, P.A. (1967). *The Art of the Soluble*. Methuen, London.

Medawar, P.A. (1974) Hypothesis and imagination. Pp. 241–273 in *The Philosophy of Karl Popper*, Vol. 1 (P.A. Schlipp, ed.). Open Court, LaSalle.

Mickevich, M.F. (1978) Taxonomic congruence. *Syst. Zool.* **27**:143–158.

Novacek, M. (1985) Cranial evidence for rodent affinities. Pp. 59–81 in *Evolutionary Relationships Among Rodents: A Multidisciplinary Analysis* (W.P. Luckett and J.-L. Hartenberger, eds.). Plenum Press, New York.

Penny, D. (1985) The evolution of meiosis and sexual reproduction. Biol. J. Linn. Soc. **25**:209–220.

Penny, D., L.R. Foulds, and M.D. Hendy. (1982) Testing the theory of evolution by comparing phylogenetic trees constructed from five different protein sequences. *Nature* **297**:197–200.

Penny, D., and M.D. Hendy. (1985a) Testing methods of evolutionary tree construction. *Cladistics* **1**:266–278.

Penny, D., and M.D. Hendy. (1985b) The use of tree comparison metrics. *Syst. Zool.* **34**:75–82.

Penny, D., and M.D. Hendy. (1986) Estimating the reliability of evolutionary trees. *Mol. Biol. Evol.* **3**:403–417.

Penny, D., M.D. Hendy, E.A. Zimmer, and R.K. Hamby. (1990) Trees from sequences: panacea or Pandora's box? *Aust. Syst. Bot.* **3**:21–38.

Popper, K.R. (1963) *Conjectures and Refutations: The Growth of Scientific Knowledge*. Routledge and Kegan Paul, London.

Popper, K.R. (1972) *Objective Knowledge: An Evolutionary Approach*. Oxford University Press, Oxford.

Popper, K.R. (1976) *Unended Quest: An Intellectual Autobiography*. Fontana, London.

Popper, K.R. (1978) Natural selection and the emergence of mind. *Dialectica* **32**:339–355.

Popper, K.R. (1984) Erkenntnis und gestaltung der wirklichkeit: die suche nach einer besseren welt. Pp. 11–40 in *Vorträge und Aufsätze Aus Dreissig Jahren*. R. Piper, Munich. English translation: (1991) Knowledge and the shaping of reality: the search for a better world. New Zealand Sci. Rev. 48 (in press).

Riddiford, A., and D. Penny. (1984) The scientific status of evolutionary theory. Pp. 1–38 in *Evolutionary Theory: Paths Into the Future* (J.W. Pollard, ed.). John Wiley and Sons, London.

Robinson, D.F., and L.R. Foulds. (1981) Comparison of phylogenetic trees. *Math. Biosci.* **53**:131–147.

Ruse, M. (1975) Darwin's debt to philosophy: an examination of the influence of the philosophical ideas of John F.W. Herschel and William Whewell on the development of Charles Darwin's theory of evolution. *Stud. Hist. Philos. Sci.* **6**:159–181.

Schweber, S.S. (1977) The origin of the *Origin* revisited. *J. Hist. Biol.* **10**:229–316.
Simberloff, D.S. (1987) Calculating probabilities that cladograms match: a method of biogeographical inference. *Syst. Zool.* **36**:175–195.
Steel, M.A. (1988) Distribution of the symmetric difference metric on phylogenetic trees. *SIAM J. Disc. Math.* **1**:541–551.
Steel, M.A. (1989) *Distributions on bicoloured evolutionary trees*. Ph.D. Dissertation, Massey University, Palmerston North.
Steel, M.A. (1992) Distributions on bicoloured binary trees arising from the principal of parsimony. *Discr. Appl. Math.* (in press).
Steel, M.A., M.D. Hendy, and D. Penny. (1988) Loss of information in genetic distances. *Nature* **336**:118.
Steel, M.A., M.D. Hendy, and D. Penny. (1991) Significance of the length of the shortest tree. *J. Classif.* (in press)
Thompson, E.A. (1975) *Human Evolutionary Trees*. Cambridge University Press, Cambridge.

10

Parsimony and Phylogenetic Inference Using DNA Sequences: Some Methodological Strategies

JOEL CRACRAFT AND KATHLEEN HELM-BYCHOWSKI

Sequence data of DNA are becoming increasingly important within systematic and evolutionary biology. With this expanding emphasis, many investigators have rising expectations that sequence variation offers an almost unlimited potential to resolve phylogenetic relationships and thereby provide us with a highly robust, if not definitive, description of the pattern of life's history (e.g., Goodman et al., 1987; Sibley and Ahlquist, 1987; Goodman, 1989). Various reasons have been given in support of this expectation. One of the most common is that because DNA is the physical information system recording all inherited attributes of organismal history, direct comparison of its structure provides the most basic of all data for phylogenetic reconstruction. Inasmuch as all historical change is encoded in DNA, it is reasoned, it makes sense to examine that change directly. A second argument is one of large numbers: we can obtain far more character-state data from DNA sequences than we can, say, from traditional morphological comparisons, and therefore, statistically, as additional data are examined, true patterns of phylogenetic relationships will eventually emerge. Furthermore, it is often claimed that because morphology is so readily subject to convergence when compared to sequence data, obtaining more and more of the latter represents our best hope for resolving the phylogenetic pattern.

Technological innovations are also fueling the rush to obtain sequence data. Comparative sequence data are much easier and less expensive to collect than they were even a year ago. Just as importantly, there has been a rapid growth in the amount of computer hardware and software that makes the analysis of comparative sequence data readily accessible to workers at all levels of expertise (Platnick, 1989; see also many of the papers in Doolittle, 1990). In particular, the availability of tree-building algorithms for microcomputers has been a contributing factor to the marked increase in the numbers of papers that use DNA sequences to answer phylogenetic questions (see summary in Swofford and Olsen, 1990). At its simplest, therefore, it is relatively straightforward to collect comparative sequence data, analyze them with a given tree-building method, and publish the results.

Yet, it is clear from the literature that phylogenetic inference using sequence data is anything but simple. There is, for example, substantial disagreement over the best method to resolve relationships from sequence data. And even if agreement could be reached on a general method, it still is not clear how that method might be applied in order to obtain a phylogenetic hypothesis that is judged reliable, or best, by some criterion. These and other problems make phylogenetic analysis of DNA sequences extremely complex.

There are three main classes of methods of phylogenetic inference applied to sequence data. First, with distance analysis, a distance (similarity/dissimilarity) coefficient is computed for all pairwise comparisons of raw sequence data, or for data corrected for multiple substitutions, and the taxa are then clustered based on the values in the distance matrix (Fitch and Margoliash, 1967; Farris, 1972; Felsenstein, 1984, 1988; Saitou and Nei, 1987). In general, the tree having the shortest overall branch length relative to that implied by the original data is chosen as the preferred tree. See Swofford and Olsen (1990) for details about optimization procedures, which differ among distance methods. Second, the method of maximum likelihood uses the original sequence data and, working from a prior evolutionary model of nucleotide substitution, computes a likelihood score for each tree given the original data (Felsenstein, 1981; Bishop and Friday, 1985; Saitou, 1988, 1990). That tree with the highest likelihood is the most preferred. The third method, cladistics or parsimony analysis, also uses the original sequence data, and from calculations of character-state changes (nucleotide substitutions) on alternative trees, the investigator chooses the one having the fewest character-state transformations, that is, the one for which homoplasies—parallelisms and reversals—are minimized (Fitch, 1971, 1977; Hendy and Penny, 1982; Cedergren et al., 1988).

There are several important reasons for preferring a parsimony approach to phylogenetic inference when using DNA sequences. First, parsimony methods are applied to the original sequence data rather than to some transformed distance metric, thus avoiding an inevitable loss of information (Farris, 1981, 1985; Penny, 1982). Second, while not entirely assumption free, parsimony methods are either less dependent upon assumptions about sequence evolution than are other methods; or they at least permit a more informed investigation of any assumptions that might be made (Cedergren et al., 1988). Third, algorithms that implement parsimony procedures are far more sophisticated than distance or maximum-likelihood alternatives and consequently allow a deeper analysis of the phylogenetic structure of one's data and of the dynamics of sequence evolution. All three of these reasons mean that parsimony provides the investigator with enhanced interpretability of the results as compared to other methods.

Nevertheless, as noted above, the application of parsimony procedures to sequence data is complex and has not yet received sufficient attention. This chapter addresses two of the more general problems involving the use of parsimony procedures in phylogenetic inference: (1) given that there are physico-chemical/functional constraints on sequence evolution, especially in sequences coding for proteins or structural RNAs, how might parsimony be applied in order to infer phylogenetic relationships, and (2) how might we judge the phylogenetic informativeness of sequence data? Our inquiry into these problems will be based on an analysis of mitochondrial DNA (mtDNA), but the findings should be applicable to other parts of the genome as well.

Although it is a simple enough procedure to obtain a global parsimony estimate of the phylogenetic structure inherent in any given data set, recognition that rates of transitions are, at least in vertebrate mtDNA, much higher than those for transversions (Brown et al., 1982; Brown, 1983, 1985; Moritz et al., 1987), that there may be biases in base composition, and that silent substitutions in the third codon position are more common than those at first and second positions, have suggested to some investigators that parsimony procedures will have difficulties when used to resolve phylogenetic relationships, especially among more distantly related taxa (Moritz et al., 1987; Hayasaka et al., 1988a; Irwin et al., 1991). If homoplasy is too extensive, it is argued, obtaining a strongly corroborated phylogenetic tree might be difficult because even a minimum-length tree will have a high number of homoplastic changes (see Archie, 1989; Smith, 1989). To the extent that this is true, it raises several important questions: How might we undertake parsimony analysis so as to reduce, or at least identify, the influence of homoplasy? Second, how do we identify those parts of the tree that seem to be less strongly supported and at the same time specify which components of the data are less informative phylogenetically? Finally, in any given study, how do we determine how much sequence must be obtained in order to reveal a reliable phylogenetic signal?

These questions speak to the notion of *reliability* and *informativeness*, terms that have been used in different ways in the literature. Some investigators speak of "reliability of methods," referring to that method which finds the best-fit tree for a given set of data. Others view reliability in terms of an "ideal" result: the ability of a method, or a set of data, to enable us to infer a tree that reflects the one true phylogeny. It is well to keep the distinction of these two views of reliability in mind, for often it is implied that once a best-fit tree is obtained, so too has an accurate estimate of the one true phylogeny. It is sometimes argued, moreover, that under a particular model of nucleotide sequence change, one or more methods will fail to be reliable; that is, it would be expected that the best-fit tree generated by those methods will not be an accurate estimate of the true phylogeny (e.g., Felsenstein, 1978). All of these views on reliability, however, are more arguments over methods, or perhaps of models of evolutionary change, inasmuch as the one true phylogeny can never be specified with certainty. Nor is it clear that a "realistic" evolutionary model can be developed without some prior hyothesis of relationships with which to investigate character-state change. Therefore, given only the data set of interest, it is difficult to say whether the best-fit tree produced by any method is or is not an accurate representation of the true phylogeny; it is simply the most economical explanation for the available data.

Ideally, "reliability" is best thought of as a parsimony problem and consequently should be judged in terms of congruence of data. "Informativeness," likewise, is also a parsimony problem. The two concepts are related. A given tree, calculated to be the most-parsimonious, or best-fit, for some set of data, can be taken to be a reliable estimate of the true phylogenetic history of the taxa if most-parsimonious trees for other, independent data are congruent with it. A given sample of sequence can be judged phylogenetically informative if it corroborates the most-parsimonious tree derived from other data. This reciprocal duality between the structure of data and what is inferred from them manifests the application of parsimony (and congruence analysis) to hypotheses at two different levels of inference: that of the data (character homology or hypotheses of synapomorphy), and that of the phylogenetic hypotheses themselves (Cracraft and Mindell, 1989).

THE DATA

This study uses a single data set, and therefore, informativeness and reliability will be assessed in relation to the "stability" seen in the cladistic signal across partitioned subsets of the data. That is, to the extent to which clades are corroborated by these different subsets, the data will be judged informative and the phylogenetic hypotheses themselves reliable. "Noise," on the other hand, will be judged to exist when the data do not corroborate a consistent (or stable) cladistic signal.

The sequence data analyzed in this study include an 898 base pair (bp) fragment of mtDNA that encompasses 458 bp of the 3′ end of the NADH-dehydrogenase subunit 4 (ND4 gene), 198 bp of three transfer RNA genes ($tRNA^{His}$, $tRNA^{Ser}$, and $tRNA^{Leu}$), and 242 bp of the 5′ end of the ND5 gene (Brown et al., 1982; Hayasaka et al., 1988a: fig. 1:629). Comparative sequences were available for 12 primate taxa: human (*Homo sapiens*), common chimpanzee (*Pan troglodytes*), gorilla (*Gorilla gorilla*), orangutan (*Pongo pygmaeus*), gibbon (*Hylobates lar*), Japanese macaque (*Macaca fuscata*), rhesus macaque (*M. mulatta*), crab-eating macaque (*M. fascicularis*), Barbary macaque (*M. sylvanus*), squirrel monkey (*Saimiri sciureus*), Philippine tarsier (*Tarsius syrichta*), and ring-tailed lemur (*Lemur catta*).

METHODS OF ANALYSIS

All analyses were undertaken using the Phylogenetic Analysis Using Parsimony (PAUP) computer program of D. L. Swofford (1990). Version 3.0 of this program for the Macintosh system is especially well suited for analysis of sequence data because it readily permits investigation of the effects of transitions versus transversions, the inclusion/exclusion of first, second, or third codon positions in coding sequences, and of nucleotide substitution parsimony of amino acid replacements. In addition, the program can compute consensus trees by different methods, perform bootstrap analyses, and calculate numerous measures of the information content for a given tree structure.

In order to investigate the effects of sample size of characters on the cladistic structure of the data, the 898 bp fragment was partitioned into four sequential subsets: (1) the first 225 bp of the ND4 gene; (2) a 458 bp segment including all of the ND4 gene that was sequenced; (3) a 656 bp fragment including the 458 bp ND4 fragment plus the 198 bp of the three tRNAs; and (4) the entire 898 bp fragment representing the ND4 fragment, tRNAs, and the 242 bp fragment of the ND5 gene. The effects of data set size were studied using this approach (rather than by random subsamples of different sizes) because systematists add sequence data sequentially and because it permitted an analysis of regions of the data set having different functional domains. In all analyses, the trees were rooted by designating the lemur as the outgroup. The results would have been the same if the tarsier had been designated the outgroup; therefore, this study is only investigating patterns of relationships within the clade consisting of the other primate species (the anthropoids; terminology of clades follows Hayasaka et al., 1988a). A nonprimate outgroup was not used in order to make our results more comparable to those of Hayasaka et al. (1988a).

Hayasaka et al. (1988a) proposed a phylogenetic hypothesis for the 12 primate taxa based upon substitution differences among those taxa and using the neighbor-joining method of Saitou and Nei (1987). A single tree was obtained (Fig. 10-1), which according to Hayasaka et al. (1988a:633), was also obtained using the unweighted pair-group method of distance analysis and by the distance-Wagner method. Hayasaka et al. concluded that "Because these three different methods give phylogenetic trees with the same topology, the phylogenetic relationships derived from these mtDNA sequence comparisons [see our Fig. 10-1] appear reliable." It is not the purpose of this chapter to speak to the issue of primate relationships *per se*; rather the goal is to examine influences on the phylogenetic structure of these mtDNA sequences, to explore the implications of that structure for assessing the "reliability" of sequence data in general, and to investigate how parsimony analysis might be effectively applied to sequence data.

Two series of analyses are used to examine and describe the extent to which the primate mtDNA data set is phylogenetically informative. The first series employs parsimony analysis on different subsets of the data with the purpose of revealing the cladistic structure of most-parsimonious and

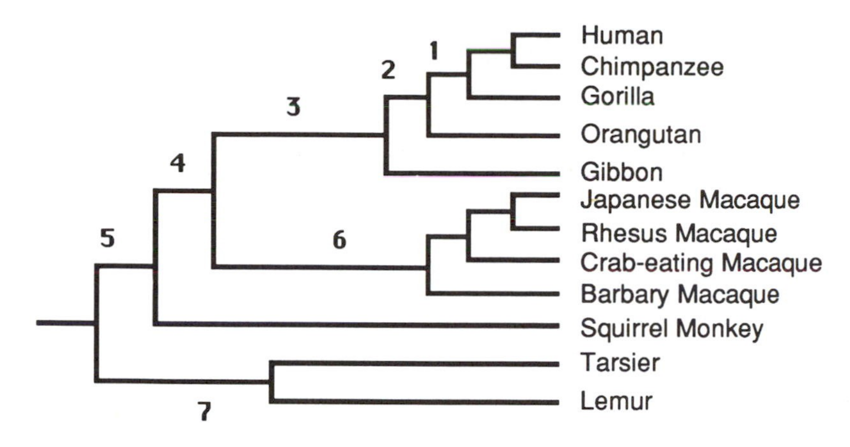

Figure 10-1 Phylogenetic hypothesis for 12 species of primates using a distance analysis (neighbor-joining method; Saitou and Nei, 1987) of an 898 bp fragment of mtDNA (from Hayasaka et al., 1988a). Names applied to clades discussed in this chapter follow the terminology of Hayasaka et al. (1988a): (1) hominines, (2) great apes, (3) hominoids, (4) catarrhines, (5) anthropoids, (6) macaques, and (7) prosimians.

near–most-parsimonious trees. The most-parsimonious tree, or collection of equally parsimonious trees, was found. Once its length was known, all trees within five steps of that length were retained, and a majority-rule consensus tree (Margush and McMorris, 1981) computed (majority-rule consensus trees were adopted because they permit an analysis of the relationship between the amount of phylogenetic signal and data set size and because they are comparable to the results of bootstrap analyses as discussed below). The collection of trees having lengths within 1% of the minimum-length tree was also examined and will be mentioned briefly.

A second series of analyses used the same subsets of data and constructed a majority-rule bootstrap tree (Felsenstein, 1985) from 100 randomly sampled (with replacement) replications of each of the subsets. The bootstrap, in this case, is not intended as a test of statistical significance but as an additional means of describing and evaluating the structure of the phylogenetic signal inherent in the data.

RESULTS

Global Parsimony Analysis

Most-Parsimonious and Near–Most-Parsimonious (Five-Step) Trees

A global parsimony analysis was undertaken using all nucleotide substitutions in each of the four subsets of the 898 bp fragment. This analysis was designed to answer two principle questions: Can a global parsimony analysis, in which transitions and transversions are equally (uniformly) weighted, reveal a stable phylogenetic signal? And, second, does the stability of that signal change as sample size is increased? Most-parsimonious trees are described but not figured. Primary attention is given to the set of trees within five steps of the most-parsimonious tree (or trees), but the collection of trees within 1% of the minimum-length tree is also examined.

1. ND4:225 bp Fragment. A parsimony analysis of the smallest data set of 225 characters yielded six equally parsimonious trees of 264 steps. These trees have only two topological ambiguities: (1) the hominine trichotomy; and (2) a trichotomy among the hominoids, macaques, and the squirrel monkey. All other relationships were as postulated by Hayasaka et al. (1988a: fig. 3:636) (see Fig. 10-1).

There are 95 trees within five steps of the shortest trees. The consensus tree exhibits relatively poor structure except within the macaques (Fig. 10-2A). There is no resolution within the great apes clade or among the basal lineages of the anthropoids.

2. ND4:458 bp Fragment. When the analysis is expanded to include all of the ND4 fragment, a single most parsimonious tree of 608 steps is found that is similar to the distance tree except for a sister-group relationship between the chimp and gorilla.

In the 458 bp data set for ND4 there are 17 trees within five steps of the single most-parsimonious tree (Fig. 10-2B). Resolution within the hominoids has improved substantially, although the human-chimpanzee-gorilla trichotomy, while recognized in 88% of the trees, is not resolved internally. The hominoid relationships of the gibbon are still somewhat ambiguous, occurring in only 53% of the trees. The additional data do not change the resolution within the macaques, but do now place them closer to the hominoids than to the squirrel monkey in 65% of the trees.

3. ND4 + tRNAs: 656 bp Fragment. A parsimony analysis of the ND4 fragment and the three tRNA genes produced a single most-parsimonious tree of 766 steps identical to that of the preceding analysis in which the gorilla is joined to the chimpanzee.

There were 9 trees within five steps of the most-parsimonious tree (Fig. 10-2C). With the addition of the 198 bp tRNA fragment to the ND4 sequence, resolution within the hominoid clade is improved, although the human-chimpanzee-gorilla trichotomy remains. Resolution in other parts of the tree remains relatively the same as for the 458 bp fragment.

4. ND4 + tRNAs + ND5: 898 bp Fragment. A parsimony analysis of the entire data set yielded two equally most-parsimonious trees (1,157 steps), one uniting chimpanzee and human, the other uniting chimpanzee and gorilla; all other relationships were as in the distance tree (Fig. 10-1).

Within five steps of the most-parsimonious trees, five trees were found.

The consensus tree (Fig. 10-2D) is fully resolved but a human-chimpanzee relationship was supported in only three of those trees (60%). Interestingly, only the branch uniting the orangutan to the hominines received less support as compared to the 656 bp fragment five-step consensus tree. Overall there was a marked improvement in resolution with the larger data set.

5. *ND4 + ND5 Fragments Compared With the tRNA Fragment.* In order to assess the relative influence on phylogenetic structure of mtDNA sequences that code for proteins versus structural RNAs, the two regions were analyzed separately. A parsimony analysis of the 700 bp encompassing the two coding regions of ND4 and ND5 taken together produced two equally parsimonious trees (999 steps) that differ only in the relationships of the chimpanzee to humans or to the gorilla; in all other respects the two trees are identical to the distance tree (Fig. 10-1).

There were five trees within five steps of the two shortest trees, and their consensus is shown in Figure 10-2E. Resolution is improved over the 656 bp fragment except in the placement of the orangutan with the hominines. Compared to the entire 898 bp data set (Fig. 10-2D), the five-step tree of ND4 + ND5 has less resolution within the hominines but the relationship between the hominoids and the macaques is more strongly supported by the ND4 + ND5 data set.

An analysis of the 198 bp fragment that includes the three tRNAs suggests that their contribution to instability is minimal. A single most-parsimonious tree was found (154 steps) that resolved relationships within the hominoids and macaques identical to those of the distance tree. This tRNA tree, however, placed the macaques outside the other taxa to yield: (macaques (tarsier (squirrel monkey + hominoids))).

There are 254 trees five steps away from the shortest tRNA tree and, not unexpectedly, the consensus tree (Fig. 10-2F) exhibits relatively little resolution.

Near–Most (1%)-Parsimonious Trees

The collection of trees within 1% of the length of the most-parsimonious tree has been used as a method to investigate the cladistic structure of sequence data (Smith, 1989) and was also examined in this study. Table 10-1 summarizes the results of this analysis.

The number of trees recovered within 1% of the shortest tree(s) did not vary noticeably among data sets despite the fact that larger data sets had trees of longer branch lengths. Only two clades, hominines and hominoids, consistently gained cladistic support as the size of the data set increased. All others showed decreasing support or no marked change. Some clades (anthropoids and macaques) were well defined no matter what size the data set. Interestingly, a global parsimony analysis shows strong resolution within the macaques across the four primary data sets, but no resolution is seen within the hominines.

95 trees (269 steps or less)
shortest tree= 264 steps (6 trees)

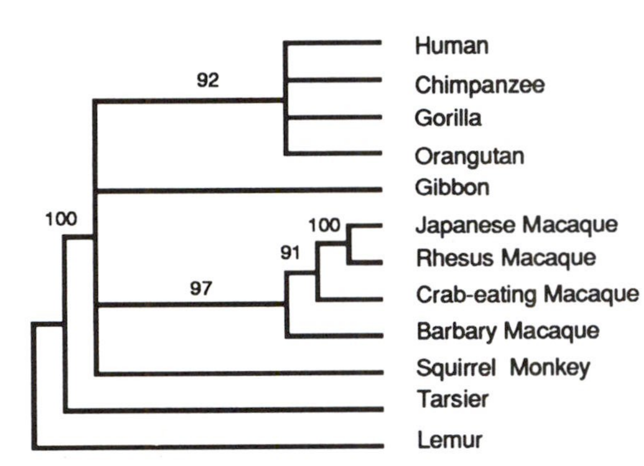

B. ND4: 458 bp

17 trees (613 steps or less)
shortest tree= 608 steps (1 tree)

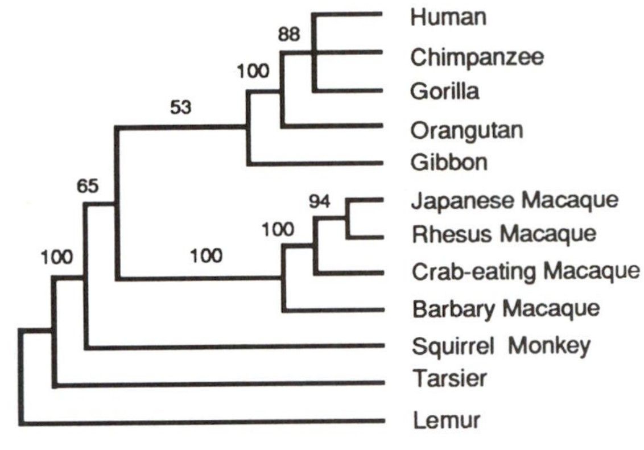

9 trees (771 steps or less)
shortest tree = 766 steps (1 tree)

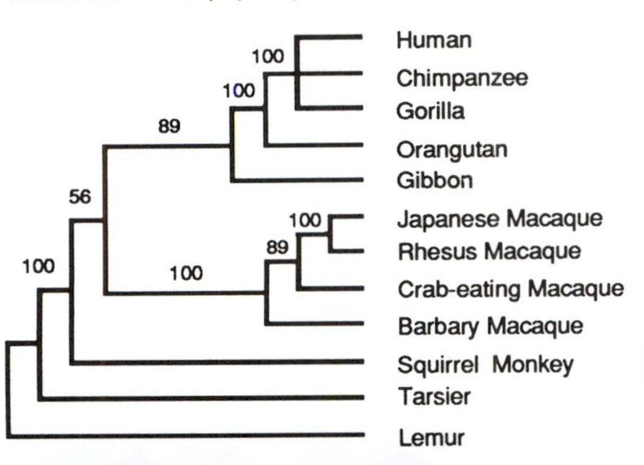

D. ND4+tRNAs+ND5: 898 bp

5 trees (1162 steps or less)
shortest tree = 1157 steps (2 trees)

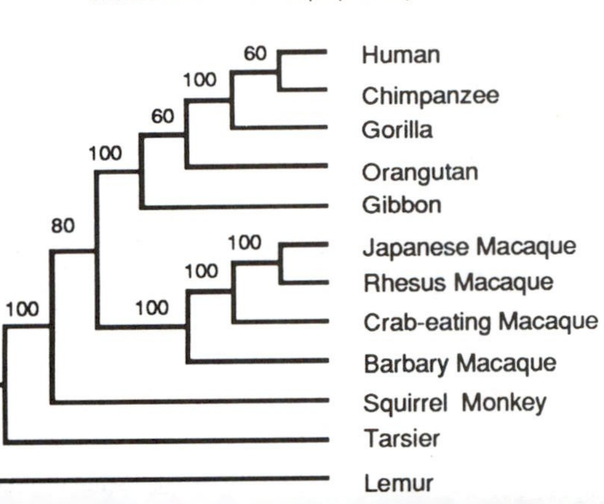

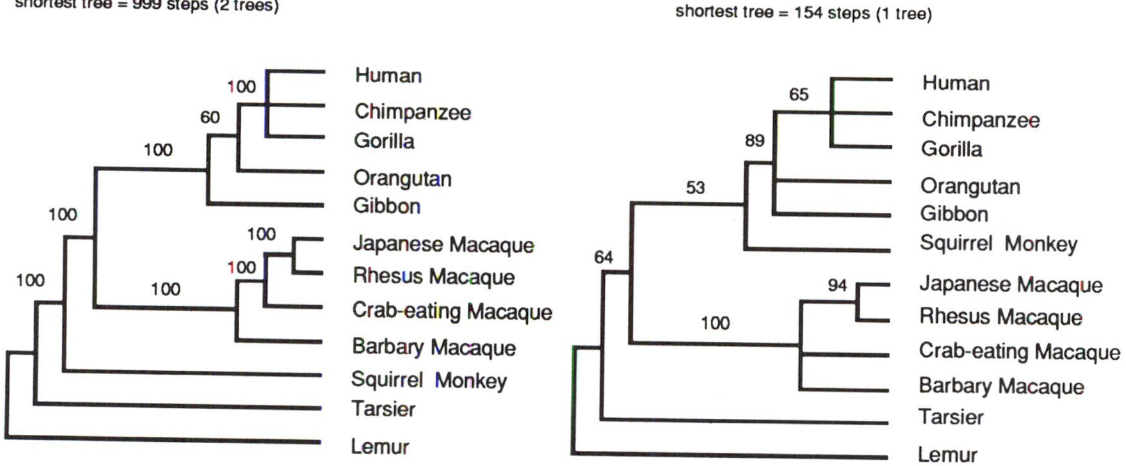

Figure 10-2 Majority-rule consensus trees of a global-parsimony analysis of all trees within five steps of the most-parsimonious solution for subsets of the ND4 + tRNAs + ND5 fragment of mtDNA, as discussed in the text. Numbers next to internal branches represent frequencies of occurrence for those clades among the most- and near–most-parsimonious solutions.

Table 10-1 Global Parsimony Analysis of Near–Most (1%)-Parsimonious Trees: Summary of Majority-Rule Consensus Trees

Fragment (no. of trees)	Clade								
	Human + Chimpanzee	Hominines	Great Apes	Hominoids	Catarrhines	Anthropoids	Japanese + Rhesus	Japanese + Rhesus + Crab-eating	Macaques
ND4: 225 bp (29)	n.r.*	52†	100	55	62	100	100	100	100
ND4:458 bp (24)	n.r.	79	96	58	71	100	92	96	100
ND4 + tRNAs: 656 bp (22)	n.r.	86	95	73	64	100	100	82	100
898 bp fragment (17)	n.r.	100	59	100	59	100	100	88	100
ND4 + ND5: 700 bp (13)	n.r.	100	69	100	69	100	85	100	100
tRNAs: 198 bp (25)	56	100	n.r.	100	n.r.‡	60	100	52	100

*n.r., clade not resolved at 50% level.
†Percentage of trees in which clade is resolved in majority-rule consensus trees.
‡Squirrel monkey clustered with the hominoids.

Summary of Global Parsimony Analysis

Numerous cladistic groupings are consistently revealed in a global parsimony analysis. With the exception of the smallest data set, the first 225 bp fragment of ND4, global parsimony analysis of each of the other data sets was able to resolve all relationships consistently except for the trichotomy within the hominines. This is also true of the combined ND4 + ND5 fragment of 700 bp, and, as noted above, even the most-parsimonious tree of the tRNAs was able to resolve most of the relationships.

These results indicate that this relatively small 898 bp fragment of mtDNA contains significant phylogenetic signal for taxa whose divergences extend as far back as 50 to 60 million years, that even "noise" produced by the high transition rate of mtDNA cannot obscure the signal, and that a global parsimony analysis is sufficiently powerful to resolve a consistent cladistic pattern even in data that are generally thought to be "noisy."

These conclusions are reinforced when the collections of near–most-parsimonious trees are examined (Fig. 10-2). In general, there is improved resolution as more data are added (Fig. 10-2A through 10-2E), but the trend is not always strong. Most importantly, a consistently strong cladistic signal is found in all but the very smallest data sets.

When the structure of trees within 1% of the most-parsimonious solution is examined, however, far fewer of the relationships are identified by a strong cladistic signal (e.g., clusters supported in 90% or more of the trees; see Table 10-1). Only the monophyly of the macaques and of the anthropoids is consistently revealed across all four primary data sets. As noted earlier, furthermore, saving 1% of the most-parsimonious trees does not produce a clear convergence in the phylogenetic signal as the data set is enlarged. This results from the fact that in each subset of the data there are many trees within 1% of the minimum-length tree, including some of extremely variant tree topology.

Transversion Parsimony Analysis

Most-Parsimonious and Near–Most-Parsimonious (Five-Step) Trees

Does the large bias for transition substitutions within mtDNA increase the "noise" and lead to instability of the phylogenetic signal within the primate data set? Can the signal be improved by examining transversions alone? And, does the stability of a phylogenetic signal inferred from a parsimony analysis of transversions improve with increasing sample size? These questions were investigated by repeating the above analysis using transversion differences.

1. ND4: 225 bp Fragment. An analysis of the first 225 bp of the ND4 fragment produced two equally parsimonious trees (79 steps). In both trees, the macaques were not fully resolved, with only the Japanese-rhesus macaque clade being held in common. The major difference between the two trees was found in the placement of *Homo*: in one tree it was with gorilla, in the other with chimpanzee.

There were 681 trees within five steps of the most-parsimonious solutions. Their majority-rule consensus (Fig. 10-3A) shows very strong support for three clades, the anthropoids, catarrhines, and macaques, but fails to resolve the relationships of any other group.

2. ND4:458 bp Fragment. A transversion parsimony analysis of the entire ND4 fragment yielded two equally parsimonious trees (196 steps). One is identical to the distance tree (Fig. 10-1), and the other failed to resolve a trichotomous relationship at the base of the macaques: (Barbary, crab-eating, Japanese + rhesus).

Although 114 trees were found within five steps of the most-parsimonious solution, the majority-rule consensus (Fig. 10-3B) shows a marked improvement in resolution compared to the 225 bp fragment. Clades are resolved within the hominoids, but not within macaques.

3. ND4 + tRNAs: 656 bp Fragment. A transversion analysis of the ND4 fragment along with the tRNAs produced two equally parsimonious trees (239 steps) that have identical structure to the two most-parsimonious trees of the preceding analysis.

There were 59 trees within five steps of the two most-parsimonious trees. Compared to the 458 bp fragment, some additional, slight resolution is apparent in the hominines and the macaques (Fig. 10-3C).

4. ND4 + tRNAs + ND5: 898 bp Fragment. A transversion parsimony analysis of the entire data set yielded a single most-parsimonious tree (381 steps). It has the same structure as the distance tree (Fig. 10-1).

For the transversions in the total data set, 46 trees were found within five steps of the most-parsimonious tree. Although there is no resolution within the hominines or macaques, compared to the previous consensus trees, there is a marked increase in support for the hominines and catarrhines (Fig. 10-3D).

5. ND4 + ND5 Fragments Compared With the tRNA Fragment. In a transversion parsimony analysis of the ND4 and ND5 fragments combined (700 bp), a single most-parsimonious tree was obtained (338 steps) that is identical to the distance tree. A transversion parsimony analysis of the three tRNAs (198 bp) found seven equally parsimonious trees (42 steps). The consensus tree for these showed the chimpanzee united with the human in 77% of the cases, but with no resolution among the other hominoids, including the gorilla. The Japanese and rhesus macaques were clustered as sister-species, but no other nodes were resolved within macaques, and the squirrel monkey was united with the macaques.

Transversion analysis of the 700 bp coding region finds 66 trees within five steps of the most-parsimonious tree. Their majority-rule consensus tree (Fig. 10-3E) is very similar to that of the entire 898 bp data set (Fig. 10-3D) except that the great apes are much less supported (76% versus 91%). Not unexpectedly, there are a large number of trees within five steps of the most-parsimonious solutions for the tRNA fragment. The majority-rule consensus for 3,000 trees (computer memory was exhausted; many more could have been found) is shown in Figure 10-3F. Only the macaques as a group are well defined.

Near–Most (1%)-Parsimonious Trees

Table 10-2 summarizes majority-rule consensus tree support for all trees with lengths within 1% of the most-parsimonious transversion solutions. A consistent pattern is observed: except for the tRNA fragment, virtually all of the "deep" cladistic events are strongly supported in each of the five primary analyses, and this also includes the shortest 225 bp fragment. In contrast, strong resolution of the "more recent" cladistic events, including those within the macaques and hominines, is lacking.

Summary of Transversion Parsimony Analysis

The most general conclusion to be drawn from a comparison of the global parsimony analysis (Fig. 10-2; Table 10-1) with the transversion parsimony analysis (Fig. 10-3; Table 10-2) is that the use of transversions, by themselves, reveals a strong cladistic signal for the relatively older clades. This is especially evident when examining those trees having a length within 1% of the most-parsimonious solutions (Table 10-2), which in most of the data sets included all those trees within two or three steps of the minimum-length solution. Since many more trees were examined in the five-step analysis (Fig. 10-3), support for some of these deeper branches declined, but in general remained fairly strong. None of these methods of analysis, when applied to transversion differences alone, were able to reveal a strong cladistic signal within the hominines or within the macaques.

The five step analysis revealed a slight increase in the strength of the cladistic signal as the data set increased in size. The 656 bp fragment (Fig. 10-3C) is not much more informative than the 458 bp fragment (Fig. 10-3B), but the total data set (Fig. 10-3D) and the ND4 + ND5 fragment (700 bp; Fig. 10-3E) exhibit stronger cladistic support within the hominoids.

Global Parsimony of First and Second Codon Positions

Does the high rate of transitions in third positions contribute noise and therefore result in decreased stability of the phylogenetic signal? This question was investigated by a global parsimony analysis of the ND4 and ND5 fragments in which the third positions were excluded. The analysis yielded a single most-parsimonious tree of 519 steps. There were only two differences from the distance tree. The first had the chimpanzee united with the gorilla, but the most unexpected result (given the preceding anal-

681 trees (84 steps or less)
shortest tree= 79 steps (2 trees)

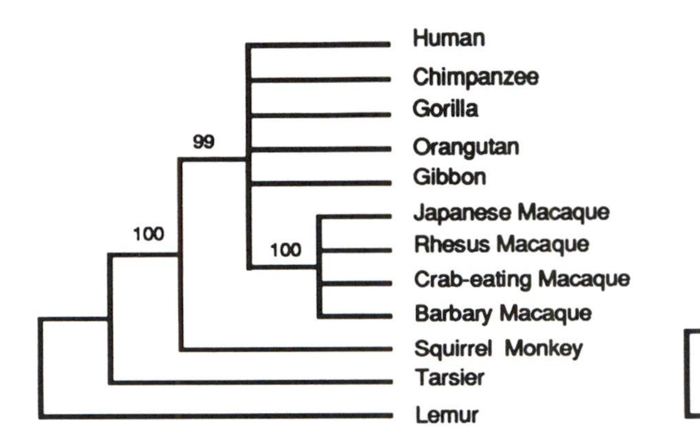

B. ND4: 458 bp

114 trees (201 steps or less)
shortest tree= 196 steps (2 trees)

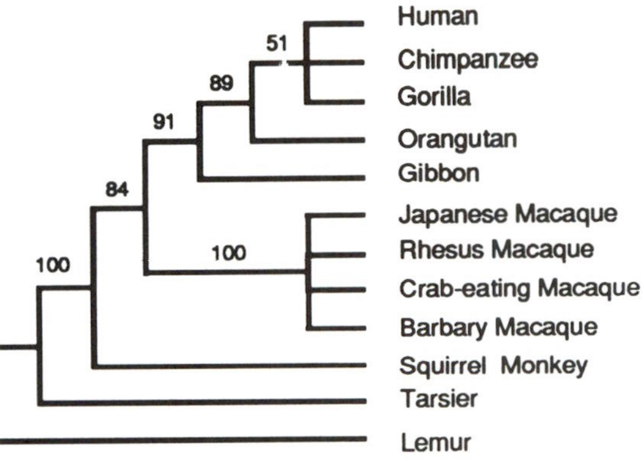

59 trees (244 steps or less)
shortest tree = 239 steps (2 trees)

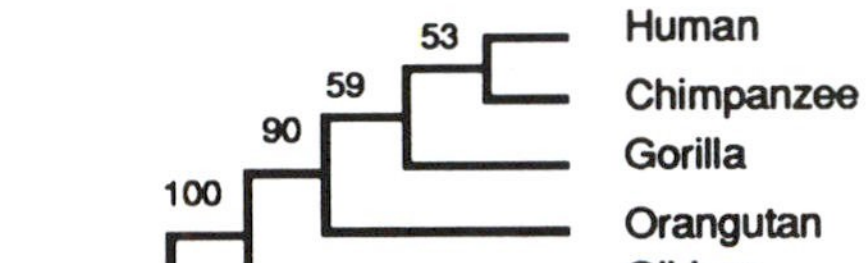

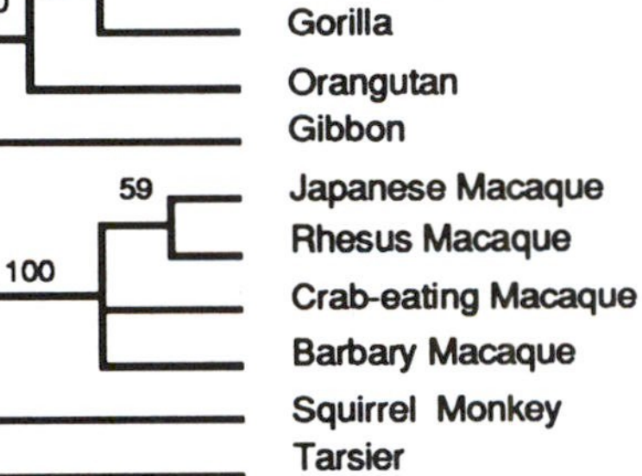

D. ND4+tRNAs+ND5: 898 bp

46 trees (386 steps or less)
shortest tree = 381 steps (1 tree)

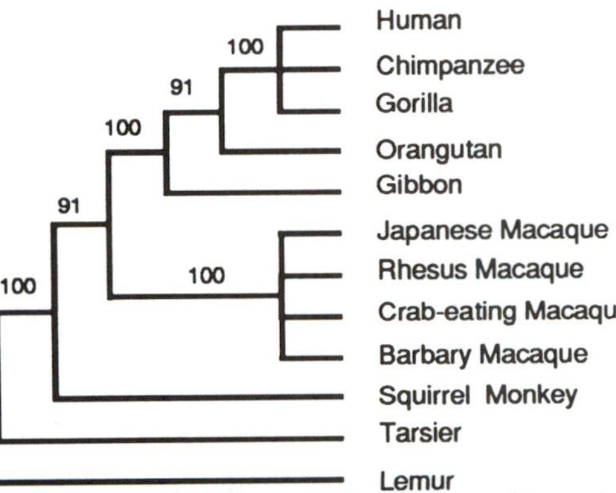

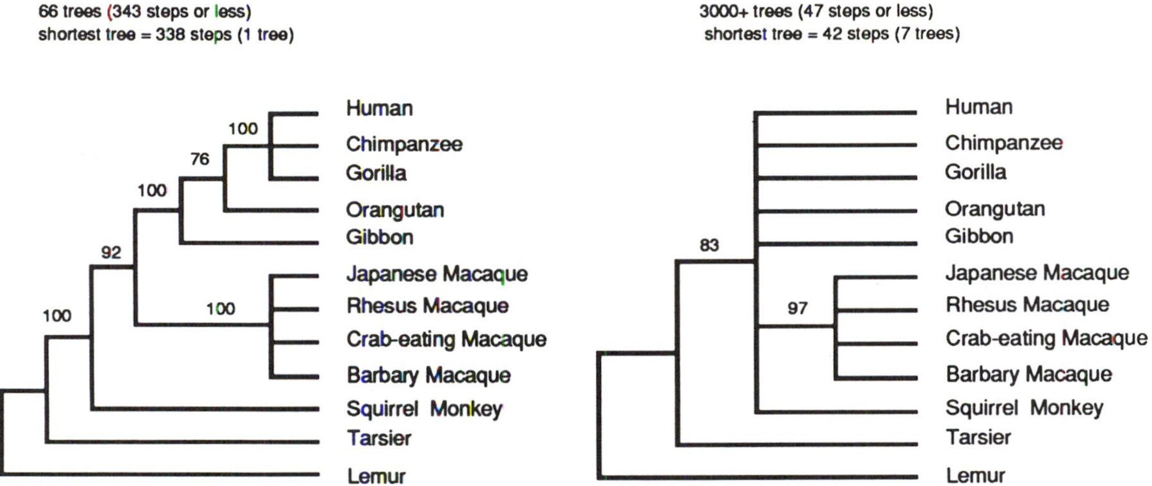

Figure 10-3 Majority-rule consensus trees of a transversion parsimony analysis of all trees within five steps of the most-parsimonious solution for subsets of the ND4 + tRNAs + ND5 fragment of mtDNA, as discussed in the text. Frequencies as in Figure 10-2.

Table 10-2 Transversion Parsimony Analysis of Near–Most (1%)-Parsimonious Trees: Summary of Majority-Rule Consensus Trees

Fragment (no. of trees)	Clade								
	Human + Chimpanzee	Hominines	Great Apes	Hominoids	Catarrhines	Anthropoids	Japanese + Rhesus	Japanese + Rhesus + Crab-eating	Macaques
ND4: 225 bp (17)	n.r.*	65†	100	100	100	100	53	n.r.	100
ND4:458 bp (14)	n.r.	100	100	100	100	100	n.r.	n.r.	100
ND4 + tRNAs: 656 bp (8)	75	100	100	100	100	100	n.r.	n.r.	100
898 bp fragment (27)	59	100	96	100	100	100	n.r.	n.r.	100
ND4 + ND5: 700 bp (38)	n.r.	100	100	100	100	100	n.r.	n.r.	100
tRNAs: 198 bp (77)	77	n.r.	n.r.	74	n.r.‡	100	64	n.r.	100

*n.r., clade not resolved at 50% level.

†Percentage of trees in which clade is resolved in majority-rule consensus trees.

‡Squirrel monkey clustered with macaques.

yses) was the closer relationship of the gibbon, rather than the orangutan, to the great apes.

There were eight trees within five steps of the shortest tree (Fig. 10-4A). The consensus tree showed little resolution within either the hominoids or the macaques, but it did resolve the monophyly of both of these clades and clustered them relative to the squirrel monkey. This analysis suggests that attempting to enhance the phylogenetic signal by eliminating transitions in the third position was unsuccessful. It had no effect on improving the resolution of the older cladistic events, which were strongly resolved by global parsimony, and not unexpectedly, it led to a decrease in the signal within the macaques (compare Fig. 10-4A with Fig. 10-2E).

Amino Acid Replacements: Substitution Parsimony

Amino acid replacements are often taken as being more informative of relationships than nucleotide substitutions, especially for more distant branching events. The structure of the phylogenetic signal of replacements was examined by parsimony analysis of the numbers of substitutions that account for amino acid differences among the taxa. A transformation matrix for vertebrate mtDNA provided by D. L. Swofford was used to convert amino acid differences to substitution differences and then each tree was evaluated.

A single most-parsimonious tree of 387 steps was found for the 231 codons of the ND4 and ND5 fragments arranged tandemly. Relationships were similar to those of the distance tree except within the hominoids where gorilla and human were united and the gibbon (not orangutan) was the sister-group of the hominines.

There were 49 trees within five steps of the most-parsimonious tree. The majority-rule consensus tree (Fig. 10-4B) shows relatively little resolution, either within the macaques or the hominoids, thus suggesting that this method is less effective in revealing phylogenetic signal than is a global parsimony analysis.

Bootstrap Analysis

Global Parsimony Bootstrap

1. ND4: 225 bp Fragment. A bootstrap analysis of the first 225 bp of ND4 produced a majority-rule consensus tree identical to the distance tree except for chimpanzee being united with gorilla (Fig. 10-5A). The latter relationship means little, however, given that all three alternatives within the hominines are about equally supported. Clades having substantial support in this data set include: (1) the macaques and their inclusive groupings; (2) the great apes; and (3) the anthropoids.

2. ND4: 458 bp Fragment. A bootstrap analysis of the entire ND4 fragment produced a tree very similar to the preceding, except that human

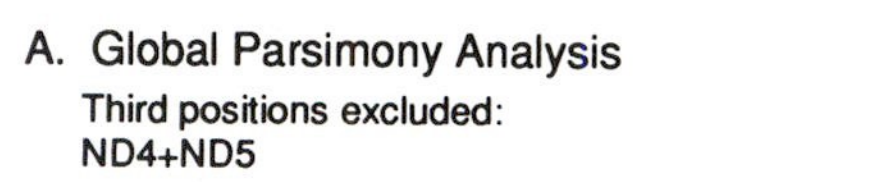

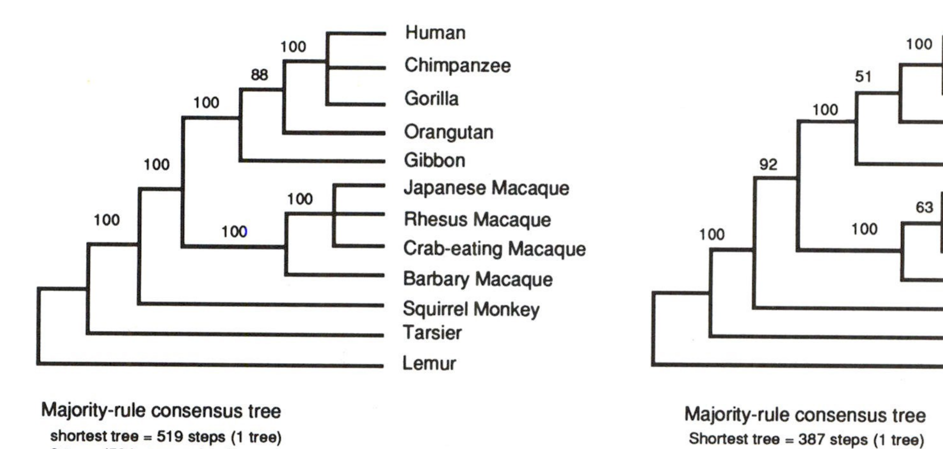

Figure 10-4 (A) A majority-rule consensus tree of a global parsimony analysis of all trees within five steps of the most-parsimonious solution for the tandemly arranged fragments of ND4 and ND5 genes of mtDNA with third codon positions excluded. (B) A majority-rule consensus tree of a global parsimony analysis of nucleotide substitutions resulting in amino acid replacements in 231 codons of the ND4 and ND5 genes of mtDNA. Frequencies in both (A) and (B) as described in Figure 10-2.

and chimpanzee are narrowly united (Fig. 10-5B). Support for each of the clades has generally improved, especially for the catarrhines, the hominines, and the macaques.

3. ND4 + tRNAs: 656 bp Fragment. The majority-rule consensus tree for this data set (Fig. 10-5C) is identical to the preceding. In general, support for each clade increases, although the catarrhine clade is found in only 50% of the replicates.

4. ND4 + tRNAs + ND5: 898 bp Fragment. The majority-rule consensus tree for all the data keeps the same pattern of relationships as the preceding, except for uniting chimpanzee and gorilla (50% of the replicates; Fig. 10-5D). Bootstrap support for the other clades has increased, sometimes substantially (e.g., hominoids, catarrhines).

5. ND4 + ND5 Fragments Compared With the tRNA Fragment. A bootstrap majority-rule consensus tree of the protein-coding region (Fig. 10-5E) shows a similar pattern to that of the distance tree. Levels of support are close to those of the 898 bp tree (Fig. 10-5D). Bootstrapping the 198 bp of the rRNAs (Fig. 10-5F) suggests that they contribute less noise to the data than might be expected, except at the base of the anthropoids where the squirrel monkey is united with the hominoids. The tRNAs, however, strongly support the monophyly of the macaques and a rhesus + Japanese macaque relationship.

Summary of the Global Parsimony Bootstrap

Not unexpectedly, the phylogenetic signal, as measured by the percentage of congruent bootstrap replicates, improves as the sample size increases. Thus, for the 898 bp fragment (Fig. 10-5D), six clades were supported more than 97% of the time. Nevertheless, if one preferred to interpret the "significance" of these results stringently by collapsing all branches that failed to meet a 90 or 95% criterion, the bootstraps of characters for each subset of the data would reveal a cladistic structure not very much different from that seen in previous analyses (Fig. 10-2).

Transversion Parsimony Bootstrap

1. ND4: 225 bp Fragment. A bootstrap of transversions contained in the first 225 bp of the ND4 fragment produces a majority-rule consensus tree much like the distance tree except for the lack of resolution at the base of the macaques (Fig. 10-6A). The monophyly of the macaques, the catarrhines, and the anthropoids is strongly supported by a bootstrap analysis of these data.

2. ND4: 458 bp Fragment. A bootstrap of the entire ND4 fragment produced a tree identical to the distance tree (Fig. 10-6B). Relative to the 225 bp fragment, the pattern of support varies. Clades within the hominoids are better supported with this expanded data set, although the monophyly

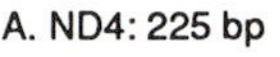

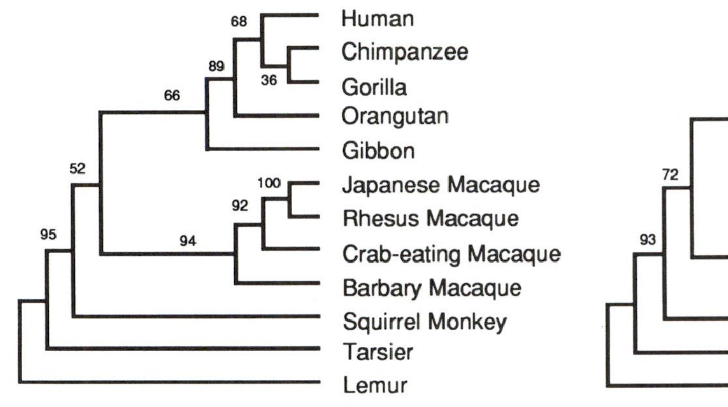

B. ND4: 458 bp

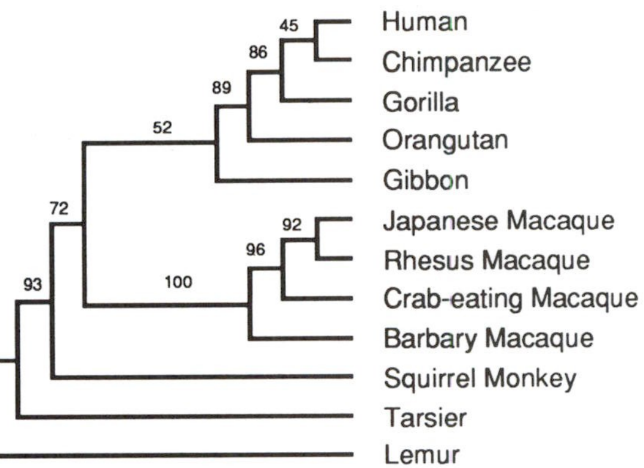

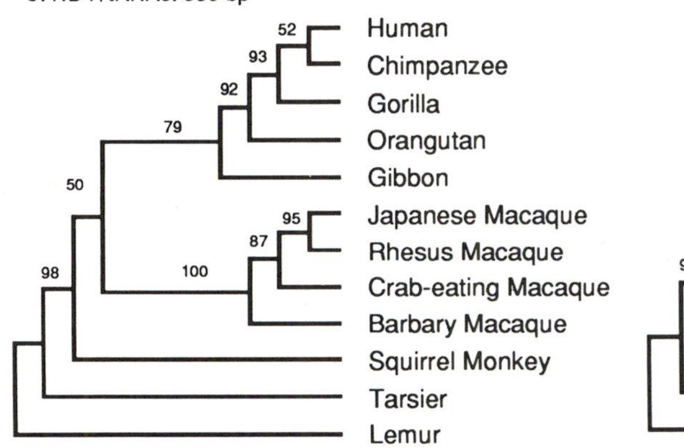

D. ND4+tRNAs+ND5: 898 bp

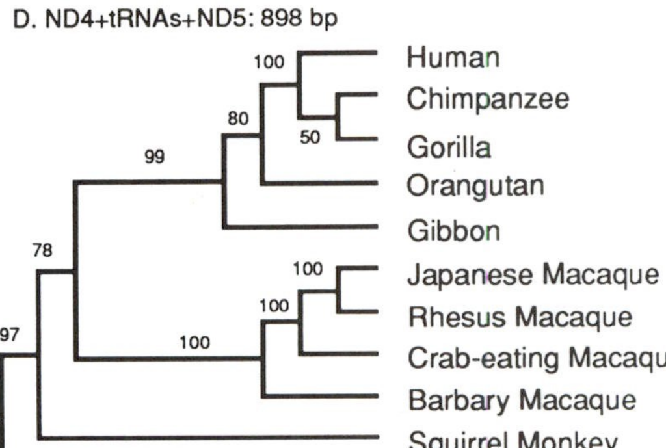

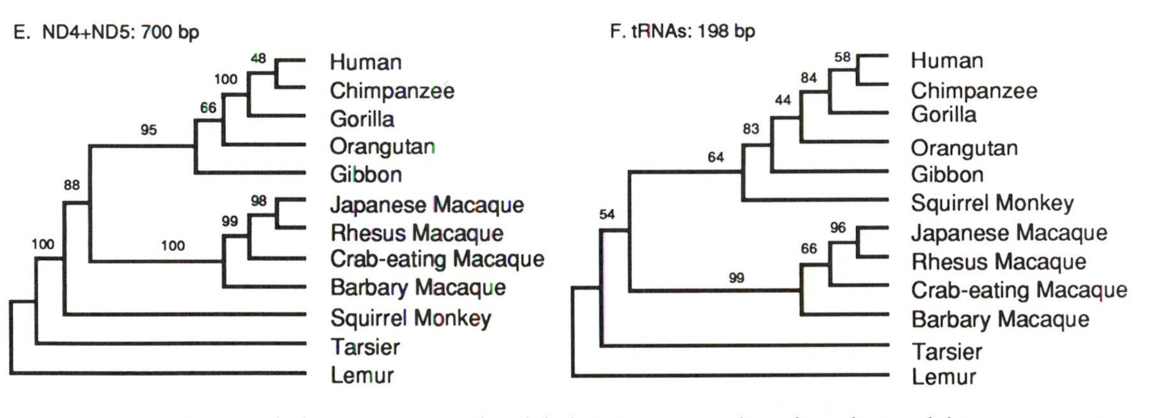

Figure 10-5 Majority-rule bootstrap trees of a global parsimony analysis for subsets of the ND4 + tRNAs + ND5 fragment of mtDNA, as discussed in the text. Frequencies of occurrence among the bootstrap trees are presented for the internal branches.

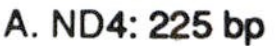
A. ND4: 225 bp

B. ND4: 458 bp

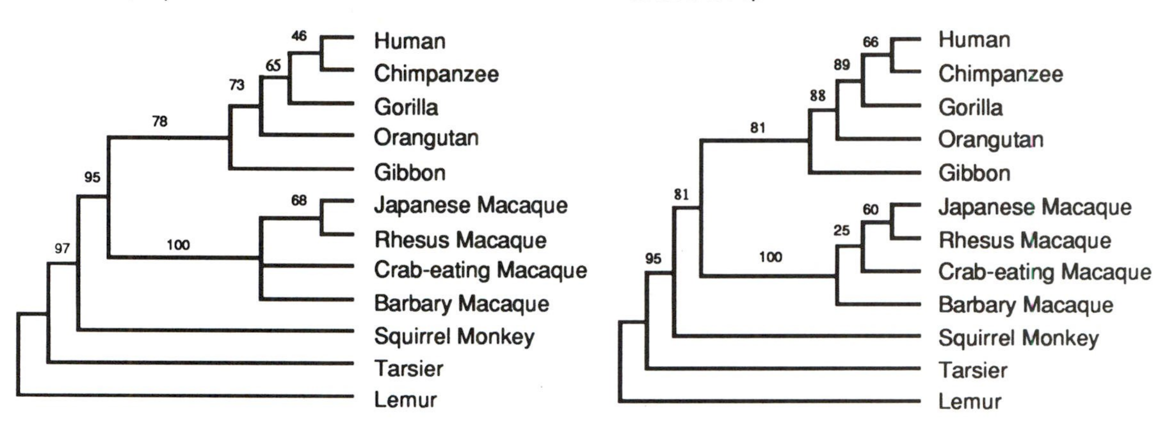
Human
Chimpanzee
Gorilla
Orangutan
Gibbon
Japanese Macaque
Rhesus Macaque
Crab-eating Macaque
Barbary Macaque
Squirrel Monkey
Tarsier
Lemur
46
65
73
78
68
100
95
97
Human
Chimpanzee
Gorilla
Orangutan
Gibbon
Japanese Macaque
Rhesus Macaque
Crab-eating Macaque
Barbary Macaque
Squirrel Monkey
Tarsier
Lemur
66
89
88
81
60
25
100
81
95

C. ND4+tRNAs: 656 bp

D. ND4+tRNAs+ND5: 898 bp

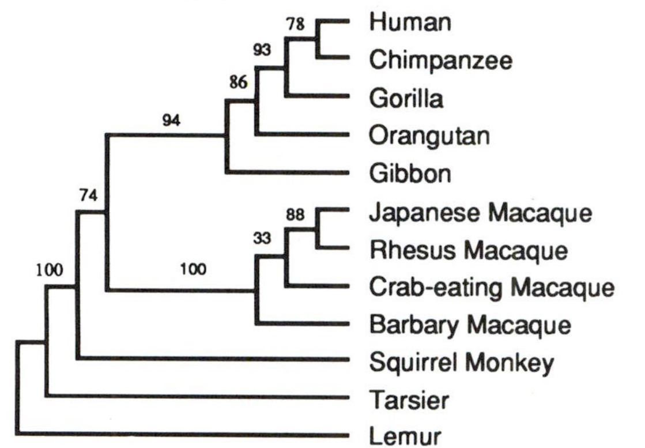
Human
Chimpanzee
Gorilla
Orangutan
Gibbon
Japanese Macaque
Rhesus Macaque
Crab-eating Macaque
Barbary Macaque
Squirrel Monkey
Tarsier
Lemur
78
93
86
94
88
33
100
74
100

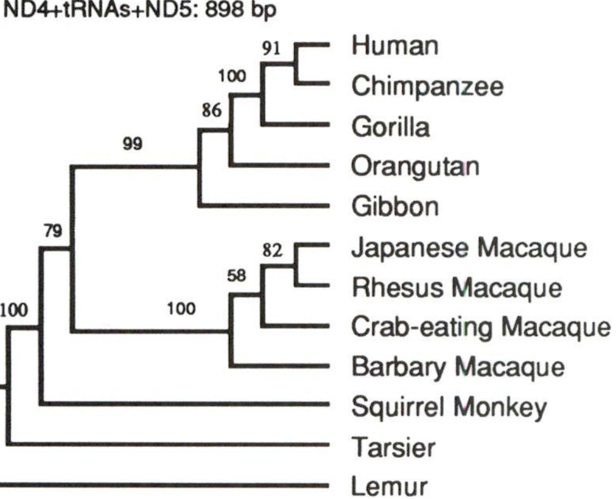
Human
Chimpanzee
Gorilla
Orangutan
Gibbon
Japanese Macaque
Rhesus Macaque
Crab-eating Macaque
Barbary Macaque
Squirrel Monkey
Tarsier
Lemur
91
100
86
99
82
58
100
79
100

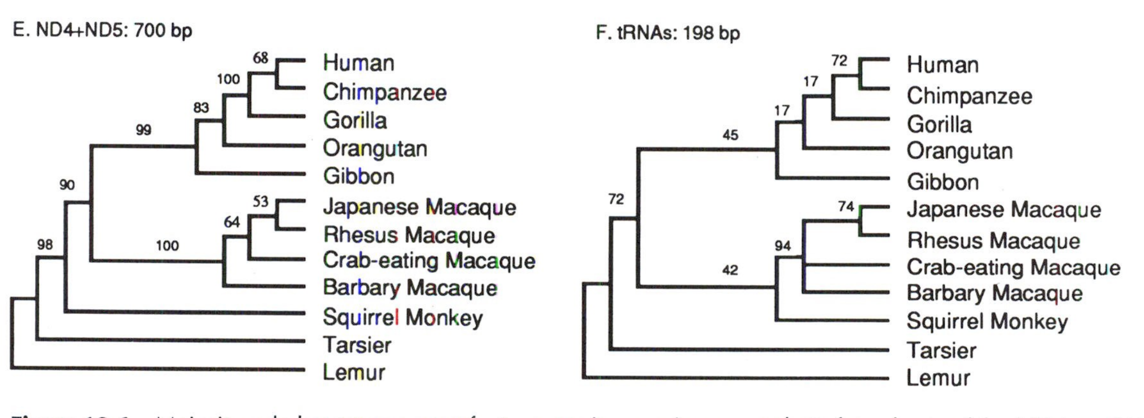

Figure 10-6 Majority-rule bootstrap trees of a transversion parsimony analysis for subsets of the ND4 + tRNAs + ND5 fragment of mtDNA, as discussed in the text. Frequencies as in Figure 10-5.

of the catarrhines is much less so. Support for a human + chimpanzee relationship has increased to 66% of the replicates.

3. ND4 + tRNAs: 656 bp Fragment. The relationships implied by this data set (Fig. 10-6C) are the same as in the preceding analysis. There is an increase in the support of all clades except that of the catarrhines.

4. ND4 + tRNAs + ND5: 898 bp Fragment. A bootstrap of the transversions in the entire data set yielded a tree showing no changes in cladistic structure from the preceding analysis, but with a general increase in the cladistic support at most nodes (Fig. 10-6D).

5. ND4 + ND5 Fragments Compared to the tRNA Fragment. A transversion bootstrap analysis of the combined ND4 and ND5 sequences once again produced the same tree as did using all the data and, in general, with approximately the same level of support (Fig. 10-6E). An analysis of transversions in the tRNAs shows that this data set retains much of the phylogenetic structure present in the protein-coding regions, but virtually all of the nodes are poorly supported (Fig. 10-6F). In the tRNA data set, the squirrel monkey clusters with the macaques, showing ambiguity at the base of the anthropoids.

Summary of the Transversion Bootstrap Analysis

As with the global parsimony bootstrap analysis (Fig. 10-5), there was a general increase in support for the nodes as sample size was increased. Neither bootstrap analysis provided significantly better support than the other. Although the transversion bootstrap produced much better resolution within the hominines, the global bootstrap analysis resolved the relationships within the macaques much more convincingly. Both did about equally well with the other parts of the tree.

The trees of Figure 10-6 are all majority-rule consensus trees, and if the level of stringency of cladistic support was taken to be 90 or 95% of the bootstrap replicates, the resulting topologies would not differ very much from those of the transversion parsimony analysis (Fig. 10-3).

DISCUSSION

Informativeness, Noise, and the Primate Data Set

A major concern when analyzing any systematic data is to assess the extent to which they are phylogenetically informative. Systematists are also interested in knowing whether a particular analytical method is effective in maximizing informativeness and thus results in phylogenetic hypotheses that are more strongly supported than those produced by other methods. These problems are especially acute with sequence data, not only because of the large numbers of characters that are normally generated, but also because the evolutionary dynamics of DNA are such that a large amount of noise (i.e., homoplasy) would not be unexpected.

How might the informativeness and noisiness of the primate data set be evaluated? One approach might examine sets of near–most-parsimonious trees and judge informativeness by the degree to which the cladistic structure is corroborated across those trees. A second method might examine consensus trees produced by resampling of the data themselves and ask to what extent each node is supported; that is, in what percentage of the trees forming the consensus does a given node occur? Both methods examine the structure of data, but in different ways.

A consideration of the amount of support for each node as seen in the majority-rule consensus trees (Figs. 10-2 and 10-3; Tables 10-1 and 10-2) permits a detailed description of the phylogenetic signal contained in the data. In the global majority-rule consensus of trees within five steps of the most-parsimonious solutions (Fig. 10-2), the relationships within the macaques are fully resolved using the largest data sets. The hominines always cluster relative to the orangutan, and these two groups cluster relative to the gibbon. The major lineages of the tree—hominoids, catarrhines, anthropoids, and macaques—are also well supported by the largest data sets. In the transversion majority-rule trees (Fig. 10-3), there is little or no phylogenetic signal within the macaques or hominines, but the data unambiguously resolve all other clades, especially using the largest data sets. Taken together, these two analyses reveal a consistent phylogenetic signal for all clades except those within the hominines.

Bootstrap analysis constitutes a second approach to characterizing the amount of phylogenetic signal in the data. Bootstrap sampling of the total data set (Fig. 10-5) reveals a strong phylogenetic signal within the macaques, and comparison with the bootstrap analysis of transversions alone (Fig. 10-6) leads to the conclusion that a substantial component of that signal is in fact coming from transition differences. Within the hominines, however, it is transversions, not transitions, that provide the most consistent signal. In both bootstrap analyses, the major lineages are all well supported using the largest data sets, although the phylogenetic position of the squirrel monkey is the least well-corroborated.

How do the two approaches compare? Not unexpectedly, the results using these two different methods parallel one another to a very great extent (e.g., compare Figs. 10-2 and 10-5; and 10-3 and 10-6). Whenever a clade is strongly supported in a near–most-parsimonious consensus tree, it is also strongly supported in the comparable bootstrap tree. Differences arise in those clades that lack strong support. Because bootstrap trees tend to be fully resolved, they have the appearance of being more informative, yet there is generally little character support underlying that increased resolution. Still, when examination of the structure of near–most-parsimonious trees often yields an unresolved polytomy, bootstrap resampling of the data seems more sensitive to finding the "correct" signal, even if that signal is not especially strong. Again, one need not interpret bootstrap results in a strict statistical sense, but simply as another method of evaluating the presence and strength of signal in the data. Thus, bootstrapping of transversions across the entire data set provides resolution of relationships within the macaques (Fig. 10-6D), whereas the majority-rule consensus analysis of near–most-parsimonious trees (Fig. 10-3D) does not. This result generally holds for most subsets of the data, not only for relationships within the macaques, but also within the hominines.

Does Phylogenetic Signal Stabilize with More Data?

A critical practical problem of phylogenetic inference using DNA sequences is deciding how much sequence must be obtained before relationships can be resolved. In attempting to answer this question, factors other than sequence length *per se* cannot be ignored. It is generally assumed, for example, that certain segments of DNA will be more informative than others and therefore less sequence will be required to reveal a phylogenetic signal. This supposition usually entails considerations of evolutionary rate, which is discussed in the next section. Here we ask whether phylogenetic signal within the primate data set is related to data set size and which method of analysis best reveals that signal.

The global parsimony analysis of near–most-parsimonious trees shows an overall increase in the strength of signal as sample size is increased (Fig. 10-2A through 10-2E). There is a marked improvement in the 458 bp tree as compared to the 225 bp tree. With respect to the 656 bp consensus tree (Fig. 10-2C), some nodes are slightly better supported compared to the 458 bp tree, others are not. Generally, the 898 bp and 700 bp trees have stronger corroboration than the 656 bp tree.

The transversion parsimony analysis of near–most-parsimonious trees also exhibits increasing strength of signal with increasing sample size (Fig. 10-3). That increase in signal, however, is confined to the deeper branches and to the hominoids. Resolution within the macaques is not affected by increases in sample size using transversions alone.

As noted earlier there is a general increase in phylogenetic signal, as determined by bootstrap support, when sample size increases. This is true for both the global parsimony and transversion parsimony analyses (Figs. 10-5, 10-6). Not all nodes show increased support between successive subsets of data, and not all nodes are well supported, but there is a definable convergence toward a single tree. These results suggest that bootstrap analysis may be somewhat more informative as a descriptor of the effects of sample size on phylogenetic signal than is examination of near–most-parsimonious trees.

Evolutionary Rate and Phylogenetic Signal

The degree to which a segment of DNA is considered likely to be informative is generally thought to be related to its rate of base substitution relative to the divergence times that the investigator is trying to resolve. Segments with high rates of change are assumed to be less informative for deeper branches because the accumulation of homoplasy (parallel and back mutations) erases any signal (Hayasaka et al., 1988a; Smith, 1989). As substitutions "saturate," genetic distances among the taxa approach an asymptote relative to divergence time (Brown, 1983; Moritz et al., 1987), and this observation has led to the conclusion that mtDNA sequence variation will be less useful in resolving relationships among relatively more distantly related taxa (Moritz et al., 1987).

To the extent that data are available (Brown, 1985), the genes examined in this study (ND4 and ND5) might be categorized as "relatively fast" compared to many other protein-coding genes, tRNAs, or ribosomal RNA genes. Hayasaka et al. (1988a:637–638) estimated that the relationship between divergence time and the number of substitutions among taxa [corrected for multiple mutations (Gojobori et al., 1982)] is linear only up to about 30 million years and that older divergence times (e.g., those deeper than the base of the hominoids) cannot be resolved "conclusively." Yet, such a conclusion is method-dependent. Even though ND4 and ND5 are "fast," the present analysis suggests their rate may have only marginal influence on the relationship between phylogenetic signal and divergence time (at least within primates). Depending upon how parsimony is applied to the data, these genes are just as capable of resolving the more ancient divergences within the primates as they are the most recent. Thus, by examining the relative influence of transitions and transversions, a reasonably consistent phylogenetic signal is revealed across the tree (relationships within the hominines constitute a slight exception; see below). These conclusions deserve to be tempered, however, by noting that transition/transversion differences do not become equal until the very most distant relationships between the prosimians and the other taxa are considered (Hayasaka et al., 1988a; see our Table 10-3). It could be argued, therefore, that relationships within the anthropoids are not affected by a saturation effect (although saturation is evident at the base of the anthropoids).

The Resolving Power of Most-Parsimonious Trees

It is often thought within molecular systematics that parsimony methods will often fail to be informative because of the high degree of homoplasy in sequence data (see next section). Therefore, it is useful to examine that supposition by asking whether the most-parsimonious trees, based on a consideration of all of the data, were in fact accurate estimates of the phylogenetic relationships of the primates, or whether the "noise" in these "fast" evolving genes made that resolution difficult or impossible.

The answer to this question may be surprising to some: the relationships of the primates were consistently revealed by the most-parsimonious trees of each of the data sets. In general, the only ambiguity involved resolution

Table 10-3 Transition/Transversion Differences Among 12 Primate Taxa (Sequence Data Derived From Hayasaka et al., 1988a)

Clade	1	2	3	4	5	6	7	8	9	10	11	12
1. Human	—											
2. Chimpanzee	74/5	—										
3. Gorilla	84/8	86/9	—									
4. Orangutan	108/35	119/34	116/33	—								
5. Gibbon	116/45	124/44	123/45	115/52	—							
6. Japanese macaque	134/74	144/73	139/72	143/75	145/77	—						
7. Rhesus macaque	131/77	145/76	135/75	141/78	133/80	32/3	—					
8. Crab-eating macaque	149/74	166/73	162/72	158/75	154/77	71/4	81/5	—				
9. Barbary macaque	155/74	149/73	146/72	137/77	139/77	103/8	100/9	104/6	—			
10. Squirrel monkey	146/98	154/99	145/96	139/111	135/106	156/102	158/104	154/102	154/102	—		
11. Tarsier	150/136	147/137	145/132	136/131	143/133	145/134	144/134	144/132	148/136	137/151	—	
12. Lemur	133/142	134/141	123/138	129/129	128/137	126/126	128/128	140/126	128/128	119/135	101/126	—

of relationships within the hominines, which has been a long-standing problem that even very large amounts of data have had difficulty in resolving. The most parsimonious solution of the smallest data set, the 225 bp fragment, also showed a trichotomy at the base of the anthropoids, yet was still remarkably effective in resolving relationships among the other taxa.

It therefore seems that a straightforward parsimony analysis of unweighted data, including both transitions and transversions, is an effective method for estimating what appears to be the "correct" tree, even in subsets of the data. It may be that because most of the data are not saturated with transition substitutions, there remains relatively little noise in the data and, therefore, parsimony is effective in revealing the signal. On the other hand, it may be that a simple parsimony approach is in fact a very powerful method and can discern phylogenetic signal, even in data having a strong component of noise [consistency indexes (Kluge and Farris, 1969) calculated after excluding uninformative characters typically had values slightly less than 0.600].

Parsimony and Phylogenetic Inference Using DNA Sequences

Many workers have long advocated parsimony analysis as a method of phylogenetic inference for DNA sequence data (Fitch, 1971, 1977; Hendy and Penny, 1982). At the same time, there has been concern within the molecular systematic community that once the algorithmic problems have been solved—and large advances have been made in recent years—there will still remain important questions as to exactly how parsimony procedures are to be applied to sequence data. The focus of most applications of parsimony has been to find the most-parsimonious tree or trees. Yet, systematic (comparative) and functional analyses of DNA and its evolution have established a body of knowledge that creates a broad spectrum of choices and challenges for the systematist. Included in this body of knowledge are the observations that the frequency of transitions and transversions are not always equal; third position changes are more frequent than first or second; and secondary structure configurations can bias the frequency and direction of nucleotide change.

The investigator is thus faced with procedural difficulties from the outset, difficulties that raise important philosophical and empirical questions. At their most basic level, these questions involve the problem of what constitutes evidence in phylogenetic inference, a subject that has received too little attention in the literature (a recent notable exception is Kluge, 1989). Do we base our best estimate of phylogenetic relationships on all the sequence evidence that is available, or, if on the basis of other evidence we have reason to think that some of the data are likely to be misleading, should those be eliminated from the analysis? If the latter, what are the criteria for choosing that portion to be eliminated? Moreover, what justifications do we advance to establish the validity of those criteria? Do we treat all characters (in this case, nucleotide positions) equally ("weighted uniformly") or should some be weighted more than others (eliminating characters might also be viewed as a form of weighting)? Do we treat all character-state transformations at any given nucleotide position equally, or are they to be weighted differently? In both cases, how is weighting itself to be justified and how might weights be determined objectively? These questions can only be a prelude to a full discussion of how parsimony is to be applied inasmuch as many other issues (e.g., identifying signal, assessing topological reliability, the notion of congruence) also have a bearing on our choices regarding systematic procedure (Cedergren et al., 1988; Kluge, 1989).

One theme of this chapter has been to ask what might be the best approach for applying parsimony procedures to sequence data. Because the problem involves so many complex issues, not the least of which are the philosophical tendencies of the investigator, it is unlikely that the systematic community will agree on a unified answer. If one sees the goal of systematic research to be the discovery of the one true history of life, and accepts the precept that we will never have certain knowledge of that history, then our only recourse would seem to be to accept the hypothesis of relationships which is most consistently revealed by the data regardless of which analytical techniques might have been employed (this does not imply, of course, that all methods of analysis are necessarily equally useful or appropriate). This is essentially an appeal for acceptance of stability of cladistic signal and to identify as noisy those cladistic patterns that are ambiguous or less consistently revealed.

There has been substantial support in the literature for weighting among nucleotide positions (e.g., within codons) or among alternative character-state transformations within positions, and many weighting procedures have been suggested (e.g., Farris, 1969; Altschul et al., 1989; Williams and Fitch, 1989, 1990). Although there is a rationale for thinking that weighting will improve the resolution of cladistic signal, what is less clear is which method of weighting is "best" in any case. In principle, at least, a tree derived from weighted data should be a better or more reliable estimate of the true phylogeny than a tree derived from unweighted data, simply because the effect of weighting should be to emphasize the best available evidence (Carpenter, 1988). Yet, a final conclusion regarding the reliability of the resulting tree can only be reached by comparison to results based on more data and an assessment of congruence of their cladistic signal. It is possible, after all, that various assumptions about the evolutionary process that may underlie a particular method of weighting are themselves incorrect and thus may have led to spurious results. An unfortunate tendency within molecular systematics has been the application of a weighting scheme and the uncritical acceptance of the resulting tree. Currently available parsimony algorithms are designed to produce a single point solution; that is, a single most-parsimonious tree or a set of equally most parsimonious trees. Yet, it is difficult to judge from a single most-parsimonious tree whether a cladistic grouping represents a strong (stable) signal found in the data or whether the data are noisy with respect to that group's resolution (e.g., within the hominines). Thus, even if the data are weighted, questions about the reliability of the result still exist in addition to those raised by any justification of the weighting methods themselves.

The analyses undertaken in this study explore ways of revealing phylogenetic signal in sequence data through parsimony procedures. Taken together, the results of this analysis suggest that *if there is any phylogenetic signal inherent in the data, parsimony methods will find it.* The ultimate arbiter of this conclusion must be congruence with phylogenetic relationships inferred from other data sets, which in the case of the primates examined here support the conclusions of this chapter (e.g., Hayasaka et al., 1988b; Miyamoto and Goodman, 1990). Taken separately, each approach to parsimony reveals a significant portion of that signal, but resolution in some parts of the tree may remain ambiguous. Most-parsimonious trees based on the entire 898 bp are essentially congruent with the distance tree of Figure 10-1, there being two equally most-parsimonious trees differing only in the placement of chimpanzee with human or with gorilla; the single most-parsimonious tree using transversions alone resolved that ambiguity in favor of chimpanzee + human. An assessment of the strength of the phylogenetic signal is best described, however, by comparison of the majority-rule consensus trees derived from near–most-parsimonious solutions using global (Fig. 10-2) and transversion (Fig. 10-3) parsimony, along with those derived from bootstrap analysis of global (Fig. 10-5) and transversion (Fig. 10-6) parsimony solutions. Indeed, *a comparison using these four methods of analysis demonstrates a strong phylogenetic signal in even the smallest subsets of the data, including the first 225 bp fragment and the 198 bp tRNA fragment.* Caution must be exercised in generalizing these results inasmuch as this is only one data set, but parsimony analyses such as these appear to constitute a powerful methodological tool for resolving phylogenetic signal even in a small amount of data. Hayasaka et al. (1988a:637) concluded that the high rate of homoplasy in mtDNA makes the more distant relationships of the tree less reliable than those that are more recent, yet parsimony analysis revealed a strong phylogenetic signal across the entire tree.

Phylogenetic Inference Using Sequence Data: Some Comments on Alternative Methods

There has been considerable discussion in recent years about which method of phylogenetic inference is the most appropriate for DNA sequence data. Efforts have been made to adjudicate this issue by simulation experiments in which a data set is generated based on a "known true" phylogeny and given assumptions as to how sequence evolution proceeds (Tateno et al., 1982; Sourdis and Krimbas, 1987; Saitou and Imanishi, 1989). Different methods are then applied to the data set to see if they recover the "true phylogeny." These simulations are perhaps useful for identifying conditions under which a given method may have limitations, but they are encumbered with philosophical and empirical problems. Because certain knowledge of phylogenetic relationships is beyond us, it is questionable whether some artificially generated "truth" can serve as the arbiter of the best method for recovering that which we cannot hope to know with certainty. The evolutionary models underlying these simulations also suffer from a critical methodological circularity: it is reasonable to assume that a detailed and empirical understanding of the mechanisms of molecular change will not be obtained unless it is based on prior hypotheses about the phylogenetic relationships of organisms. Consequently, to incorporate assumptions about the evolutionary process into a simulation procedure that will be used to choose a method of phylogenetic inference may bias our recovery of those relationships, which themselves will bias and confound our understanding of processes of molecular evolution. Perhaps as important, these simulation experiments generally take as their model of evolutionary change one that most closely applies to noncoding, randomly evolving DNA. In practice, however, the majority of molecular systematists are attempting to infer relationships from sequences that will seldom, if ever, evolve under the constraints specified by those models.

In real-world cases, we might expect that when there is a strong phylogenetic signal inherent in the data, virtually any method will recover it. When data are extremely noisy, say as a result of too little data being used to resolve very close internodal distances or of too much homoplasy, then perhaps no method would be expected to resolve a reliable signal. Somewhere in between these extremes we might expect that different methods will sometimes yield different phylogenetic hypotheses for a given set of data. Choice among those hypotheses should not, however, be based on an appeal to method, but on comparison and congruence with phylogenetic inferences derived from other data.

At this time, three primary methods of phylogenetic inference are applied to sequence data; namely, parsimony analysis, distance analysis, and maximum likelihood. Distance analysis has had both its defenders (Felsenstein, 1984; Nei, 1987) and critics (Farris, 1981, 1985). Tree-building algorithms that work on distance matrices are necessary for certain kinds of data, such as DNA-DNA hybridization or immunological distances, but it is questionable whether they should be applied to discrete character data such as sequences. At issue is not whether distance analysis of sequences is able to reveal phylogenetic structure, for cleary it can. More to the point is the inevitable loss of information in converting sequence data to distances (Farris, 1981, 1985; Penny, 1982), their sensitivity to saturation effects so that they must be corrected, the loss of flexibility to examine the deep structure of the data underlying the results, and the inability of most distance algorithms to examine the structure of the large suite of trees having near minimum-length fit. Taking the distance analysis of the primate data set as an example, Hayasaka et al. (1988a) did not cluster using original distances but corrected them for multiple substitutions (Gojobori et al., 1982) and then clustered on those new distances. A single tree was produced by the neighbor-joining method (Saitou and Nei, 1987) (Fig. 10-1) but without an assessment of whether it was the best-fit tree. That tree, moreover, was not compared quantitatively or qualitatively to other near–minimum-length trees. Many different models have been proposed to "correct" for multiple substitutions. Yet, even though substitution probabilities can be estimated from observed nucleotide frequencies, there is no guarantee that those probabilities actually reflect the dynamics of processes that produced the observed nucleotide variation. That being the case, it remains to be seen to what extent these different models that are being applied to sequence data will influence inferred phylogenetic structure, not only of the minimum-length tree, but of those trees close to that solution.

Maximum-likelihood methods are more developed in theory than in practice (Felsenstein, 1981; Bishop and Friday, 1985; Saitou, 1988, 1990; Fukami and Tateno, 1989), and we will not discuss them extensively here. A choice among trees using maximum-likelihood methods depends upon accepting an underlying model of evolutionary change. Often, that model incorporates equal probabilities of substitutional change from one nucleotide to another, but some have been developed to account for differences in rate between transitions and transversions (Saitou, 1990). Calculation of likelihood functions for trees is computationally difficult, and the method cannot be extended easily beyond cases for five or six taxa. It seems that if more realistic, and thus more complicated, models of evolutionary change are to be incorporated, computational difficulty is sure to increase.

Cedergren et al. (1988:102) note that parsimony methods have virtue in phylogenetic inference using sequence data because they result "in the most economical reconstruction of mutational history, with no assumptions and with the minimum of coincidence and unobserved changes. . . ." Although it is arguable whether parsimony is assumption free (all scientific methods designed to discover pattern involve some assumptions), the strengths of parsimony analysis over other methods of analyzing sequence data are easy to enumerate. With parsimony, it is possible to use a minimum of assumptions about the processes governing molecular evolution and still obtain interpretable results. Weighting of characters or character-states involves assumptions, but that too can be accomplished conservatively such that narrowly constrained assumptions about molecular processes are avoided. Today's parsimony algorithms are fast and efficient at searching among all possible trees, thus guaranteeing that not only will the most-parsimonious tree be found, but that the cladistic structure of near–most-parsimonious trees can be examined. Moreover, with parsimony it is possible to examine and evaluate the data supporting each cladistic component on a tree. For these reasons alone, parsimony is the preferred method for analyzing DNA sequences.

ACKNOWLEDGMENTS

We would like to thank T. E. Dowling, D. M. Hillis, T. Jones, D. Mindell, and especially M. M. Miyamoto for their helpful comments on the manuscript. We thank E. Chipouras and M. M. Miyamoto for providing the calculations for Table 10-3. This research was supported by the National Science Foundation through grants BSR-8805957 and BSR-9007652.

REFERENCES

Altschul, S. F., R. J. Carroll, and D. J. Lipman. (1989) Weights for data related by a tree. *J. Mol. Biol.* **207**:647–653.

Archie, J. W. (1989) A randomization test for phylogenetic information in systematic data. *Syst. Zool.* **38**:239–252.

Bishop, M. J., and A. E. Friday. (1985) Evolutionary trees from nucleic acid and protein sequences. *Proc. R. Soc. London* **226B**:271–302.

Brown, W. M. (1983) Evolution of mitochondrial DNA. Pp. 62–88 in *Evolution of Genes and Proteins* (M. Nei and R. K. Koehn, eds.). Sinauer Associates, Sunderland.

Brown, W. M. (1985) The mitochondrial genome of animals. Pp. 95–130 in *Molecular Evolutionary Genetics* (R. J. MacIntyre, ed.). Plenum Press, New York.

Brown, W. M., E. M. Prager, A. Wang, and A. C. Wilson. (1982) Mitochondrial DNA sequences of primates: tempo and mode of evolution. *J. Mol. Evol.* **18**:225–239.

Carpenter, J. M. (1988) Choosing among multiple equally parsimonious cladograms. *Cladistics* **4**:291–296.

Cedergren, R., M. W. Gray, Y. Abel, and D. Sankoff. (1988) The evolutionary relationships among known life forms. *J. Mol. Evol.* **28**:98–112.

Cracraft, J., and D. P. Mindell. (1989) The early history of modern birds: a comparison of molecular and morphological evidence. Pp. 389–403 in *The Hierarchy of Life. Molecules and Morphology in Phylogenetic Analysis* (B. Fernholm, K. Bremer, and H. Jörnvall, eds.). Elsevier Science Publishers B. V., Amsterdam.

Doolittle, R. F. (1990) (ed.). *Molecular Evolution: Computer Analysis of Protein and Nucleic Acid Sequences*. Methods in Enzymology, vol. 183. Academic Press, New York.

Farris, J. S. (1969) A successive approximations approach to character weighting. *Syst. Zool.* **18**:374–385.

Farris, J. S. (1972) Estimating phylogenetic trees from distance matrices. *Am. Nat.* **106**:645–668.

Farris, J. S. (1981) Distance data in phylogenetic analysis. Pp. 2–23 in *Advances in Cladistics. Proceedings of the First Meeting of the Willi Hennig Society*. (V. A. Funk and D. R. Brooks, eds.). New York Botanical Garden, Bronx.

Farris, J. S. (1985) Distance data revisited. *Cladistics* **1**:67–85.

Felsenstein, J. (1978) Cases in which parsimony or compatibility methods will be positively misleading. *Syst. Zool.* **27**:401–410.

Felsenstein, J. (1981) Evolutionary trees from DNA sequences: a maximum likelihood approach. *J. Mol. Evol.* **17**:368–376.

Felsenstein, J. (1984) Distance methods for inferring phylogenies: a justification. *Evolution* **38**:16–24.

Felsenstein, J. (1985) Confidence limits on phylogenies: an approach using the bootstrap. *Evolution* **39**:783–791.

Felsenstein, J. (1988) Phylogenies from molecular sequences: inference and reliability. *Ann. Rev. Genet.* **22**:521–565.

Fitch, W. M. (1971) Toward defining the course of evolution: minimum change for a specific tree topology. *Syst. Zool.* **20**:406–416.

Fitch, W. M. (1977) On the problem of discovering the most parsimonious tree. *Am. Nat.* **111**:223–257.

Fitch, W. M., and E. Margoliash. (1967) Construction of phylogenetic trees. *Science* **155**:279–284.

Fukami, K., and Y. Tateno. (1989) On the maximum likelihood method for estimating molecular trees: uniqueness of the likelihood point. *J. Mol. Evol.* **28**:460–464.

Gojobori, T., K. Ishii, and M. Nei. (1982) Estimation of average number of nucleotide substitutions when the rate of substitution varies with nucleotide. *J. Mol. Evol.* **18**:414–423.

Goodman, M. (1989) Emerging alliance of phylogenetic systematics and molecular biology: a new age of exploration. Pp. 43–61 in *The Hierarchy of Life. Molecules and Morphology in Phylogenetic Analysis* (B. Fernholm, K. Bremer, and H. Jornvall, eds.). Elsevier Science Publishers B. V., Amsterdam.

Goodman, M., M. M. Miyamoto, and J. Czelusniak. (1987) Pattern and process in vertebrate phylogeny revealed by coevolution of molecules and morphology. Pp. 141–176 in *Molecules and Morphology in Evolution: Conflict or Compromise*? (C. Patterson, ed.). Cambridge University Press, Cambridge.

Hayasaka, K., T. Gojobori, and S. Horai. (1988a). Molecular phylogeny and evolution of primate mitochondrial DNA. *Mol. Biol. Evol.* **5**:626–644.

Hayasaka, K., S. Horai, T. Gojobori, T. Shotake, K. Nozawa, and E. Matsunaga. (1988b) Phylogenetic relationships among Japanese, rhesus, Formosan, and crab-eating monkeys, inferred from restriction-enzyme analysis of mitochondrial DNAs. *Mol. Biol. Evol.* **5**:270–281.

Hendy, M. D., and D. Penny. (1982) Branch and bound algorithms to determine minimal evolutionary trees. *Math. Biosci.* **59**:277–290.

Irwin, D. M., T. D. Kocher, and A. C. Wilson. (1991) Evolution of the cytochrome *b* gene of mammals. *J. Mol. Evol.* **32**:128–144.

Kluge, A. G. (1989) A concern for evidence and a phylogenetic hypothesis of relationships among *Epicrates* (Boidae, Serpentes). *Syst. Zool.* **38**:7–25.

Kluge, A. G., and J. S. Farris. (1969) Quantitative phyletics and the evolution of anurans. *Syst. Zool.* **18**:1–32.

Margush, T., and F. R. McMorris. (1981) Consensus *n*-trees. *Bull. Math. Biol.* **43**:239–244.

Miyamoto, M. M., and M. Goodman. (1990) DNA systematics and evolution of primates. *Ann. Rev. Ecol. Syst.* **21**:197–220.

Moritz, C., T. E. Dowling, and W. M. Brown. (1987) Evolution of animal mitochondrial DNA: relevance for population biology and systematics. *Ann. Rev. Ecol. Syst.* **18**:269–292.

Nei, M. (1987) *Molecular Evolutionary Genetics*. Columbia University Press, New York.

Penny, D. (1982) Towards a basis for classification: the incompleteness of distance measures, incompatibility analysis and phenetic classification. *J. Theor. Biol.* **96**:129–142.

Platnick, N. I. (1989) An empirical comparison of microcomputer parsimony programs. II. *Cladistics* **5**:145–161.

Saitou, N. (1988) Property and efficiency of the maximum likelihood method for molecular phylogeny. *J. Mol. Evol.* **27**:261–273.

Saitou, N. (1990) Maximum likelihood methods. Pp. 584–598 in *Molecular Evolution: Computer Analysis of Protein and Nucleic Acid Sequences* (R. F. Doolittle, ed.). Methods in Enzymology, vol. 183. Academic Press, New York.

Saitou, N., and T. Imanishi. (1989) Relative efficiencies of the Fitch-Margoliash, maximum-parsimony, maximum-likelihood, minimum-evolution, and neighbor-joining methods of phylogenetic tree construction in obtaining the correct tree. *Mol. Biol. Evol.* **6**:514–525.

Saitou, N., and M. Nei. (1987) The neighbor-joining method: a new method for reconstructing phylogenetic trees. *Mol. Biol. Evol.* **4**:406–425.

Sibley, C. G., and J. E. Ahlquist. (1987) Avian phylogeny reconstructed from comparisons of the genetic material, DNA. Pp. 95–121 in *Molecules and Morphology in Evolution: Conflict or Compromise*? (C. Patterson, ed.). Cambridge University Press, Cambridge.

Smith, A. B. (1989) RNA sequence data in phylogenetic reconstruction: testing the limits of its resolution. *Cladistics* **5**:321–344.

Sourdis, J., and C. Krimbas. (1987) Accuracy of phylogenetic trees estimated from DNA sequence data. *Mol. Biol Evol.* **4**:159–166.

Swofford, D. L. (1990) *PAUP: Phylogenetic Analysis Using Parsimony*, Version 3.0. Illinois Natural History Survey, Champaign.

Swofford, D. L., and G. J. Olsen. (1990) Phylogeny reconstruction. Pp 411–501 in *Molecular Systematics* (D. M. Hillis and C. Moritz, eds.). Sinaurer Associates, Sunderland.

Tateno, Y., M. Nei, and F. Tajima. (1982) Accuracy of estimated phylogenetic trees from molecular data. I. Distantly related species. *J. Mol. Evol.* **18**:387–404.

Williams, P. L., and W. M. Fitch. (1989) Finding the minimal change in a given tree. Pp. 453–470 in *The Hierarchy of Life. Molecules and Morphology in Phylogenetic Analysis* (B. Fernholm, K. Bremer, and H. Jörnvall, eds.). Elsevier Science Publishers B.V., Amsterdam.

Williams, P. L., and W. M. Fitch. (1990) Phylogeny determination using dynamically weighted parismony method. Pp. 615–626 in *Molecular Evolution: Computer Analysis of Protein and Nucleic Acid Sequences* (R. F. Doolittle, ed.). Methods in Enzymology, vol. 183. Academic Press, New York.

11

Evolutionary Analysis of Length-Variable Sequences: Divergent Domains of Ribosomal RNA

ALLAN LARSON

It is important in molecular phylogenetic studies to match the timescale of the evolutionary divergence events being studied to the evolutionary rate and pattern of the molecules used to resolve them. In many taxa, and especially in vertebrates, divergence events occurring approximately 70 to 300 million years (Myr) before present have been particularly difficult to resolve. Comparatively rapidly evolving molecules whose study has clarified phylogenetic relationships on a more recent timescale (mitochondrial DNA, serum albumin, various proteins compared electrophoretically) are often too highly saturated with change to be useful for resolving divergence events in this time range. The highly conserved sequences that have been used to investigate much older divergences [cytochrome *c*, small-subunit ribosomal RNA (rRNA)] usually do not offer sufficient variation to resolve divergence events within the 70 to 300 Myr range. The "divergent domains" of the large subunit of nuclear-encoded rRNA (Hassouna et al., 1984) constitute a group of sequences whose evolutionary rate and pattern facilitate phylogenetic resolution in this time range (Qu et al., 1988; Larson and Wilson, 1989; Larson, 1991), and also occasionally for more recent phylogenetic divergences (Gonzalez et al., 1990). These are the most highly variable segments of rRNA. Divergent domains may differ extensively both in sequence and length among even closely related genera, but they occupy homologous positions relative to the more conserved parts of the large ribosomal subunit (Hassouna et al., 1984; Gerbi, 1985; Clark, 1987).

Neither the higher-order structures nor the cellular functions of the divergent domains are well documented. Larson and Wilson (1989) found indirect estimates of higher-older structures of divergent domains to be highly conflicting, and the secondary structures of these sequences are likely to be quite variable on an evolutionary timescale (Gonzalez et al., 1985; Gorski et al., 1987). Divergent domains have been hypothesized to be remnants of mobile elements that inserted into the ribosomal genes (Clark et al., 1984), or alternatively, the remnants of linkers that connected different "functional segments" during the evolutionary assembly of the ribosome, and which were subsequently eliminated from all but the nuclear-encoded eukaryotic ribosomes (Clark, 1987; see also Gonzalez et al., 1985; Spencer et al., 1987). Larson and Wilson (1989) concluded, however, that the evolutionary patterns of at least four of these domains are inconsistent with the hypothesis that they are evolving free of functional constraint.

Because the ribosomal divergent domains are potentially of great importance for molecular systematic studies, it is necessary to investigate in detail the evolutionary pattern demonstrated by these sequences and to evaluate whether this pattern is prone to support or to violate the requirements of commonly used molecular phylogenetic methods. The superposition of information contained in base substitutions with information from length mutations is particularly important for analyzing these sequences. In this chapter, I present an investigation of the evolutionary properties of 11 of the 12 divergent domains of the large ribosomal subunit using comparisons of salamanders and several closely-related outgroup taxa. A schematic diagram of the positions of these divergent domains in the large ribosomal subunit of the frog *Xenopus laevis* is shown in Figure 11-1. The data used in this analysis (Fig. 11-2) represent a subset of the information presented in my recent phylogenetic studies of salamanders (Larson and Wilson, 1989; Larson, 1991), aligned with sequences published elsewhere for *Mus domesticus* (Hassouna et al., 1984) and *Xenopus laevis* (Ware et al., 1983).

This evaluation of evolutionary pattern requires partitioning the changes

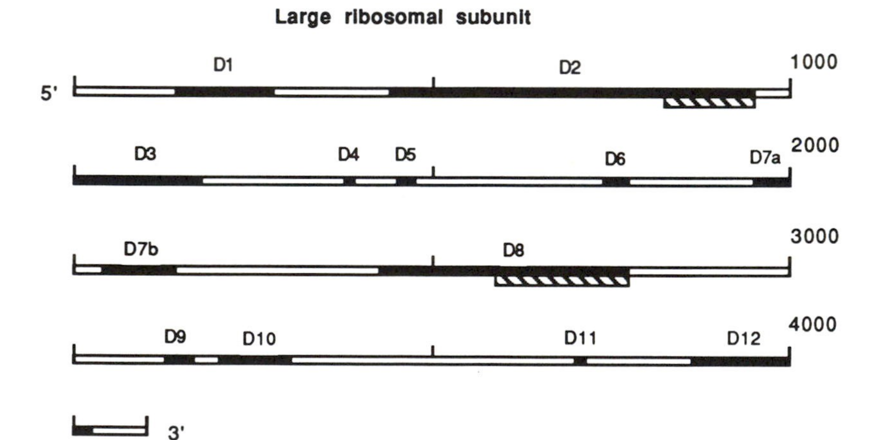

Figure 11-1 Schematic diagram of the large ribosomal subunit of *Xenopus laevis* showing positions of divergent domains 1–12 (solid lines). For domains D2 and D8, only the segments marked with diagonal bars are included in this study.

```
                                       D1------------------------------->
                                                130       140       150
                                                  +         +         +
1. Aneides flavipunctatus              CCCCG--CGUCCGGCG-GGCGCGGGAAAUGUG
2. Desmognathus ochrophaeus            .....--.........-...............
3. Amphiuma means                      .....--.........-...............
4. Rhyacotriton olympicus              .....--.........-...............
5. Siren intermedia                    .....UC..CGU.C..-...............
6. Andrias davidianus                  .....--.........-.........C.....
7. Hynobius sp.                        .....--.........-...............
8. Necturus beyeri                     -....UC.........-....U..........
9. Notophthalmus viridescens           .....--..-G.....-...............
10. Ambystoma californiense            -....UC..-G.....-...............
11. Dicamptodon sp.                    .....--..-G.....-...............
12. Typhlonectes compressicauda        .....--...G.....-...............
13. Xenopus laevis                     -....--..C...C..-.........CG....
14. Mus domesticus                     .....-C..CG..U..C....U..........
15. Scaphirhynchus platorynchus        -....-C...GU.C.AC...............

     D1-------------------------------------------------------------->
          160       170       180            190       200       210
            +         +         +              +         +         +
1.   GCGUACAGAAGACCG--CCUCCCCGGCGUCGCUCA---GGGGCCCAAGUCCUUCUGAUCGAGG
2.   ...............--..................---.........................
3.   ...............--.................---G.C.......................
4.   ...............--..................---.........................
5.   .....U.........--...........n....--UGC.........................
6.   ...............--...........C......---.........................
7.   ...............--...........Cnn....---.........................
8.   ...............--..................---.........................
9.   ...............-A..................--G.........................
10.  ...............-A..................--G.........................
11.  ...............-Annn.........nn....--G.........................
12.  ......G........--..C.U......Cnn...---G.......G.................
13.  ......G.G......GA..C........CG....---G.........................
14.  ......G.......C-A.-.........C.....GUGG.........................
15.  ...............AA--....U.....nn...---G.........................

     D1-------------------------------------------------------------->|
         220       230       240       250       260       270
           +         +         +         +         +         +
1.   CCCAnCCCGCGGACGGUGUUAGGCCGGU-AGCGGCCCCGGCGCGCCGGGAUCAGGUCUUCUCG|
2.   ....n.......................-..................................|
3.   .U..n.......................-..................................|
4.   .U..n...U...................-..................................|
5.   .U..n...U...................-..................................|
6.   .U..n...U...................-..................................|
7.   .U..n...U...................-..................................|
8.   .U..n...U...................-..................................|
9.   .U..C.......................-.......................G..........|
10.  .U..C...U...................-.......................G..........|
11.  .U..n...U...................-.......................G..........|
12.  .U..n.......................-....................CC.G..........|
13.  ....G......................GG.....................C.C.....C....|
14.  ....G...U.........G.........-....................C..G.......C..|
15.  .U..n...A...................-......U...U.........G...........n.|
```

Figure 11-2 Aligned sequences for 11 divergent domains of the large subunit of rRNA (after Larson, 1991). Samples include 11 salamanders (1–11), a caecilian (12), a frog (13), a mammal (14), and a fish (15). Numbers at the tops of the columns are the standard position numbers for the published *Xenopus laevis* (13) sequence (Ware et al., 1983). Positions not present in the *Xenopus* sequence are denoted alphabetically. Dots denote identity to the sequence in row 1; dashes denote gaps in the aligned sequence. Divergent domains are marked above the aligned sequences (D1→|, D2→|, etc., to D12). Vertical bars mark breaks in the reported sequence. Base designations follow IUB recommendations: A, adenine; C, cytosine; G, guanine; U, uracil; S, A or C; W, G or U; n, any or unknown. Lower case base letters (a, g, u) denote uncertain positions in the published *Xenopus* sequence (Ware et al., 1983). Large inserts in the aligned sequences are denoted by letters J, K, L, M,

```
     D2-------------------------------------------------------------->
             830       840       850       860       870       880
               +         +         +         +         +         +
1.   CCCCCCGCCCCCGACGCGACUGUCGACCGGAGCGGACUGUCCUCAGU-CGCCCCGACCGC
2.   ....---.......................G................G...G........
3.   ........U.....................G.........G......-............
4.   .....U..U....G.............U..G.........G......-............
5.   .....U..U....G.............U..G.........G......G............
6.   .....U..U....G..........A..U..G................-............
7.   .....U.UU---.G..........A.....G................G............
8.   .....U..U....G.............U..G.........U......G.........U..
9.   .....U..U....G.............U..G.........G.....C-............
10.  .....U..U....GU............U..G.U..............-.....A......
11.  .....U..U....G.............U..G................G-----.......
12.  .............G..........A.....G................G............
13.  ........U....G..........A.....G........C..C....G.......U....
14.  U.U...A.....--------------J...G...........C....G.......G--..
15.  ........U....G.............U..GC...............GG...U.......
```

J = *CUCCGUCGCCUCUCUCGGGGCCCGGUGGGGGG*

```
     D2-------------------------------------------------------------->
             890       900          920       930       940       950
               +         +            +         +         +         +
1.   GUCGCGCCGCCG-GGCGGGGA|   GGG-CGCCAGGGGUCUGCGGCGAUGUCGGUGACCCACCCG|
2.   ............-........|   ...-....................................|
3.   ............-........|   ...-....................................|
4.   ............-........|   ...-..........................U.........|
5.   ....U...CGG.C...--...|   ...-.A..................................|
6.   .........U..-........|   ...-...U.......................G........|
7.   ...........--........|   ...-...........................G........|
8.   ............-........|   ...-U..U......................U.........|
9.   ....U.......-........|   ...-..........................U.........|
10.  ............-........|   ...-................U.........U.........|
11.  ............-........|   ...-................U.........U.........|
12.  ............-........|   ...-...........G...............G........|
13.  .C..........A...-....|   .A.C.....A.....C...............U........|
14.  ....UCG.....K..-.....|   .A.-...AC......G.............CU.........|
15.  ......UU....-.......-|   ...-...........G.............CU.....U...|
```

K = *UCGGGUCCC*

```
     D3-------------------------------------------------------------->
            1000      1010        1020      1030       1040      1050
               +         +           +         +          +         +
1.   GAGUCGGAGGG---CCGUGCGAAC-CC-CCGUGGCGCAAUGAAGGUGA-GGGCC---GCGCG
2.   ...........---.A.C......-..-....................-.....---.....
3.   ...........---...C......-..-.U..................-.....---.....
4.   ...........---..........-..-.U..................-.....---.....
5.   ...........---..U.U....A-..-.U..................-.....---.....
6.   ...........---.........A-..-.U..................-.....---.....
7.   ...........---.........A-..-.U..................-....----.....
8.   ...........---.........A-..-.U..................-.....---.....
9.   ...........---.........A-..-.U..................-.....---.....
10.  ...........---...AA....A-..-.U..................-.....---.....
11.  ...........---...A.....A-..-.U..................-.....---.....
12.  U..........---...C.A...A-..-.U.C................-.....---.....
13.  ...........ACU.U.C.....A-..-.U..................-.....GGG....C
14.  .....--....GCU.-..C....AG..G....................A.....CCGC.CG.
15.  .....A.C...GCU.-U..U...A-..-....................-.....---.....
```

Figure 11-2 (*continued*).

N, and O and are identified below the appropriate segment. Alignments are tentative for sites 2664–2688. Sequences reported for domains D2 and D8 are partial sequences (see Fig. 11-1) and the sequence reported for domain D5 includes a short segment outside the 3′ end of that domain (as identified by Hassouna et al., 1984). (*Figure continues.*)

```
    D3----------------------------------------------------------------->
          1060            1120      1130      1140      1150      1160
             +               +         +         +         +         +
1.  CCGGCUGAGGUGGGA| GCGGGCGCACCACCGGCCCGUCUCGCCCGCUCCGUCGGGGAGGUGGA
2.  ...............| ..A............................................
3.  ...............| ...............................................
4.  ...............| ...............................................
5.  ...............| ..............................C................
6.  ...............| ................................U..............
7.  ...............| ...............................................
8.  ...............| ................................U..............
9.  ...............| ................................U..............
10. ...............| ................................U..............
11  ...............| ................................U..............
12. ...............| ..............................C................
13. ...............| ..............................C................
14. GG.C.C.........| .A............................CG..C............
15. ...............| ........................A.....A................

    D3-------->|   D4------------------>|   D5----------------------->
          1170     1360          1370                 1460      1470
             +        +             +                    +         +
1.  GCAUGAGCGC|   GGAAACCC--------CAGU|   GAA---CGGC---UCGCUGGCCUGG
2.  ..........|   ........--------....|   ...---....---............
3.  ..G.......|   ...C....--------....|   ..----....ACG........G...
4.  .........U|   ........--------....|   ..----....ACG........G...
5.  .........U|   ..CCG...--------....|   ...GCC....---........G...
6.  .........U|   ..CC....--------....|   ..----....-CG........G...
7.  .........U|   ..CC....--------....|   ..----....-CG........G...
8.  .........U|   ..-C....--------....|   ..----....ACG........G...
9.  .........U|   ....C...--------....|   ..----....ACG........G...
10. .........U|   ........--------....|   ..----....ACG........G...
11. .........U|   ........--------....|   ..----....ACG........G...
12. ..........|                   ....|   ------...GA--........U...
13. ..G.......|   --G-U.----CGU-CG....|   ------...GA--........U...
14. ...C.....U|   UCGCU...GACGUACG....|   ...GCC....---........G...
15. .........U|   U-G-U...-------G....|   ...GCC....---........U...

    D5---------------------------->| D6---------------------------------->
             1480      1490      1500 1730      1740      1750      1760
                +         +         +    +         +         +         +
1.  AGCC-GGGCGUGGAAUGCG-AGCCCGC---| GGCCGUCGCCGGCGCU-GAG-AGCCCGCGGGGGCU
2.  ....-..............-.......---| ................-...-..............
3.  ....-..............-.......---| ................-...-..............
4.  ....-....A.........-..-U-..C-U| ................-...-..............
5.  ....GC.............-...U...---| .............A..G...-.....A........
6.  ....-....A.........-..-U-..C-U| .....G..........-...-..............
7.  ....-....A.........-..-U-..C-U| ................-...-..............
8.  ....-....A.........-..-U-..C-U| ................-...-..............
9.  ....-....A.........-..-U-..C-U| ................-...-..............
10. ....-....A.C.......-...U...---| ................-...-..............
11. ....-....A.C.......-...U...---| ................-...-..............
12. ....-..............-...--..C-U| .........U......-.G.-..............
13. ....g..............n...A...CAU| ................G.GUC..U...........
14. ....-..............-..-U-..C-U| ........L.....UCG.GCC---......A.C..
15. ....-..............-..---CUCGC| ...........C.A.-G.G--........AA....
```

L = *CCCCCGCUUGGGCCGCGCGCCUCCCCUCCGCCCCCUGCCCGGGGGCGGUGCGUGGGGGCUG*
GGGCUCUCUGCGCGUGGGCGCCGGCGAGGGCAGGGCAAGGCAAGGCUGACGCC

Figure 11-2 *(continued)*.

```
     D6-------------->| D7a---------------------------------------->
          1770          1950      1960      1970      1980      1990
             +           +         +         +         +         +
1.   AAGCCGCGAC-GAGU| UAGGCGAGCGCCGUUCGGAAGGGACGGGCGAUGGCCUCCGU-GC
2.   ..........-....| .........................................-..
3.   .G........-....| .........................................UC.
4.   ..........-....| .........................................U..
5.   .G........-....| .G.......................................U..
6.   ..........-....| .........................................U..
7.   ..........-....| .........................................U..
8.   ..........-....| .........................................U..
9.   ..........-....| .G.......................................U..
10.  ..........-....| .G.......................................U..
11.  ..........-....| .G.......................................U..
12.  ..........-....| .........................................UC.
13.  .G........U....| .G.......................................C..
14.  .C........-....| .G......-----.G.C........................U..
15.  ..........-....| .......AU.......U...A....................C..

     D7a-------------->|  D7b---------------------------------->|
        2000                  2050  2100      2110      2120
           +                     +   +         +         +
1.   CCUCGGCCGAUCGAAAG|  GAGACGGG| GGCGUCCAGUGCGGUAACGCGACCGAUC|
2.   .................|  ........| ............................|
3.   .................|  ........| ....................A.......|
4.   ....A............|  ........| ....................A.......|
5.   ....A............|  ........| ....................A.A.....|
6.   ....A............|  ........| ..............CG....A.......|
7.   ....A............|  ........| ..............CG....A.......|
8.   ....A............|  ........| ....................A.......|
9.   ....A............|  ........| ....................A.A.....|
10.  ....A............|  ........| ....................A.A.....|
11.  ....A............|  ........| ....................A.A.....|
12.  .................|  ........| ............................|
13.  .................|  ........| ..............CG............|
14.  .................|  ....U...| C.A.G.......................|
15.  ..C....U.........|  ........| ....................A.......|

     D8---------------------------------------------------------------->
      2600      2610      2620      2630      2640      2650      2660
         +         +         +         +         +         +         +
1.   AGCGGCGGCGACUCUGGACGUGCGCCGGGCCCUUCCUGUGGAUCGCCUCAGCUGCGGCG-GUCGCC
2.   ...........................................................-......
3.   G..........................................................-.G....
4.   G..........................................................-......
5.   G...................C.............................---..U.G.-......
6.   G...................C......................................-.G....
7.   G...................C.A....................................-.--...
8.   G.....................A..................................U.-......
9.   G...................C....................................U.-......
10.  G...................C.A..................................U.-......
11.  G...................C.A..................................U.-......
12.  G...................C......................................C...-U.
13.  G...................C..........................C...........-CG....
14.  G...................C.A.............C......................-.G..--
15.  G............U......C.A........................C...........-UG..U|
```

Figure 11-2 *(continued)*.

on a phylogenetic tree topology using parsimony (Fitch, 1971). The effects of varying tree topology are evaluated to assess the inaccuracies that may result from basing the study on an incorrect tree topology. The three topologies investigated correspond to: (1) the topology favored by the parsimony analysis of Larson (1991); (2) a tree representing an alternative hypothesis of salamander relationships (Duellman and Trueb, 1986) but retaining the same topology as tree 1 for outgroup taxa (the fish *Scaphirhynchus*, the mammal *Mus*, the frog *Xenopus*, and the caecilian *Typhlo-*

```
      D8------------------------------------------------------------->
                2670      2680      2690      2700      2710      2720
                   +         +         +         +         +         +
 1.  UGGUCCCCGGC--------------CGCCUCCUCGCCGG-GAGG-UCGGGGC---------
 2.  C...G-.....--------------..............-....-C......---------
 3.  .C-........--------------.C.U..GC.------.G..-.......---------
 4.  .C-.......---------------.C....AC.C-----.G..-.......---------
 5.  .C-.......UCCCGU---------.U..G.UC..G...G.C..-C......---------
 6.  .C-.......---------------.C.....--------.G..-.......---------
 7.  .C------..-----------UCCCU......--------.G..-.......---------
 8.  .C-.......-----------------.....C.------.G..-.......---------
 9.  .C-.......U--------------.......C.------.G..CG......---------
10.  .C-..U....U--------------U-.....C.------.G..U.......---------
11.  .C-..-----U--------------...U...--------.G..-.......---------
12.  .C-.....-----------------.......C.C....-.G..-G......---------
13.  .C-.....C..GCCGUCCCCCUCCUG......C.C..UCA.G..-A......GCGWGCSGC
14.  .C..--------------------MU-..G..C.C....-|
```

M = GGGCCGCCCCCUUGUGGCUGCGCCUCCAAGGGGGGUGUGUGCGCGCCCUGUGCGGGCGGGCGGG GGCGGUGCGUGGAGCCCUCUCUGCUACCCCCCCACCUCGCUCCGGGGCGCCCCUCCUUUGGGCGCGGC CGUGGGCGGCCCGAGGGGCCCUCGCCGGC

```
      D8------------------------------------------------------->|
              2730      2740      2750      2760      2770
                 +         +         +         +         +
 1.  -----------CGGCG-GUCCG-CCUCGGCCGGCGCCUAGCAGCUGACUUAGAACUGG|
 2.  -----------.....-.....-...................................|
 3.  -----------.....-C....-...................................|
 4.  -----------.....-.....-...................................|
 5.  ------------....---.U.-...................................|
 6.  -----------.....-U..-.-...................................|
 7.  -----------.....-U..-.-...................................|
 8.  -----------.....-C..U.-...................................|
 9.  -----------.....-U..U.-...................................|
10.  -----------.....-U..U.-.........A.........................|
11.  -----------.....-U..U.-.........A.........................|
12.  -----------....nnnn...-...........A.......................|
13.  GGGGCGGCCGG....G.C...G...................................|
14.           CG.....CCC...-.....................C.............|
15.                        ...................................|
```

```
     D9--------------------------->|   D10------------------------>
     3140      3150      3160            3210      3220      3230
        +         +         +               +         +         +
 1.  ----------------CCU--CGGGCCG|   CCGGUGAGGCGGGGGGCGAGCCCUGU
 2.  ----------------...--.......|   ..........................
 3.  ----------------...--.......|   .......................C.A
 4.  ----------------...--.......|   ..........................
 5.  ----------------...--.......|   .......................C.A
 6.  ----------------...--.......|   ..........................
 7.  ----------------...--.......|   ..........................
 8.  ----------------.U.--.......|   ..........................
 9.  ----------------...--.......|   .......................C..
10.  ----------------...--.......|   .......................C..
11.  ----------------...--.......|   .......................C..
12.  ----------------...--.......|   .......................C.A
13.  CCCCCGCGCUCGUCGCAAAGGG.C....|   .......................C.A
14.  CCCCCGNGCACGCCGG...CG.......|   .......................C.A
15.  ----------------...--.......|   .......A.....A......U..C.A
```

N = GCGCCCGGCCCCGUCCUCGCGUCGGGUCGGG

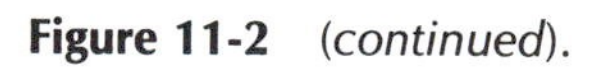

Figure 11-2 (*continued*).

```
    D10--------------------------------------------->|
           3240      3250          3280      3290
             +         +             +         +
1.  GGGGCUCUCGCUUCUGGCUCCAAGC|  CGGGCGCGACCCGCUCCGGGG|
2.  .........................|  .....................|
3.  .........................|  .....................|
4.  .........................|  .....................|
5.  .........................|  .....................|
6.  .........................|  ....U................|
7.  .........................|  .....................|
8.  .........................|  .....................|
9.  .........................|  .....................|
10. .........................|  .....................|
11. .........................|  .....................|
12. .........................|  .....................|
13. ................AC.......|  ..................A..|
14. .................G.......|  .....................|
15. .C........A.....UG.......|  .....................|
```

```
    D12----------------------------------------------------------->
    3860        3870      3880      3890   |      3920      3930
       +          +         +         +    |        +         +
1.  CCGCAGCGCCG-UGGAGCCUCGGUUGGCCUGGGA-UA-C|  GCCG-GUGCGUAGAGCC
2.  ...........-......................-..G.|  ....-............
3.                                                   ..........
4.  ...........-......................-..--|  ....-............
5.  ...........-...G.............AAC..-..-.|  ....-.....C......
6.  ...........-......................-..-.|  ....-............
7.  ...........-......................-..-.|  ....-............
8.  ......U....-......................-..--|  ....-............
9.  --....U....-C.....................-..G.|  ....-............
10. nnn......U.-..................A...-..--|  ....-............
11. -........U.-......................-..--|  ....-............
12. .U.........-C....-...........GA...-..G.|  ....C.A...-......
13. ...........-C....-......C.....C...U..G.|  ....-.C.G.C......
14. .G.........AA................CC...-..G.|  O...-.....G......
15. ..nn......CA...CU..GU.............-..G.|  .U..-...U.G..U...
```

O = GGU

```
    D12---------------------------------------------------->
            3940      3950       3960      3970       3980
              +         +          +         +           +
1.  GCUCG-CCUCGGGACCGGAGUGCGGACAGAA-GGGGGCCGCCUCUCUCUCCCGU
2.  .....-.........................-......................
3.  .....-.........................-......................
4.  .....-.........................-......................
5.  U....-................U.......U-...C.........--.....CG
6.  .....-.......G........U....G...-...C...A.....--.......
7.  .....-.......G........U....G...-...C...A.....--.......
8.  .....-................U........-......................
9.  .....-.U.G..........C......G...-............CA..G...U.
10. .....-.........UC.....U........-.......A........G.....
11. .....-.........UC.....U........-.......A........G.....
12. .....-..............C..........-.............--.......
13. .....-..............C......G-..A.............--......G
14. .U...U.U.G..A.A...G......C.G...A..........-.....G.....
15. A....-UGA.U..UU.....A.........UGCAU.U.....--....A..UU.
```

Figure 11-2 (*continued*).

```
     D12---------------------------------------------->|
                 3990       4000       4010       4020
                    +          +          +          +
1.   -A-GCGCACCGCAUGUUCGUGGGGAACCUGGUGCUAAAUCAUUCG|
2.   -.-..........................................|
3.   -.-..........C...............................|
4.   -.-..........C...............................|
5.   -G-.U........C...............................|
6.   -.-..........................................|
7.   -.-..........................................|
8.   -.-..........C...............................|
9.   -.-.U.................C..G..C................|
10.  -.-.U........................................|
11.  -.-.U........................................|
12.  -.-..........C...............................|
13.  -.-..........C...............................|
14.  C.C.UUG.A....C.......U.........C......C......|
15.  U.--.................UU...................A..|
```

Figure 11-2 *(continued)*.

nectes); and (3) a tree retaining the salamander relationships of tree 1 but shuffling the relationships of outgroups (see Fig. 11-3).

MOLECULAR EVOLUTIONARY PATTERN AND PHYLOGENETIC RECONSTRUCTION

The methods used to infer phylogenies from molecular sequence data are reviewed by Felsenstein (1988). Despite repeated criticism, the most prevalent criterion used for constructing molecular phylogenetic trees is parsimony, which favors the topology that minimizes the total amount of homoplastic change in the characters being studied. A related criterion, character compatibility, favors the topology specified by the largest group of nonconflicting characters. Felsenstein (1988) notes that both of these methods are subject to statistical inconsistency under certain conditions; when these conditions occur, the parsimony and compatibility criteria will favor an incorrect topology more strongly as the size of the data set increases. Inequality of rates of molecular evolution among different lineages coupled with high rates of molecular evolution will promote inconsistency. What constitutes a high rate of change is relative to the divergence events being investigated. A molecule that evolves too rapidly for investigating one set of evolutionary branching events may be found to evolve conservatively when evaluating more recent ones.

In the context of parsimony, compatibility, and alternative criteria, it is important to investigate the following parameters of molecular evolutionary pattern: (1) variance of evolutionary rate among lineages; (2) distribution of change among sites in the molecular sequence; (3) evidence for saturation of substitution (multiple substitutions occurring at the same site between lineages being compared); (4) ratio of transitions to transversions; (5) evolutionary rate for base substitutions versus length mutations; (6) overall evolutionary rate; and (7) base composition.

Variance of evolutionary rate among lineages is assessed using the index

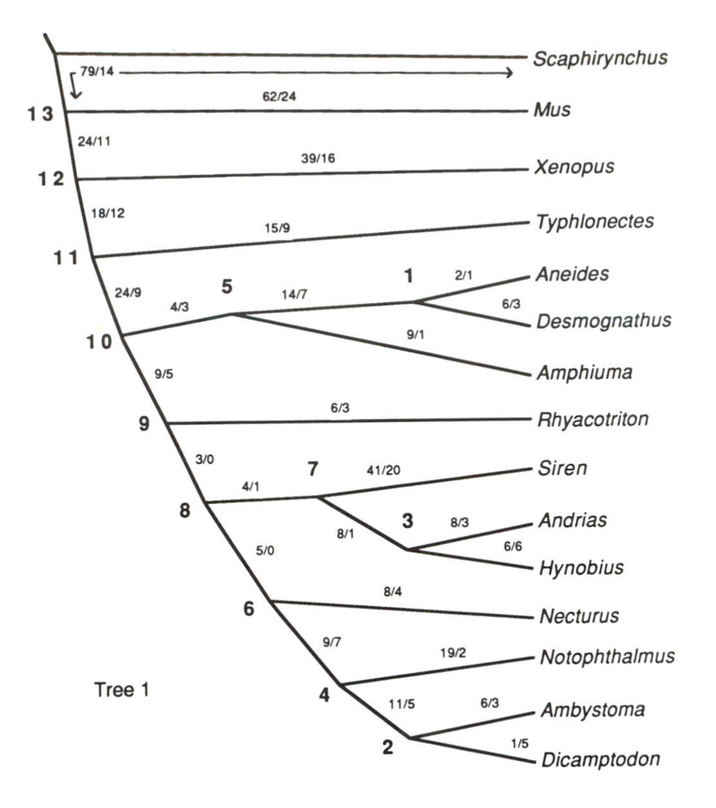

Figure 11-3 Three tree topologies used to investigate molecular evolutionary patterns. Tree 1 is the topology favored by a parsimony analysis of rRNA sequence variation by Larson (1991). Tree 2 is the topology presented by Duellman and Trueb (1986) for salamander samples with the same configuration of outgroups used in tree 1. Tree 3 is the same configuration of salamander samples used in tree 1 with an alternative topology for outgroup taxa. Nodes are numbered (1–13) for reference to Table 11-2. For each branch, the inferred number of substitutions is given followed by the inferred number of length mutations. (*Figure continues.*)

of dispersion (R = variance/mean rate; see Gillespie, 1986a) for all paired sister lineages in the tree. If the molecular clock is a Poisson process, then $R = 1$, although the values measured for the proteins used in molecular clock studies are usually higher (approximately 2 to 3; Kimura, 1983), and values exceeding 30 have been reported in more extensive surveys of protein and nucleic acid evolution (Gillespie, 1986b, 1989). Interpretations of this outcome regarding the clocklike nature of molecular evolution are highly controversial. Compare, for example, Kimura (1983) and Gillespie (1986a, 1986b, 1989).

The density distribution of changes among sites in the molecular sequence is investigated using the negative binomial (Bliss and Fisher, 1953;

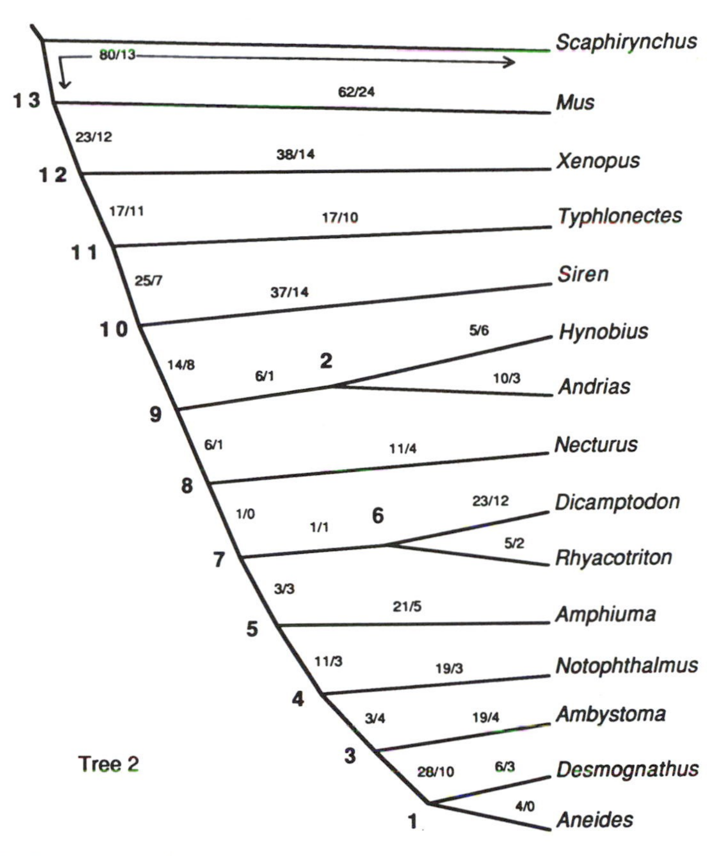

Figure 11-3 *(continued).*

Uzzell and Corbin, 1971). The negative binomial distribution is defined by the mean (m) number of changes per site (as measured on the phylogenetic tree) and $k = m^2/(s^2 - m)$ where s^2 is the variance. Values of k are very large if the distribution of substitutions is Poisson; when k is small, some molecular sites are much more likely than others to undergo evolutionary change. A similar parameter, *alpha* ($= m^2/s^2$) has been used for estimating molecular similarity of homologous sequences from restriction map comparisons (Nei and Li, 1979).

Saturation of sites in molecular sequence comparisons can be investigated using patterns in the ratio of transitions (substitutions replacing purines with purines and pyrimidines with pyrimidines) to transversions (substitutions replacing pyrimidines with purines and purines with pyrimidines). Comparison of recently diverged molecular sequences demonstrates an approximately twofold excess of transitions over transversions for cellular genomes (Fitch, 1980; Jukes, 1980; Nichols et al., 1980), and the transition bias is even larger for some organellar genomes (see review by Moritz et al., 1987). As more distantly related (i.e., more extensively saturated) comparisons are made, the measured transition/transversion ratio drops

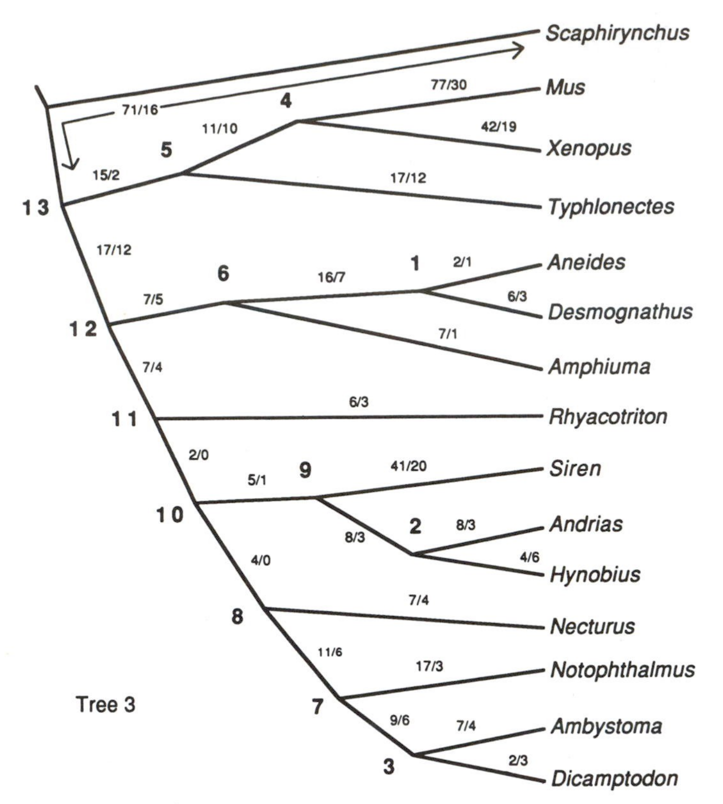

Figure 11-3 *(continued).*

(DeSalle et al., 1987). The expected equilibrium ratio depends upon the base composition of the molecular sequence and can be evaluated using the equations of Holmquist (1983). Comparisons of observed and expected transition/transversion ratios are used to evaluate the degree of saturation of ribosomal divergent domain sequences among species separated by approximately 70 to 200 Myr.

Evolutionary rates of the divergent domains are estimated using comparisons for which approximate evolutionary divergence dates are available from paleontological data and previous molecular studies. The paleontology of salamanders is reviewed by Estes (1981), and its relevance to the molecular phylogeny of salamanders is discussed by Larson (1991). Fossil evidence suggests that the separation of *Ambystoma* and *Dicamptodon* was complete by 65 Myr before present. Maxson and Wilson (1979) placed the split between *Ambystoma* and the amphiumid and plethodontid lineages (represented here by *Amphiuma*, *Aneides*, and *Desmognathus* samples) at 200 Myr ago based upon comparative immunological results. According to the molecular phylogeny of Larson (1991), this comparison spans the oldest divergence event in the history of the extant salamander families.

Relative rates of substitutions and length mutations are evaluated without exact dates by comparing the numbers of changes inferred for each category.

Where the rate assumptions of parsimony methods are violated, alternative approaches that are rate-invariant can be used. The invariant method of Lake (1987) is designed for use in sequence comparisons where rates are likely to be unequal among lineages and where saturation is a problem. Its application requires larger amounts of evolutionary change than are available in the data set on which this study is based, and it is less likely than parsimony to be informative under conditions where parsimony analysis is appropriate (Prager and Wilson, 1988). Felsenstein (1988) notes that Lake's invariant method contains the assumption that the alternative outcomes of transversion events occur with equal probability (that G to C changes and G to U changes are equally likely, etc.) and that the method is subject to error when this assumption is violated. This assumption can be tested using the data for ribosomal divergent domains evaluated here, and it is relevant for the possible use of the divergent domains for evaluating evolutionary divergences older than the ones considered here.

Compositional statistics is a rate-invariant method presented by Sidow and Wilson (1990) as an alternative to Lake's evolutionary parsimony. It shares with evolutionary parsimony the potential to resolve deep branches in the tree of life using sequence comparisons that demonstrate substitutional saturation and variable rates of evolution. Unlike evolutionary parsimony, however, it utilizes information on base composition and does not assume balanced transversions. The evolution of base composition within the divergent domains is analyzed here with respect to the application of compositional statistics.

RESULTS

The boundaries of the divergent domains used in this analysis correspond to those identified by Hassouna et al. (1984) (see Fig. 11-1). A short region located immediately 3′ to domain 5 shows extensive length variation in salamanders and is included with domain 5 for this analysis. Molecular changes are partitioned on the alternative tree topologies using the computer program, Phylogenetic Analysis Using Parsimony (PAUP) (Swofford, 1984). The types of molecular evolutionary change inferred for each topology in Figure 11-3 are summarized in Table 11-1.

Indices of Dispersion

Indices of dispersion are presented for each of the 13 nodes of each tree depicted in Figure 11-3 (Table 11-2). Because of the relatively small numbers of changes recorded within the individual domain segments, R values are calculated for all divergent domains together, including base substi-

Table 11-1 Summary of Evolutionary Changes for Each Divergent Domain*

Domain†	T	V	I	Del	L	Total	N	Total/N
1 (122a–274)	39	17	15	11	1	83	160	0.5
2 (821–950)	44	23	6	14	0	87	121	0.7
3 (992–1170)	24	23	4	4	1	56	132	0.4
4 (1358a–1370)	2	7	6	5	1	21	20	1.1
5 (1449–1500)	5	8	13	13	3	42	53	0.8
6 (1727–1775)	13	5	3	3	3	27	50	0.5
7a (1947–2007)	12	5	0	2	0	19	61	0.3
7b (2043–2122)	9	5	0	0	0	14	36	0.4
8 (2596–2777)	32	32	17	19	1	101	152	0.7
9 (3137–3164)	2	8	0	2	2	14	28	0.5
10 (3204–3291)	10	8	0	0	0	18	73	0.2
12 (3857–4023)	55	52	8	16	2	133	155	0.9
Total Tree 1	247	193	72	89	14	615	1,041	0.6
Total Tree 2	283	212	81	83	14	673	1,041	0.6
Total Tree 3	234	192	89	79	16	610	1,041	0.6

*Changes are inferred from the lineages on tree topology 1 (Fig. 11-3) for each of the divergent domains. Total changes (all domains summed) are given for tree topologies 1–3 (Fig. 11-3) at the bottom of the table.

Abbreviations used are: T, transitions; V, transversions; I, insertions; Del, deletions; L, unpolarized length changes; Total, sum of mutational events; N, number of bases per segment.

†Domains are numbered as recognized by Hassouna et al. (1984). Numbers in parentheses denote the sites scored (as numbered in Fig. 11-2). Only partial sequences are given for domains 2 and 8 (see Fig. 11-1) and sequences reported for domain 5 include a short segment located 3′ to this domain as recognized by Hassouna et al. (1984).

tutions and length mutations. For each tree, *R* ranges from values much lower to values much larger than the value of 1 expected if molecular change through time is Poisson distributed. On tree 1 (the topology favored by Larson, 1991), the median value (1.5) is similar to observations for protein sequences on which molecular clock studies are based (Wilson et al., 1977; Kimura, 1983). The mean value (3.65 ± its standard error of 1.16) exceeds the median because a few nodes have values that are much higher than the others. Use of the alternative tree topologies increases the median value of *R* (2.6 for tree 2, and 2.3 for tree 3).

Indices of dispersion greater than 1 indicate some degree of clumping of changes on the phylogenetic tree. There are only a few examples where particular taxa are associated consistently with rate inequalities on all three topologies tested. The terminal lineages leading to *Amphiuma*, *Necturus*,

Table 11-2 Indices of Dispersion (*R*) Representing the Variance/Mean Ratios of the Inferred Numbers of Changes Occurring on Sister Lineages of Each of the Tree Topologies Shown in Figure 11-3 and the Median (*M*)*

	Node													
	1	2	3	4	5	6	7	8	9	10	11	12	13	M
Tree 1	1.5	0.3	0.0	0.1	3.9	6.9	10.1	1.6	10.2	1.4	9.5	1.5	0.5	1.5
Tree 2	1.0	0.0	3.4	2.8	2.6	10.7	2.2	3.7	1.6	0.1	13.6	0.0	3.2	2.6
Tree 3	1.5	0.0	1.1	6.3	21.6	6.0	0.1	7.6	9.5	2.3	9.8	0.2	1.7	2.3

*Nodes are as numbered on Figure 11-3.

Rhyacotriton, and *Typhlonectes* are short relative to their sister lineages on all three trees. The lineage leading to *Siren* is anomalously long on trees 1 and 3, but nearly identical in length to its sister lineage on tree 2. A partitioning of changes for each divergent domain individually on tree 1 (not shown) reveals that most domains have changes distributed throughout the tree rather than being anomalously clumped on one or a few lineages. There are, however, a few apparent exceptions to this generalization. Five of the 11 domains examined (1, 5, 6, 8, and 12) have a somewhat higher than expected total amount of change on the lineage leading to *Siren*. Domains 1 and 4 show a somewhat high concentration of change on the lineage immediately ancestral to all salamanders. Domain 12 also shows a higher than expected amount of change on the *Notophthalmus* lineage, and for domain 9 the majority of the observed changes are on the *Xenopus* lineage.

Density Distribution of Substitutions and Length Mutations

For each divergent domain, the negative binomial parameters (m, k, and *alpha*) are evaluated separately for base substitutions (Table 11-3), length mutations (Table 11-4), and for all changes together (Table 11-5). For making these calculations, a "site" is defined as a column of single-base width in the block of aligned sequences (although the individual sequences in the alignment will not always have a base present at each site; see Fig. 11-2). Multiple-base deletions are counted separately for each site covered in the sequence alignment. Multiple-base insertions are counted as single events and the position of the insertion is treated as a single site for this analysis. The total number of sites counted for evaluating base substitutions alone (Table 11-3) does not include "sites" produced by insertions unique to single sequences and therefore differs from the totals used to evaluate length mutations (Table 11-4) and total changes (Table 11-5).

For tree 1, the overall mean number of base substitutions per molecular site (m) is 0.44. Estimates for tree 2 are generally higher (mean = 0.49) reflecting the fact that it is less parsimonious overall. Estimates derived from tree 3, which rearranges only the outgroup taxa, are closer overall to values obtained with tree 1 (mean = 0.42). The mean number of length changes per molecular site (Table 11-4) is less than the mean number of substitutions by a factor of approximately 1.5. Overall, the mean number of changes per site (substitutions and length mutations) is 0.68 for tree 1, with the other trees differing only slightly. Mean values estimated for m, k, and *alpha* (Tables 11-3–11-5) differ by no more than a factor of 1.5 when different tree topologies are used. For length mutations, large differences in the estimates of these parameters are seen for some individual domains when tree topology 3 is compared to topologies 1 and 2 (see, for example, domains 6 and 9 in Table 11-4). This results mainly from differences in partitioning a few large length mutations occurring in some of the smaller domains among the caecilian, frog, salamander, and mammalian

Table 11-3 Density Distribution of Base Substitutions Per Site Within Divergent Domains of the Large Ribosomal Subunit Estimated Separately Using Tree Topologies 1, 2, and 3 (Fig. 11-3)

		Domain*												
		1	2	3	4	5	6	7a	7b	8	9	10	12	Mean†
Total sites		152	119	128	12	52	47	61	36	147	28	73	151	83.83
Mean	(1)	0.37	0.56	0.37	0.75	0.25	0.38	0.28	0.39	0.44	0.36	0.25	0.71	0.44
	(2)	0.39	0.61	0.43	0.67	0.31	0.38	0.34	0.44	0.56	0.36	0.23	0.78	0.49
	(3)	0.34	0.55	0.38	0.75	0.21	0.36	0.26	0.36	0.43	0.36	0.26	0.66	0.42
k	(1)	0.40	0.94	0.76	0.93	0.48	0.76	0.36	0.52	0.39	—	0.25	1.18	0.63
	(2)	0.31	0.73	0.32	1.14	0.33	0.76	0.21	0.38	0.29	—	0.28	0.96	0.47
	(3)	0.42	0.93	0.53	0.93	0.41	0.62	0.54	0.64	0.34	—	0.23	1.39	0.63
Alpha	(1)	0.19	0.35	0.25	0.42	0.16	0.25	0.16	0.22	0.20	0.42	0.12	0.44	0.26
	(2)	0.17	0.33	0.18	0.42	0.16	0.25	0.13	0.20	0.19	0.42	0.13	0.43	0.24
	(3)	0.19	0.35	0.22	0.42	0.14	0.21	0.18	0.23	0.19	0.42	0.12	0.45	0.26

*Calculations for domain 2 omit one region of the *Mus* sequence in Figure 11-2 that is not alignable with the others. Values of k are omitted for domain 9 because very low values of the variance to the mean produce negative values of k.

†Values of parameters for the individual domains are weighted by the length of the domain to obtain mean values.

Table 11-4 Density Distribution of Sites Covered by Length Mutations Within Divergent Domains of the Large Ribosomal Subunit Estimated Separately Using Tree Topologies 1, 2, and 3 (Fig. 11-3)*

		Domain†												
		1	2	3	4	5	6	7a	7b	8	9	10	12	Mean‡
Total sites		160	121	132	20	53	50	61	—	152	28	—	155	93.20
Mean	(1)	0.18	0.24	0.12	0.95	0.64	0.22	0.10	—	0.39	1.29	—	0.19	0.29
	(2)	0.21	0.22	0.12	0.95	0.72	0.22	0.10	—	0.39	1.29	—	0.19	0.29
	(3)	0.20	0.24	0.14	0.65	0.79	0.14	0.10	—	0.41	0.07	—	0.21	0.26
k	(1)	0.10	0.22	0.48	—	0.57	1.53	—	—	0.42	—	—	0.26	0.38
	(2)	0.09	0.43	0.48	—	0.45	1.53	—	—	0.45	—	—	0.26	0.41
	(3)	0.10	0.22	0.26	—	0.61	0.20	—	—	0.47	—	—	0.24	0.28
Alpha	(1)	0.06	0.12	0.10	1.39	0.30	0.19	0.11	—	0.20	1.80	—	0.11	0.21
	(2)	0.06	0.15	0.10	1.39	0.28	0.19	0.11	—	0.21	1.80	—	0.11	0.21
	(3)	0.07	0.12	0.09	0.80	0.34	0.08	0.11	—	0.22	0.08	—	0.11	0.14

*No Length Mutations Were Recorded in Domains 7b and 10.

†Calculations for domain 2 omit one region of the *Mus* sequence in Figure 11-2 that is not alignable with the others. Values of *k* are omitted for domains 4, 7a, and 9 because very low values of the variance to the mean produce negative values of *k*.

‡Values of parameters for the individual domains are weighted by the length of the domain to obtain mean values.

Table 11-5 Density Distribution of Changes (Length Mutations and Base Substitutions) Within the Divergent Domains of the Large Ribosomal Subunit Estimated Separately Using Tree Topologies 1, 2, and 3 (Fig. 11-3)

		Domain*												
		1	2	3	4	5	6	7a	7b	8	9	10	12	Mean†
Total sites		160	121	132	20	53	50	61	36	152	28	73	155	86.75
Mean	(1)	0.53	0.78	0.48	1.40	0.89	0.54	0.38	0.39	0.82	1.68	0.25	0.88	0.68
	(2)	0.58	0.82	0.54	1.35	1.02	0.54	0.44	0.44	0.92	1.68	0.23	0.95	0.73
	(3)	0.52	0.79	0.51	1.10	0.98	0.48	0.36	0.36	0.83	0.43	0.26	0.85	0.64
k	(1)	0.48	1.38	0.94	8.17	0.98	1.55	0.57	0.52	0.66	—	0.25	1.79	1.10
	(2)	0.36	1.33	0.47	10.27	0.75	1.55	0.33	0.38	0.49	—	0.28	1.47	0.95
	(3)	0.19	1.45	0.71	1.75	1.00	0.92	0.79	0.64	0.65	0.47	0.23	2.34	0.96
Alpha	(1)	0.25	0.50	0.32	1.20	0.47	0.40	0.23	0.22	0.37	3.02	0.12	0.59	0.46
	(2)	0.22	0.51	0.25	1.19	0.43	0.40	0.19	0.20	0.32	3.02	0.13	0.58	0.43
	(3)	0.14	0.51	0.30	0.68	0.50	0.32	0.25	0.23	0.37	0.23	0.12	0.62	0.36

*Calculations for domain 2 omit one region of the *Mus* sequence in Figure 11-2 that is not alignable with the others. Some values of *k* are omitted for domain 9 because very low values of the variance to the mean produce negative values of *k*.

†Values of parameters for the individual domains are weighted by the length of the domain to obtain mean values.

lineages. This has a relatively small impact, however, on the estimated negative binomial parameters for all domains combined (Tables 11-4 and 11-5).

The values of k and *alpha* estimated from all three trees are too small for the density distribution of changes (for both site substitutions and length mutations) to be Poisson. Some sites are more likely than others to evolve. This replicates the observation of Larson and Wilson (1989) that some segments of sequence within the divergent domains are relatively highly conserved, whereas others undergo extensive change.

Transition/Transversion Ratios

Estimates of the ratio of transitions to transversions are given for the 11 divergent domains in Table 11-6. The results obtained using different tree topologies are within the same order of magnitude, generally differing by not more than a factor or two. Four domains (1, 2, 6, and 7a/7b) have values of transitions/transversions very close to 2, consistent with prior empirical comparisons of nuclear-encoded sequences where site saturation has not occurred. Four additional domains (3, 8, 10, and 12) have values closer to 1. All of these are well above the expected equilibrium value of approximately 0.4 calculated using the method of Holmquist (1983). Three small domains (4, 5, and 9) have values that approach the expected ratios for saturated sequences. The transversions observed in these three small domains occur mainly in comparisons to the *Mus* sequence, however, which is relatively distantly related to the amphibians and therefore more likely to demonstrate substitutional saturation. Additional sampling of the mammalian lineage may clarify the substitutional changes that separate the amphibian and mouse sequences. With this exception, it appears that molecular sites in the divergent domains do not demonstrate substitutional saturation within the scope of this study.

Bias in Transversion Substitutions

Polarized transversion substitutions were tallied for the divergent domains and adjacent regions reported by Larson (1991) for tree topology 1 (Fig. 11-3). For 16 observed transversions replacing an A base, 12 replaced it with C and only four replaced it with U. For 76 transversions replacing a G base, 51 replaced it with C and 25 replaced it with U. For 29 transversions replacing a U base, 23 replaced it with G and only six replaced it with A. Using a χ^2 goodness-of-fit test, all three of these cases are significantly different from the prediction of balanced transversions, at least at the 0.05 level. Only the transversions replacing the C base occur with approximately equal frequencies; of 61 changes, 32 replace it with G and 29 replace it with A. Overall, there is a more than twofold preference for replacing a purine with C rather than U, and an almost twofold preference for replacing a pyrimidine with G rather than with A.

Table 11-6 Ratios of Transition/Transversion Substitutions Measured According to Three Alternative Tree Topologies (Fig. 11-3), Compared With the Expected Equilibrium Ratio (Calculated According to Holmquist, 1983)

		Domain												
		1	2	3	4	5	6	7a	7b	8	9	10	12	Mean*
Observed	(1)	2.3	1.9	1.0	0.3	0.6	2.6	2.4	1.8	1.0	0.3	1.3	1.1	1.4
	(2)	2.2	2.1	1.2	0.1	0.9	3.3	3.2	1.3	1.0	0.3	1.4	1.1	1.5
	(3)	2.0	1.8	1.1	0.3	0.6	2.8	2.2	1.6	0.9	0.3	1.1	1.0	1.4
Expected		0.4	0.3	0.4	0.5	0.4	0.4	0.4	0.5	0.4	0.2	0.3	0.4	0.4
Substitutions†														
	(1)	56	67	47	9	13	18	17	14	64	10	18	107	36.67
	(2)	60	74	55	8	17	17	21	16	82	10	17	118	41.25
	(3)	51	66	49	9	11	19	16	13	63	10	19	100	35.50

*Mean values of the transition/transversion ratio are calculated by weighting the values for individual domains by the inferred number of base substitutions.

†The number of base substitutions inferred for each divergent domain for each of the tree topologies (Fig. 11-3).

Rates of Change

Approximately 1,040 bases of sequence are tallied for the divergent domains, although this is not a complete assessment of the divergent domain sequences in rRNA. Calculation of the rate of divergence between lineages (counting changes accumulated on both lineages descended from a node) per molecular site per million years ago (Mya) is done using 65 Mya as the estimated separation time of *Ambystoma* and *Dicamptodon* based on the fossil record (see review by Estes, 1981). Fifteen total changes are observed on these lineages, giving an estimated divergence rate of 0.02% change/site/Mya for the pair of lineages. The 65 Mya divergence is a minimum estimate of divergence time, making the rate estimate a maximum one. Within the resolution of the phylogenetic study of Larson (1991), the divergence of the amphiumids, plethondontids, and remaining extant salamanders constitutes an unresolvable three-way split. The divergence of the amphiumids and plethodontids from *Ambystoma* is estimated by Maxson and Wilson (1979) to be 200 Mya. Using this as a date for the hypothethical three-way split, the following rates of change are estimated: (1) amphiumids (*Amphiuma*) to plethodontids (*Aneides* and *Desmognathus*) (0.0002 changes/site/Mya); (2) amphiumids to ambystomatids (*Ambystoma* and *Dicamptodon*) (0.0002 changes/site/Mya); (3) plethodontids to ambystomatids (0.0003 changes/site/Mya). In nearly all cases, the changes measured here are single base-length mutations or substitutions, thereby representing approximately 0.02% divergence/site/Mya. This is comparable to the value estimated by Larson and Wilson (1989) for evolutionary change in the divergent domains for salamanders. Larson and Wilson (1989) interpreted these values, which were based on comparisons within plethodontids, to be underestimates for salamanders as a whole, but the results given here indicate that they are probably accurate. Values calculated by Larson and Wilson (1989) and Schmickel et al. (1990) for mammals are approximately eight to ten times larger. A possible acceleration of evolutionary rate in mammals may help to explain why the saturation inferred from three divergent domains discussed above (based on transition/transversion ratios) occurs mainly in comparisons to the *Mus* sequence.

The ratio of base substitutions to length mutations for the divergent domains (based on tree topology 1) is approximately 2.5:1 (440/175 mutations). This difference is judged not to be great enough to warrant separate weighting of the two kinds of events for phylogenetic reconstruction.

Evolution of Base Composition

For applying the rate invariant phylogenetic method of Sidow and Wilson (1990), and for estimating the expected equilibrium ratio of transitions/transversions for a molecular sequence comparison (Table 11-6), it is necessary to measure base compositions. Base compositions are determined for the 11 divergent domains for samples from the caecilian, fish, frog, salamander, and mammal lineages (Table 11-7). The mean G + C content

Table 11-7 Nucleotide Base Compositions of Divergent Domains*

		Domain											
	Base	1	2	3	4	5	6	7a	7b	8	9	10	12
Salamander	A	13	10	14	33	14	13	17	19	8	0	7	14
	C	37	42	32	33	32	33	30	28	38	50	33	32
	G	36	37	44	25	43	46	40	42	38	40	46	36
	U	14	10	10	8	11	9	13	11	16	10	14	18
Caecilian	A	9	8	15	25	12	11	17	19	8	0	8	16
	C	37	42	32	25	26	30	31	28	43	50	35	33
	G	40	40	43	25	45	48	38	42	35	40	46	37
	U	14	9	10	25	17	11	15	11	15	10	11	14
Frog	A	9	11	14	8	16	8	15	17	6	11	11	14
	C	40	45	31	33	29	31	31	31	44	46	35	35
	G	41	36	44	33	41	47	41	44	38	36	44	38
	U	11	8	11	25	14	14	13	8	12	7	9	13
Mammal	A	10	8	15	15	13	9	16	22	8	4	8	16
	C	37	41	32	40	28	34	29	25	43	56	35	31
	G	39	40	45	25	45	48	41	42	37	37	47	37
	U	13	11	8	20	13	9	14	11	12	4	9	16
Fish	A	14	8	16	10	13	18	20	22	11	0	13	16
	C	32	38	30	40	30	36	30	28	35	50	32	28
	G	36	40	42	20	43	40	34	39	37	40	43	31
	U	18	14	13	30	15	7	16	11	18	10	13	25
Mean	A	11.0	9.0	14.8	18.2	13.6	11.8	17.0	19.8	8.2	3.0	9.4	15.2
	C	36.6	41.6	31.4	34.2	29.0	32.8	30.2	28.0	40.6	50.4	34.0	31.8
	G	38.4	38.6	43.6	25.6	43.4	45.8	38.8	41.8	37.0	38.6	45.2	35.8
	U	14.0	10.4	10.5	21.6	14.0	10.0	14.2	10.4	14.6	8.2	11.2	17.2
Standard deviation	A	2.3	1.4	0.8	4.2	1.5	4.0	1.9	2.2	1.8	4.8	2.5	1.1
	C	2.9	2.5	0.9	6.2	2.2	2.4	0.8	2.1	3.9	3.6	1.4	2.6
	G	2.3	1.9	1.1	4.7	1.7	3.3	2.9	1.8	1.2	1.9	1.6	2.8
	U	2.5	2.3	1.8	8.4	2.2	2.6	1.3	1.3	2.6	2.7	2.2	4.8

*All values are given as percentages. Values are based on the sequences presented for *Aneides flavipunctatus* (salamander), *Typhlonectes compressicauda* (caecilian), *Xenopus laevis* (frog), *Mus domesticus* (mammal), and *Scaphirhynchus platorynchus* (fish).

of the divergent domains ranges from approximately 60 to 90%. Three small domains (4, 6, and 9) have an atypically large variance of base composition.

DISCUSSION

Assessment of molecular evolutionary pattern is important for deciding which analytical method will extract the maximum amount of phylogenetic information from comparative molecular sequence data without systematically incurring errors. The divergent domains of the large ribosomal subunit are perhaps unusual in the degree to which base-substitutional and length-mutational changes are superimposed. These sequences are very important for phylogenetic studies, however, because of their potential for resolving evolutionary branching events on an intermediate timescale and because they can be studied in all eukaryotic taxa. The results of this study are important not only for interpreting direct sequence comparisons involving the divergent domains, but also for evaluating phylogenetic studies based upon restriction mapping of nuclear ribosomal genes. Conclusions regarding the utility of different methods of phylogenetic analysis are expected to be of general significance for studies of ribosomal evolution.

The evolutionary pattern revealed by the divergent domains does not meet the assumptions of parsimony analysis exactly, but the approximation is judged to be close enough that statistical tests based upon parsimony should be robust for the evolutionary timescale being investigated. The molecular clock provides a good expectation for the distribution of molecular evolutionary changes on the phylogenetic tree. It is anticipated, however, that a few lineages will show substantially larger or smaller amounts of change than predicted by the molecular clock in studies of rRNA. For a strictly stochastic clock, the index of dispersion (R) should equal 1, whereas the median value measured for the divergent domains in this study is 1.5. Some parts of the tree have a much smaller variance/mean ratio, but others are anomalously high. The use of incorrect tree topologies will inflate estimates of R, thereby overestimating the deviations from clocklike evolution.

When rate variation produces inconsistency of the parsimony criterion, lineages demonstrating unusually large amounts of change will tend to cluster together, thereby obscuring the true relationships of the species being studied. Larson (1991) notes that the lineages demonstrating relatively large amounts of change in this data set are not grouped by parsimony. Although the observed rate asymmetry appears not to produce inconsistency of the parsimony criterion on the timescale investigated, it cautions against the use of parsimony analysis of divergent domains to investigate divergences much older than these. The observed rate asymmetry also cautions against making phylogenetic inferences from unweighted pair group clustering of taxa (UPGMA; see Felsenstein, 1988), which is more sensitive than parsimony to rate asymmetry. The UPGMA analysis is discouraged, furthermore, by the difficulty of converting to distance values comparisons of paired sequences that differ by both length mutations and site substitutions. The presence of length mutations that affect more than a single site has the consequence that not all evolutionary events produce equivalent amounts of divergence among paired sequences.

The observation that the overall mean number of changes inferred per site is substantially less than one (Tables 11-3–11-5) supports the conclusion that the evolutionary saturation of sites, which also can produce inconsistency of parsimony analysis, is not a problem here. This conclusion is compromised somewhat by the observation that the density distribution of changes along the molecular sequence is highly nonrandom. For example, a few small domains average more than one change per site (Tables 11-4–11-5). Furthermore, values of the negative binomial parameter, k, and the related parameter, *alpha* (Tables 11-3–11-5) show that evolutionary change in the divergent domains is not Poisson distributed, but that some sites are much more likely than others to undergo evolutionary change. If changes along the molecular sequence were Poisson distributed, evolutionary saturation of sites would occur less rapidly than it does when substitutions are concentrated in a subset of the sites. This problem can be avoided somewhat by using as taxonomic characters only the more conservatively evolving sites in the alignment. Alternatively, rate-invariant approaches (Lake, 1987; Sidow and Wilson, 1990) could be used for making phylogenetic inferences from heavily substituted sequences, although these methods are applicable only to sites that have not undergone length mutation.

The magnitude of *alpha* is important for estimating sequence divergence or distance from restriction map data (Nei and Li, 1979). Values obtained for protein sequences (1–2) are too high for use with divergent domains. This would overestimate the total divergence among paired sequences by inferring that evolutionary changes are more evenly distributed among sites than they are. The values reported here (0.46 for all mutations, 0.26 for base substitutions alone) are more compatible with the values of 0.45–0.50 used for DNA sequences by Carr et al. (1987) and Wilson et al. (1989). Even these are too high, however, for some divergent domains.

The observed ratio of transitions to transversions in sequence comparisons is also useful for assessing substitutional saturation of molecular sites. Except for three very small domains (4, 5, and 9), transition/transversion ratios exceed those expected for saturated sequences and approximate the value of 2, characteristic of recently diverged molecular sequences compared in earlier empirical studies (Table 11-6). These studies of molecular evolutionary pattern suggest that the divergent domains meet the criterion of parsimony that sequences are not heavily saturated. This result suggests also that the rate-invariant methods developed for analysis of highly saturated sequences (Lake, 1987; Sidow and Wilson, 1990) are less likely than parsimony to be informative for phylogenetic analysis of ribosomal divergent domains on the timescale addressed here.

To resolve divergence events much older than the ones addressed in this study using the ribosomal divergent domains requires a method that accommodates rate variation, substitutional saturation of sites, and length mutations. Lake's invariant method (Lake, 1987) meets the first two requirements but not the third, and its assumption of balanced transversions is violated by the pattern observed here for the divergent domains. The latter problem is remedied by the compositional statistics of Sidow and Wilson (1990). Neither of these methods, however, is able to utilize information from the length-variable sites for phylogenetic inference.

Further theoretical work is needed to make more precise quantitative statements regarding the degree to which violations of the assumptions of parsimony will promote errors in phylogenetic inference. For evaluating the usefulness of the parsimony criterion, it is particularly important to measure the effects on phylogenetic reconstruction of deviations from the following optimal conditions: (1) the median index of dispersion (R) should equal 1; (2), the mean density of substitutions per molecular site should not exceed 1; (3) the negative binomial parameter k should be large; and (4) the ratio of transition to transversion substitutions should be approximately 2:1. It is suggested that future studies address the interactions of these factors with respect to consistency of the parsimony criterion.

SUMMARY

The molecular evolutionary patterns of 11 divergent domains of the large ribosomal subunit are assessed to evaluate the appropriateness of these sequences for phylogenetic analyses using parsimony and other criteria. The taxa studied represent nine families of salamanders covering approximately 70 to 200 million years of divergence, and four vertebrate outgroups. The rate of evolution of the divergent domain sequences is estimated to produce approximately 0.02% sequence divergence between lineages per site per Mya. Consistency of the parsimony criterion for phylogenetic inference requires that rates of molecular evolution be stochastically constant. Variance of evolutionary rates is assessed using the index of dispersion (R = variance/mean rate of change along branches of the phylogenetic tree). The median value of R is 1.5, which indicates a higher than stochastic variance (for which R = 1.0), but which is well within the range used in earlier molecular clock studies, in which R may be as large as 2 or 3.

Parsimony analysis requires also that the sequences being compared are not heavily saturated with multiple substitutions at the same sites. The average total number of changes per molecular site is 0.68 for the ribosomal divergent domains. Substitutional saturation therefore appears not to be a problem at the evolutionary timescale studied, although the distribution of changes along the molecular sequence is nonrandom, thereby increasing the likelihood of substitutional saturation at the most rapidly evolving sites. The average transition/transversion ratio for the divergent domains is 1.9:1, approximating the value observed empirically for comparisons of relatively recently diverged molecular sequences (2.0) and substantially exceeding the value predicted for highly saturated sequences (0.4).

Parsimony analysis of the ribosomal divergent domains appears to be justified for investigating phylogenetic divergence events occurring approximately 70 to 200 Mya. Rate-invariant methods are preferable for investigating much older divergences. The method of compositional statistics is recommended because it is able to accommodate an observed bias in transversion substitutions. To extract the maximum amount of phylogenetic information from these sequence comparisons, however, the superposition of base substitutions and length mutations also must be addressed.

ACKNOWLEDGMENTS

I thank Norma Gieg-Sinclair for collecting data and Chris DeHaan for making calculations and figures. Joel Cracraft, Michael M. Miyamoto, Ellen Prager, Elizabeth Zimmer, and an anonymous reviewer provided helpful suggestions and discussion. Financial support was provided by NSF grant BSR-8708393 and BRSG grants from Washington University.

REFERENCES

Bliss, C. I., and R. A. Fisher (1953) Fitting the negative binomial distribution to biological data. *Biometrics* **9**:176–200.

Carr, S. M., J. A. Brothers, and A. C. Wilson. (1987) Evolutionary inferences from restriction maps of mitochondrial DNA from nine taxa of *Xenopus* frogs. *Evolution* **41**:176–188.

Clark, C. G. (1987) On the evolution of ribosomal RNA. *J. Mol. Evol.* **25**: 343–350.

Clark, C. G., B. W. Tague, V. C. Ware, and S. A. Gerbi. (1984) *Xenopus laevis* 28S ribosomal RNA: a secondary structure model and its evolutionary and functional implications. *Nucleic Acids Res.* **12**:6197–6220.

DeSalle, R., T. Freedman, E. M. Prager, and A. C. Wilson. (1987) Tempo and mode of sequence evolution in mitochondrial DNA of Hawaiian *Drosophila. J. Mol. Evol.* **26**:157–164.

Duellman, W. E., and L. Trueb. (1986) *Biology of Amphibians*. McGraw-Hill, New York.

Estes, R. (1981) Gymnophiona, Caudata. *Handb. Paläherpetol.* **2**:1–115.

Felsenstein, J. (1988) Phylogenies from molecular sequences: inference and reliability. *Ann. Rev. Genet.* **22**:521–565.

Fitch, W. M. (1971) Toward defining the course of evolution: minimum change for a specific tree topology. *Syst. Zool.* **20**:406–416.

Fitch, W. M. (1980) Estimating the total number of nucleotide substitutions since the common ancestor of a pair of homologous genes: comparison of several methods and three beta hemoglobin messenger RNAs. *J. Mol. Evol.* **16**: 153–209.

Gerbi, S. A. (1985) Evolution of ribosomal DNA. Pp. 419–517 in *Molecular Evolutionary Genetics* (R. J. MacIntyre, ed.). Plenum Press, New York.

Gillespie, J. H. (1986a) Statistical aspects of the molecular clock. Pp. 255–272 in *Evolutionary Process and Theory* (S. Karlin and E. Nevo, eds.). Academic Press, London.

Gillespie, J. H. (1986b) Variability of evolutionary rates of DNA. *Genetics* **113**: 1077–1091.

Gillespie, J. H. (1989) Lineage effects and the index of dispersion of molecular evolution. *Mol. Biol. Evol.* **6**:636–647.

Gonzalez, I. L., J. L. Gorski, T. J. Campen, D. J. Dorney, J. M. Erickson, J. E. Sylvester, and R. D. Schmickel. (1985) Variation among human 28S ribosomal RNA genes. *Proc. Natl. Acad. Sci. USA* **82**:7666–7670.

Gonzalez, I. L., J. E. Sylvester, T. F. Smith, D. Stambolian, and R. D. Schmickel. (1990) Ribosomal RNA gene sequences and hominoid phylogeny. *Mol. Biol. Evol.* **7**:203–219.

Gorski, J. L., I. L. Gonzalez, and R. D. Schmickel. (1987) The secondary structure of human 28S rRNA: the structure and evolution of a mosaic rRNA gene. *J. Mol. Evol.* **24**:236–251.

Hassouna, N., B. Michot, and J.-P. Bachellerie. (1984) The complete nucleotide sequence of mouse 28S rRNA gene: implications for the process of size increase of the large subunit rRNA in higher eukaryotes. *Nucleic Acids Res.* **12**:3563–3583.

Holmquist, R. (1983) Transitions and transversions in evolutionary descent: an approach to understanding. *J. Mol. Evol.* **19**:134–144.

Jukes, T. H. (1980) Silent nucleotide substitutions and the molecular evolutionary clock. *Science* **210**:973–978.

Kimura, M. (1983) *The Neutral Theory of Molecular Evolution*. Cambridge University Press, Cambridge.

Lake, J. A. (1987) A rate-independent technique for analysis of nucleic acid sequences: evolutionary parsimony. *Mol. Biol. Evol.* **4**:167–191.

Larson, A. (1991). A molecular perspective on the evolutionary relationships of the salamander families. *Evol. Biol.* **25**:211–277.

Larson, A., and A. C. Wilson. (1989) Patterns of ribosomal RNA evolution in salamanders. *Mol. Biol. Evol.* **6**:131–154.

Maxson, L. R., and A. C. Wilson. (1979) Rates of molecular and chromosomal evolution in salamanders. *Evolution* **33**:734–740.

Moritz, C., T. E. Dowling, and W. M. Brown. (1987) Evolution of animal mitochondrial DNA: relevance for population biology and systematics. *Ann. Rev. Ecol. Syst.* **18**:269–292.

Nei, M., and W.-H. Li. (1979) Mathematical model for studying genetic variation in terms of restriction endonucleases. *Proc. Natl. Acad. Sci. USA* **76**: 5269–5273.

Nichols, B. P., G. F. Miozzari, M. Van Cleemput, G. N. Bennett, and C. Yanofsky. (1980) Nucleotide sequences of the trp G regions of *Escherichia coli, Shigella dysenteriae, Salmonella typhimurium*, and *Serratia marcescens*. *J. Mol. Biol.* **142**:503–517.

Prager, E. M., and A. C. Wilson. (1988) Ancient origin of lactalbumin from lysozyme: analysis of DNA and amino acid sequences. *J. Mol. Evol.* **27**: 326–335.

Qu, L.-H., M. Nicoloso, and J.-P. Bachellerie. (1988) Phylogenetic calibration of the 5′ terminal domain of large rRNA achieved by determining twenty eucaryotic sequences. *J. Mol. Evol.* **28**:113–124.

Schmickel, R. D., J. Sylvester, D. Stambolian, and I. L. Gonzalez. (1990) The ribosomal DNA sequence for the study of the evolution of human, chimpanzee, and gorilla. Pp. 11–32 in *DNA Systematics Volume III: Human and Higher Primates* (S. K. Dutta and W. P. Winter, eds.). CRC Press, Boca Raton.

Sidow, A., and A. C. Wilson. (1990) Compositional statistics: an improvement of evolutionary parsimony and its application to deep branches in the tree of life. *J. Mol. Evol.* **31**:51–68.

Spencer, D. F., J. C. Collings, M. N. Schnare, and M. W. Gray. (1987) Multiple spacer sequences in the nuclear large subunit ribosomal RNA gene of *Crithidia fasciculata*. *EMBO J.* **6**:1063–1071.

Swofford, D. L. (1984) *Phylogenetic Analysis Using Parsimony*. Illinois Natural History Survey, Champaign.

Uzzell, T., and K. W. Corbin. (1971) Fitting discrete probability distributions to evolutionary events. *Science* **172**:1089–1096.

Ware, V. C., B. W. Tague, C. G. Clark, R. L. Gourse, R. C. Brand, and S. A. Gerbi. (1983) Sequence analysis of 28S ribosomal DNA from the amphibian *Xenopus laevis*. *Nucleic Acids Res.* **11**:7795–7817.

Wilson, A. C., S. S. Carlson, and T. J. White. (1977) Biochemical evolution. *Ann. Rev. Biochem.* **46**:573–639.

Wilson, A. C., E. A. Zimmer, E. M. Prager, and T. D. Kocher. (1989) Restriction mapping in the molecular systematics of mammals: a restrospective salute. Pp. 407–419 in *The Hierarchy of Life. Molecules and Morphology in Phylogenetic Analysis*. (B. Fernholm, K. Bremer, and H. Jörnvall, eds.) Elsevier Science Publishers B.V., Amsterdam.

12

Statistical Methods for Testing Molecular Phylogenies

WEN-HSIUNG LI AND MANOLO GOUY

Although numerous methods have been developed for reconstructing phylogenetic trees, there exist few methods for evaluating the statistical confidence of an inferred phylogeny or for testing whether one phylogeny is significantly better than another. In fact, the statistical methodology for testing phylogenies is in a rather primitive state. This has occurred for two reasons. First, although phylogenetic reconstruction has long been recognized as a problem in statistical inference (Edwards and Cavalli-Sforza, 1964), few authors have formulated the problem in a statistical framework. Indeed, most current methods yield one or a few trees and do not provide information concerning the confidence level of estimated phylogenies. Second, the problem is extremely complex, largely because the number of possible alternative trees is large even when only a moderate number of taxa are involved. For this reason, most current statistical tests are heuristic when the number of taxa involved is five or larger. Fortunately, the rapid accumulation of DNA sequence data has stimulated a strong trend to make phylogenetic reconstruction more statistical.

Statistical tests can be classified as analytical or resampling. Resampling methods (e.g., bootstrapping, jacknifing) resample the data to infer empirically the variability of the estimate obtained by a tree-making method. For a recent review of resampling methods, see Felsenstein (1988). In this chapter, we discuss only analytical methods. Analytical tests can be based on parsimony methods, distance methods, likelihood methods, or invariant methods. The last approach, which includes the evolutionary parsimony method, has been reviewed in Felsenstein (1988). We discuss only the other approaches.

PARSIMONY TESTS

This analytic approach was initiated by Cavender (1978, 1981). He studied the confidence limits that can be placed on a phylogeny inferred from a four-species data set. The results were extended by Felsenstein (1985), who obtained narrower confidence limits by adding the assumption of a constant rate of evolution (i.e., an evolutionary clock). For ease of representation, we shall discuss Felsenstein's results before discussing Cavender's. We shall also discuss the case of more than four species.

Four Species With a Molecular Clock

We assume that species 4 is a known outgroup, so that it can be used as a reference to infer the branching order of the other three species. The three possible rooted trees are shown in Figure 12-1a–12-1c. In terms of parsimony, a nucleotide site is informative (useful) for choosing among the three trees only if it is in the same state in two of the four species, and in another state in the other two species. For example, if the site has the configuration AAGG among species 1–4, respectively, then it supports tree 1a because this tree requires (at least) only one nucleotide substitution to explain the configuration, whereas trees 1b and 1c each require two substitutions. As another example, if the configuration is ATGG, then it is not informative because each of the three trees requires two substitutions.

We assume that all informative sites are equivalent and that each informative site supports tree 1a with probability p, tree 1b with probability q, and tree 1c with probability r. Then the probability that among n random informative sites there are i sites supporting tree 1a, j sites supporting tree 1b, and k sites supporting tree 1c is trinomial, that is,

$$\text{Prob}(i, j, k) = \frac{n!}{i!j!k!} p^i q^j r^k. \quad (1)$$

For the moment, let us assume that tree 1a is the true tree. Under the assumption of rate-constancy, an informative site has a higher probability of supporting the true tree than supporting an incorrect tree. So, $p \geq q = r$. The probability that among n random informative sites, the number (C) of sites supporting a particular incorrect tree is m or larger is given by

$$\text{Prob}(C \geq m) = \sum_{i=m}^{n} \frac{n!}{i!(n-i)!} q^i (p+q)^{n-i}, \quad (2)$$

which is obtained by expanding $[q + (p + q)]^n$. Since $p + 2q = 1$ and since $q \leq p$, formula (2) assumes the maximum value when $p = q = ⅓$, that is, if the three species represent a true trichotomy (Fig. 12-1d).

Now suppose that we do not know which tree is the true tree. Suppose further that in a sequence data set with n informative sites, the best supported tree is favored by m sites. Is this tree the true tree? Let us assume that this tree is incorrect and has by chance been supported by so many sites. The probability for a particular incorrect tree to be supported by m or more sites is given by equation (2). Therefore, the probability for one of the two incorrect trees to be supported by m or more sites is

$$P = 2\,\text{Prob}(C \geq m). \quad (3)$$

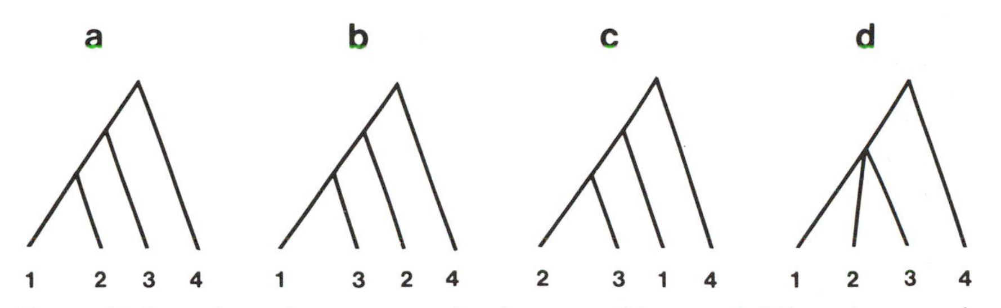

Figure 12-1 a, b, and c represent the three possible rooted, bifurcating trees for three taxa with one outgroup and d represents the case of a trichotomy. (From Li and Gouy, 1990.)

If P is smaller than α, then the support for the best tree is significant at the level of α.

In practice, p is unknown, and it is necessary to use the least favorable value, $1/3$. Table 12-1 shows the minimum m value required for equation (2) to be smaller than or equal to a given probability (δ), assuming that n informative sites are observed and that $q = p = 1/3$. It can be used to determine the significance level. For example, if $n = 10$ and $m = 8$, then the significance level is approximately 0.5% from Table 12-1 and $2 \times 0.5\% = 1\%$ from formula (3), and so the inferred tree is significant at the 1% level. If n is greater than 100, equation (2) can be computed by assuming that m/n follows a normal distribution with mean equal to $1/3$ and variance equal to $(1/3 \times 2/3)/n$; the approximation is very rough, however.

In addition to the above test, Felsenstein (1985) has also developed a statistic for testing whether the best supported tree is significantly better than the next best tree. Suppose that these two trees are supported by n_1 and n_2 sites, respectively. Let

$$S = n_1 - n_2.$$

Then the probability that the best supported tree is at least s steps better is given by

$$\text{Prob}\,(S \geq s) = \sum_{\substack{(i,j,k) \\ j - \max(i,k) \geq s \\ \text{or } k - \max(i,j) \geq s}} \text{Prob}(i,j,k). \tag{4}$$

If Prob $(S \geq s)$ is smaller than α, then the best supported tree is significantly better than the next best tree at the α level. Again we do not know the p value and so we accept the worst-case analysis assuming $p = q = r = 1/3$. Table 12-2 shows some results computed from this analysis. For example, if there are $n = 10$ informative sites, then the best supported tree must be at least five steps better than the next best tree in order to be significantly better at the 5% level.

The test given by formula (4) is computationally somewhat more complicated than that given by formula (3), but the powers of the two tests

Table 12-1 Minimum m Value Required for Prob($C \geq m$) in Equation (2) to be Smaller Than or Equal to δ*

	δ(%)						δ(%)				
n†	0.05	0.30	0.50	1.00	2.50	n	0.05	0.30	0.50	1.00	2.50
10	9	9	8	8	7	34	22	20	20	19	18
11	10	9	9	8	8	36	23	21	20	20	19
12	11	10	9	9	8	38	24	22	21	21	19
13	11	10	10	9	9	40	24	23	22	21	20
14	12	11	10	10	9	43	26	24	24	23	22
15	12	11	11	10	10	47	28	26	25	24	23
16	13	12	11	11	10	50	29	27	26	26	24
17	13	12	12	11	11	53	30	28	28	27	26
18	14	13	12	12	11	57	32	30	29	28	27
19	14	13	13	12	11	60	33	31	31	30	28
20	15	14	13	13	12	63	35	33	32	31	29
21	15	14	14	13	12	67	36	34	34	33	31
22	16	15	14	14	13	70	38	35	35	34	32
23	16	15	15	14	13	73	39	37	36	35	33
24	17	16	15	15	14	77	41	38	38	36	35
25	17	16	16	15	14	80	42	40	39	38	36
26	18	16	16	15	14	83	43	41	40	39	37
27	18	17	17	16	15	87	45	42	42	40	39
28	19	17	17	16	15	90	46	44	43	42	40
29	19	18	17	17	16	93	47	45	44	43	41
30	20	18	18	17	16	97	49	46	46	44	43
32	21	19	19	18	17	100	50	48	47	46	44

*It is assumed that $q = p = 1/3$.

†n = number of informative sites observed.

(Taken from Li and Gouy, 1990.)

Table 12-2 Difference (*S*) in the Number of Supporting Sites Required for Significance at the 5% Level

Sites	Informative (Clock)	All Sites
2	—	2
3	—	3
4	4	3
5	5	3
10	5	5
13	5	6
15	6	6
20	6	8
25	7	9
30	8	10
40	9	13
50	9	15
100	13	26
200	17	48
500	27	109
1,000	—	209
2,000	—	405
5,000	—	984
10,000	—	1,940

(From Felsenstein, 1988.)

are about the same (Felsenstein, 1985). This is not surprising, because if the best supported tree is significant, it is likely to be significantly better than the next best tree (and also the worst tree), and *vice versa*.

Although the above results were derived under the assumption that one of the four species is known to be an outgroup, they apply equally well in the absence of this knowledge if we consider unrooted trees and if the rate-constancy assumption holds.

Four Species Without a Molecular Clock

Although in the above example we have assumed a molecular clock, the analysis was made under the worst situation, namely that the probability for an informative site to support an incorrect tree is ⅓ rather than $< \frac{1}{3}$. Thus, the results are applicable even under nonconstant rates as long as the rates among the three lineages are not too different or the sequences have not diverged greatly so that the probability for an informative site to support an incorrect tree is $\leq \frac{1}{3}$. However, if this condition does not hold, then a different analysis is necessary. Cavender (1978) has carried out such an analysis, considering morphological characters with two states. Felsenstein (1983) extended the calculation to the case of four states, such as nucleotide sequences.

Cavender's formulation is as follows. Suppose that the tree in Figure 12-2 is the true tree. Assume that all characters are equivalent and can be characterized by two states, 0 and 1. Assume further that a change of a 0

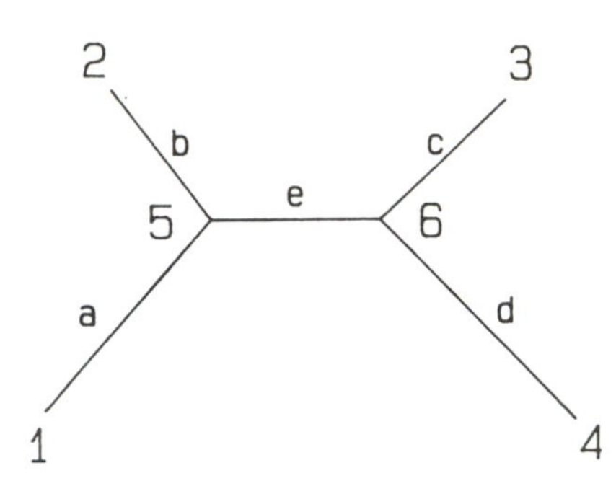

Figure 12-2 A hypothetical tree for four sequences.

to a 1 over the course of time may be followed by a change back to 0 and that a change one way is just as probable as a change the other way. In Figure 12-2, the branch length a denotes the probability that a character at vertex 1 differs in state from that at vertex 5; the other branch lengths are defined in the same manner. In the worst-case analysis, Cavender assumes that the rate of evolution is rapid in both lineages 1 and 4 so that their characters have been randomized, whereas the rates in the other lineages are extremely slow so that no change has occurred. That is, $b = c = e = 0$ and $a = d = 1/2$. If we assume that the state is 0 at vertices 2 and 3, then for the site to be informative, the state should be 1 at both vertices 1 and 4. This occurs with probability $1/2 \times 1/2 = 1/4$, and in this case the site supports a wrong tree. Thus, in the worst case, one-fourth of all characters can support an incorrect tree with none supporting the correct tree. For this reason one can conclude in favor of the most-parsimonious tree only if the evidence is stronger than that.

In the case of nucleotide sequences, with a simple symmetric model of base change, Felsenstein (1983) showed that $3/16$ of all characters can be "phylogenetically informative" and favor a wrong tree. Therefore, in this case one asks whether the proportion of sites supporting a tree is significantly greater than $3/16$ of all sites. This probability can be obtained by a binomial expansion of $(13/16 + 3/16)^n$. The third column of Table 12-2 gives the difference in steps required for significance at the 5% level.

More Than Four Species

In theory, the above analysis can be extended to the case of more than four species. However, the situation now becomes very complicated because the number of possible alternative trees increases rapidly with the number of species; for example, for five species there are 15 possible unrooted trees. Moreover, the above formulation is based on the worst-case analysis, so when the number of possible alternative trees is large, the test developed is likely to have very low power.

In view of the complexity of the problem, Kishino and Hasegawa (1989) developed a heuristic test in a manner similar to that of Templeton (1983). They considered whether two trees X and Y differ significantly in terms of the number of substitutions. Assume that all nucleotide sites are independent and identical. Let the minimum number of base substitutions at the i-th site for tree X be X_i and let

$$X = X_1 + X_2 + \dots\dots + X_n. \tag{5}$$

We define Y in the same manner. Let $Z = X - Y$. The mean of Z can be estimated by the inferred values of X_i and Y_i. The variance of Z can be estimated by the sample variance over sites, that is,

$$V(Z) = \frac{N}{N-1} \sum_{i=1}^{N} \left(Z_i - \frac{1}{N} \sum_{k=1}^{N} Z_k \right)^2, \tag{6}$$

where $Z_i = X_i - Y_i$ and N is the total number of sites on the sequence When N is large, $[Z - E(Z)]/V(Z)$ approaches the standard normal distribution and the standard normal test can be used to test the significance of Z. In practice it is difficult to know how large N should be for Z to be close to a normal distribution. We note that actually the number of informative sites is more relevant than the total number of sites.

It should also be noted that when one tree is significantly more parsimonious than another tree, it does not mean that it is the most parsimonious tree. To draw this conclusion, one must show that the tree is more parsimonious than all of the other possible trees. In practice, however, when the number of possible trees is large, we cannot compare all of them and so instead we usually compare only those that seem to be most plausible. In this respect, one should note that two trees usually differ only in a small part of the whole topology, and in comparing two trees, we implicitly assume that the other part of the tree topology is correct. Snce this assumption may not be true, it can complicate the test.

TESTS OF INTERNAL BRANCH LENGTHS

Instead of testing the significance of an inferred phylogeny, it is simpler to test the significance of estimated internodal distances. As examined below, in the case of four taxa significance of the internodal distance can be taken as significance of the inferred phylogeny. When the number of taxa under study is more than four, the two problems are no longer equivalent and the requirement of all internodal distances being significantly greater than 0 seems to be a too stringent test for the significance of the inferred branching order. A simple way to test the significance of internodal distances is to study their variances.

Variances of Internodal Distances

Mueller and Ayala (1982) proposed to compute these variances by the jackknife method, while Nei et al. (1985) derived analytic formulas for the case of a UPGMA tree. That is, a tree estimated by the unweighted pair-group method (UPGMA) (Sneath and Sokal, 1973). The UPGMA method assumes a constant rate of evolution, but there is now strong evidence that this assumption is often violated (e.g., Li et al., 1987a). It is therefore desirable to consider an approach that does not make this assumption. Li (1989) proposed a two-step approach. The first step is to infer the branching order. One can use the transformed distance method (Farris, 1977; Klotz et al., 1979; Li, 1981), the neighbor-joining method (Saitou and Nei, 1987), or any other method that does not assume rate-constancy and that has been shown to be effective for obtaining the correct tree. The second step is to estimate the branch lengths by the least-squares method (Cavalli-Sforza and Edwards, 1967; Chakraborty, 1977). The variances of internodal distances are then obtained from the equations derived from the least-squares method. The variances for the cases of four and five taxa are given below.

Four Taxa

Suppose that the inferred tree topology is as shown in Figure 12-3a. The branch lengths estimated by the least-squares method are

$$a = \tfrac{1}{2}d_{12} + \tfrac{1}{4}(d_{13} - d_{23} + d_{14} - d_{24}), \tag{7}$$

$$b = d_{12} - a, \tag{8}$$

$$c = \tfrac{1}{4}(d_{13} + d_{23} + d_{14} + d_{24}) - \tfrac{1}{2}(d_{12} + d_{34}), \tag{9}$$

$$d = \tfrac{1}{2}d_{34} + \tfrac{1}{4}(d_{13} + d_{23} - d_{14} - d_{24}), \tag{10}$$

$$e = d_{34} - d, \tag{11}$$

and the variance of c is

$$\begin{aligned} V(c) = {} & 1/16\,[V(d_{13}) + V(d_{23}) + V(d_{14}) + V(d_{24}) \\ & + 2V(d_{16}) + 2V(d_{26}) + 4V(d_{56}) + 2V(d_{53}) + 2V(d_{54})] \\ & - \tfrac{1}{2}[V(d_{15}) + V(d_{25}) + V(d_{63}) + V(d_{64})] \\ & + \tfrac{1}{4}V(d_{12}) + \tfrac{1}{4}V(d_{34}), \end{aligned} \tag{12}$$

where $V(d_{ij})$ denotes the variance of the estimate of the distance between sequences i and j.

First, consider protein sequence data. The mean and variance of the number (d_{ij}) of amino acid replacements per site between sequences i and j can be estimated by

$$d_{ij} = -\,b\,\ell n(1 - p/b), \tag{13}$$

$$V(d_{ij}) = p(1 - p)/[N(1 - p/b)^2], \tag{14}$$

where $b = 19/20$, p is the proportion of different amino acids between the two sequences, and N is the number of residue sites compared.

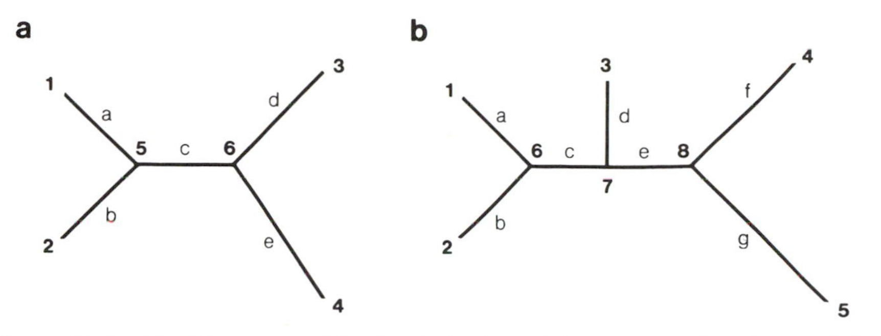

Figure 12-3 Model trees used in the derivation of the mean and variance of the branch lengths. (From Li, 1989.)

Next, consider nucleotide sequence data. Under the assumption of random substitution among the four types of nucleotides (i.e., the one-parameter model), the mean and variance of the number of substitutions per nucleotide site between sequences i and j are also given by (13) and (14), except that now $b = 3/4$, p is the proportion of different nucleotides between the two sequences, and N is the number of nucleotide sites compared (Jukes and Cantor, 1969; Kimura and Ohta, 1972). Under the two-parameter model (Kimura, 1980), the formulas corresponding to (13) and (14) are

$$d_{ij} = A + B, \tag{15}$$

$$V(d_{ij}) = [x^2P + z^2Q - (xP + zQ)^2]/N, \tag{16}$$

where P and Q are, respectively, the proportions of transitional and transversional differences between sequences i and j, $A = \frac{1}{2}\ell n(x) - \frac{1}{4}\ell n(y)$ is the number of transitional substitutions per site, $B = \frac{1}{2}\ell n(y)$ is the number of transversional substitutions per site, $x = 1 / (1 - 2P - Q)$, $y = 1 / (1 - 2Q)$, and $z = (x + y)/2$.

Five Taxa

Suppose that the inferred branching order is as shown in Figure 12-3b. The branch lengths are then given by

$$a = \tfrac{1}{2}d_{12} + 1/6\,[d_{13} - d_{23} + d_{14} - d_{24} + d_{15} - d_{25}], \tag{17}$$

$$b = d_{12} - a, \tag{18}$$

$$c = \tfrac{1}{4}(d_{13} + d_{23}) + 1/8\,(d_{14} + d_{24} + d_{15} + d_{25}) - \tfrac{1}{4}(d_{34} + d_{35}) - \tfrac{1}{2}d_{12}, \tag{19}$$

$$d = \tfrac{1}{2}(d_{13} + d_{23}) - \tfrac{1}{2}d_{12} - c, \tag{20}$$

$$e = \tfrac{1}{4}(d_{34} + d_{35} - d_{13} - d_{23}) + 1/8\,(d_{14} + d_{24} + d_{15} + d_{25}) - \tfrac{1}{2}d_{45}, \tag{21}$$

$$f = d_{34} - d - e, \tag{22}$$

$$g = d_{45} - f. \tag{23}$$

If two of the five taxa, say taxa 4 and 5, are known to be outgroups, then one needs to obtain only the variance of c.

$$\begin{aligned} V(c) = {} & 1/64\,[V(d_{14}) + V(d_{15}) + V(d_{24}) + V(d_{25})] \\ & + 1/32\,[V(d_{18}) + V(d_{28}) + V(d_{46}) + V(d_{56})] \\ & + 1/16\,[V(d_{13}) + V(d_{23}) + V(d_{34}) + V(d_{35}) \\ & + V(d_{68})] \\ & + 1/8\,[V(d_{17}) + V(d_{27}) + V(d_{36}) + V(d_{38}) \\ & - V(d_{47}) - V(d_{57})] \\ & + 1/4\,[V(d_{12}) + V(d_{67}) - V(d_{78})] \\ & - 1/2\,[V(d_{16}) + V(d_{26}) + V(d_{37})]. \end{aligned} \tag{24}$$

If only one or no outgroup is available, then one needs also to obtain the variance of e.

$$\begin{aligned} V(e) = {} & 1/64\,[V(d_{14}) + V(d_{15}) + V(d_{24}) + V(d_{25})] \\ & + 1/32\,[V(d_{18}) + V(d_{28}) + V(d_{46}) + V(d_{56})] \\ & + 1/16\,[V(d_{13}) + V(d_{23}) + V(d_{34}) + V(d_{35}) \\ & + V(d_{68})] + 1/8\,[- V(d_{17}) - V(d_{27}) + V(d_{36}) \\ & + V(d_{38}) + V(d_{47}) + V(d_{57})] + 1/4\,[V(d_{45}) \\ & - V(d_{67}) + V(d_{78})] - 1/2\,[V(d_{58}) \\ & + V(d_{48}) + V(d_{37})]. \end{aligned} \tag{25}$$

Computer programs for the above formulas are available upon request by sending an IBM PC-compatible floppy disk to the authors.

Test of Significance of an Inferred Phylogeny

In the case of three taxa with one or two outgroups, the above results can be used to test the significance of an inferred phylogeny. Since in this case there is only one internal branch (i.e., branch c), testing the significance of the internal branch is equivalent to testing the significance of the inferred phylogeny. More explicitly, the null hypothesis is that the true phylogeny is a trichotomy (i.e., the three taxa diverged at the same time). This hypothesis is the same as the hypothesis of $c = 0$. Therefore, if the estimated c is significantly greater than 0, the null hypothesis of a trichotomy is rejected and the inferred branching order can be taken as statistically significant. If one considers unrooted trees, the same argument applies to the case of four taxa with no outgroup, because in this case there is also only one internal branch (Fig. 12-3a).

The above formulas for the mean and variance of c were derived under the assumption that the inferred branching order of the three taxa was ((1, 2), 3), which means that lineage 3 branched off earlier than did lineages 1 and 2. If, instead, the inferred branching pattern is ((1, 3), 2), then the subscripts 2 and 3 in the above formulas should be exchanged, and if the inferred branching pattern is ((2, 3), 1), subscripts 1 and 3 should be exchanged. Under the null hypothesis of a trichotomy, the three branching patterns ((1, 2), 3), ((1, 3), 2), and ((2, 3), 1) occur with equal probability. However, in each data set only one pattern can occur and only one c can be positive and is tested for significant deviation from 0. Regardless of which pattern occurs, the probability for c to assume a particular (nonnegative) value is the same. If the distribution of c is the same as the distribution of $|x|$ where x is a standard normal random variate, then the standard statistical test based on the standard normal distribution can be applied. In particular, the estimated c is significantly greater than 0 at the 5% level if the ratio of mean to standard error is ≥ 2, and is significant at the 1% level if the ratio is ≥ 2.6. Obviously, the case of four taxa with no outgroup can be treated in the same manner, if one considers unrooted trees. A computer simulation (Li, 1989) for the case of three taxa with one outgroup shows that the significance level defined by the above criteria is generally applicable, though the distribution of c is not strictly normal.

When the number of taxa under study is more than four, the situation becomes complicated. For example, in the case of five taxa there are two internal branches (Fig. 12-3b), and the probability for (only) one of them to become by chance significantly greater than 0 at the level of $\alpha = 5\%$ is $2\alpha = 10\%$. Thus, in this case one cannot reject the null hypothesis that all the internal branches have 0 length (i.e., all the taxa diverged at the same time point and form a "star" phylogeny); of course, this null hypothesis can be rejected if $\alpha \leq 2.5\%$. On the other hand, the probability for both internal branches to be by chance significantly greater than 0 at the level of $\alpha = 5\%$ is only approximately $\alpha^2 = 0.0025$ (it is approximate because the two internodal distances are not estimated independently). Hence the requirement of all internal branches being significant seems to be a too stringent test for the significance of the inferred topology. However, one cannot draw a conclusion about the significance of an inferred tree topology as long as one or more of the internodal distances are nonsignificant; of course, the uncertainty can be restricted to a subset of taxa.

Numerical Examples

The following examples assume that the rate of nucleotide substitution is constant over time and that the observed number of substitutions between each pair of sequences is equal to the expected value.

Table 12-3 shows the case of three species with an outgroup (taxon 4) (Fig. 12-3a). In the table, α denotes the proportion of transitional changes; $\alpha = 1/3$ if substitutions occur randomly. The standard error (SE), which is the square root of $V(c)$, is larger for $\alpha = 2/3$ than for $\alpha = 1/3$. Since transitional changes generally occur more often than transversional changes (Brown et al., 1982; Li et al., 1984) the two-parameter model is more realistic than the one-parameter model; the former is applicable to all α values whereas the latter only to $\alpha = 1/3$.

The ratio c/SE can be used to test whether c is significantly greater than 0. A ratio of 2 can be taken as significant at the 5% level. All the values in Table 12-3 were obtained for $N = 1{,}000$. When $c = 0.01$, the ratio is 2 or larger if $\alpha \leq 2/3$. Thus, this case requires only a small amount of sequence data to resolve the branching order of the three species. When c is 0.005, then the ratio is considerably smaller than 2 (e.g., the ratio is 1.28 for $\alpha = 1/3$). Formulas (14) and (16) imply that $V(c)$ is inversely proportional to N. Therefore, for the ratio to increase from 1.28 to 2, the N value should increase from 1,000 to $N' = 1{,}000 \times (2/1.28)^2 = 2{,}500$ approximately. The other N' values in Table 12-3 were obtained in the same manner. If c is 0.001, then the number of nucleotide sites needed to be studied is rather large, greater than 50,000. Saitou and Nei (1986) have considered this problem from a different angle. They studied the probability of obtaining the correct topology as a function of the number of nucleotides studied under various tree-making methods.

Table 12-4 shows the amount of reduction in $V(c)$ when a second outgroup (taxon 5) is added (Fig. 12-3b); the $V(c)$ value for the case of one outgroup is denoted by $V_1(c)$ and that for the case of two outgroups by

Table 12-3 Standard Error (SE) of the Estimate of the Length of Branch c in Figure 12-3a*

c	α†	SE‡	c/SE	L'§
0.010	⅓	0.0046	2.17	850
	⅔	0.0050	2.00	1,000
0.005	⅓	0.0039	1.28	2,500
	⅔	0.0043	1.16	3,000
0.001	⅓	0.0032	0.31	41,000
	⅔	0.0037	0.27	54,000

*The branch lengths in Figure 12-3a are $a = b = 0.05$, $d = 0.05 + c$, $e = 0.15 - c$.

†α = proportion of transitional substitutions.

‡SE was computed under the assumption that the number of nucleotide sites (L) studied is 1,000.

§L' is the number of nucleotide sites required for the ratio c/SE to be 2 or larger (i.e., to be approximately 5% significant).

(Taken from Li, 1989.)

$V_2(c)$. The reduction increases as $V_1(c)$ becomes larger. Since $V(c)$ is inversely proportional to N, a reduction in $V(c)$ can also be achieved by increasing N. Is it more advantageous to increase N or to add a second outgroup? The total number of nucleotides sequenced is $4N$ for the case of one outgroup and $5N$ for the case of two outgroups, the latter being 1.25 times the former. Therefore, if the same total number of nucleotides are to be sequenced, it is less advantageous to add a second outgroup than to increase N if $V_1(c)/V_2(c)$ is smaller than 1.25, whereas the reverse is true if the ratio is larger than 1.25. In Table 12-4 the ratio is smaller or equal to 1.25 for the first six cases and is larger than 1.25 for the last six cases. Since the ratio tends to increase with $V_1(c)$, in general it is more advantageous to increase N if $V_1(c)$ is relatively small, but more advantageous to add a second outgroup if $V_1(c)$ is relatively large. In all the cases in Table 12-4, the distances from sequences 4 and 5 to the other three are the same (i.e., $g = f$ in Fig. 12-3b), so that the fifth sequence is as good a reference as the fourth one. If the fifth is more distantly related to the other three than the fourth sequence is, then the reduction in $V(c)$ is expected to be smaller than those shown in Table 12-4. Further, the effect will also be reduced if sequences 4 and 5 are closely related to each other.

NEIGHBORS RELATION TESTS

Let us first define the concept of "neighbors." In an unrooted tree, two operational taxonomic units (OTUs) are said to be neighbors if they are connected through a single internal node. For example, in Figure 12-4, OTUs 1 and 2 are a pair of neighbors, and OTUs 5 and 6 are another pair. In comparison, neither OTUs 1 and 3 nor OTUs 2 and 3 are neighbors. However, if we combine OTUs 1 and 2 into a composite OTU, then the combined OTU (12) and OTU 3 become a new pair of neighbors.

We explain below how the neighbors relation concept can be used to reconstruct phylogenetic trees. We then propose methods for testing the significance of neighbor pairs.

Four Taxa

Assume that the tree shown in Figure 12-3a is the true tree. The branch lengths are said to be additive, if the following equalities hold.

$$d_{13} + d_{24} = d_{14} + d_{23} = a + b + 2c + d + e$$

$$= d_{12} + d_{34} + 2c, \quad (26)$$

where d_{ij} is the distance (number of substitutions) between OTUs i and j estimated from sequence data while a, b, c, d, and e are the inferred branch lengths. Therefore, under additivity the following two conditions hold:

$$d_{12} + d_{34} < d_{13} + d_{24} \quad (27a)$$

$$d_{12} + d_{34} < d_{14} + d_{23}. \quad (27b)$$

Table 12-4 Variance ($V(c)$) of the Estimate of the Length of Branch c in Figure 12-3b

$a=b$	d	$f=g$	e	c	α*	$V_1(c)$† ($\times 10^{-4}$)	$V_2(c)$ ($\times 10^{-4}$)	$V_1(c)/V_2(c)$
0.020	0.025	0.050	0.025	0.005	⅓	0.072	0.066	1.09
					⅔	0.079	0.071	1.11
			0.029	0.001	⅓	0.028	0.023	1.22
					⅔	0.033	0.027	1.22
0.050	0.055	0.100	0.045	0.005	⅓	0.153	0.126	1.21
					⅔	0.188	0.150	1.25
			0.049	0.001	⅓	0.104	0.079	1.32
					⅔	0.135	0.100	1.35
0.100	0.110	0.150	0.040	0.010	⅓	0.445	0.351	1.27
					⅔	0.577	0.444	1.30
			0.045	0.005	⅓	0.376	0.287	1.31
					⅔	0.501	0.374	1.34

*α = proportion of transitional substitutions.

†$V_1(c)$ was obtained under the assumption that the second outgroup (taxon 5 in Fig. 12-3b) is not available.

(Taken from Li, 1989.)

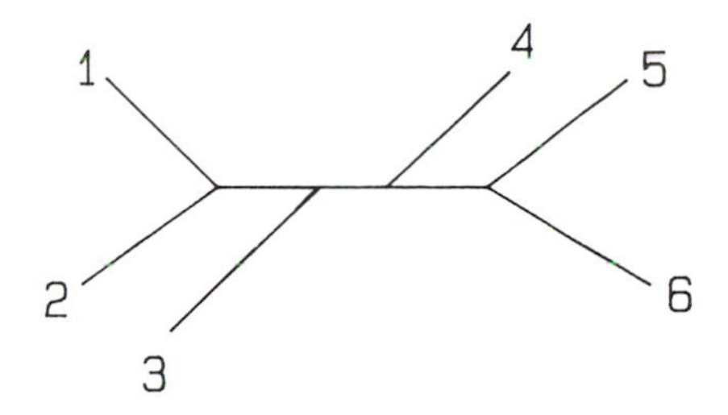

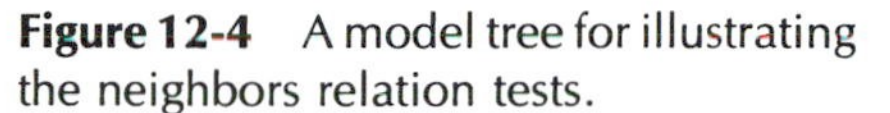
Figure 12-4 A model tree for illustrating the neighbors relation tests.

These two conditions are known as "the four-point condition" (Buneman, 1971). They may hold even if the additivity holds only approximately.

Conversely, for four OTUs with unknown relationships, if the above two conditions hold, then we may assume that OTUs 1 and 2 are neighbors, and so are OTUs 3 and 4. The question is then whether the topology inferred is statistically significant.

A simple way to test the significance of the topology is to test whether both x and y are significantly greater than 0, where

$$x = (d_{13} + d_{24}) - (d_{12} + d_{34}) \tag{28a}$$

$$y = (d_{14} + d_{23}) - (d_{12} + d_{34}). \tag{28b}$$

However, our computer simulation shows that this test is too stringent, and a more reasonable test is to consider whether $z = x + y$ is significantly greater than 0.

If the degree of sequence divergence is large, one should use the corrected number of nucleotide substitutions per site for d_{ij} (Saitou and Imanishi, 1989; Jin and Nei, 1990). In this case, one can compute the variance of z and use it to test the significance of z in a manner similar to that in the preceding section.

However, if the degree of sequence divergence is not large, say 50% or less, then it is slightly better to use the observed proportion of differences for d_{ij} (Saitou and Nei, 1987). Assume that all nucleotide sites are identical and independent. Let z_k be the z value for the k-th site and $Z = \Sigma z_k$; the d_{ij} values used are the observed numbers of differences between sequences i and j at site k. Then the variance of $z = Z/N$ can be estimated by

$$V(z) = V(Z)/N^2 \approx \frac{1}{N^2} \sum_{k=1}^{N} (z_k - \bar{z})^2 \tag{29}$$

where $\bar{z} = \Sigma z_k/N$, and N is the number of nucleotide sites on the sequence. This formulation is computationally simple and can be easily extended to the case of many OTUs (see below).

Instead of formula (29), the variance of Z can be computed as follows. Since $Z = d_{13} + d_{24} + d_{14} + d_{23} - 2(d_{12} + d_{34})$,

$$\begin{aligned} V(Z) = {} & V(d_{13}) + V(d_{24}) + V(d_{14}) + V(d_{23}) + 4V(d_{12}) \\ & + 4V(d_{34}) + 2\mathrm{Cov}(d_{13},d_{24}) + 2\mathrm{Cov}(d_{13},d_{14}) \\ & + 2\mathrm{Cov}(d_{13},d_{23}) - 4\mathrm{Cov}(d_{13},d_{12}) - 4\mathrm{Cov}(d_{13},d_{34}) \\ & + 2\mathrm{Cov}(d_{24},d_{14}) + 2\mathrm{Cov}(d_{24},d_{23}) - 4\mathrm{Cov}(d_{24},d_{12}) \\ & - 4\mathrm{Cov}(d_{24},d_{34}) + 2\mathrm{Cov}(d_{14},d_{23}) - 4\mathrm{Cov}(d_{14},d_{12}) \\ & - 4\mathrm{Cov}(d_{14},d_{34}) - 4\mathrm{Cov}(d_{23},d_{12}) - 4\mathrm{Cov}(d_{23},d_{34}) \\ & + 8\mathrm{Cov}(d_{12},d_{34}). \end{aligned} \tag{30}$$

Let p_{ij} be the proportion of different nucleotides between sequences i and j. Then $V(d_{ij}) = p_{ij}(1 - p_{ij})/N$, which is the binomial variance. For the covariances we note that (with m and n referring to the other two sequences)

$$\begin{aligned} \mathrm{Cov}(d_{ij},d_{mn}) &= E(d_{ij}d_{mn}) - E(d_{ij})E(d_{mn}) \\ &= E(d_{ij}d_{mn}) - p_{ij}p_{mn}. \end{aligned}$$

The term $E(d_{ij}d_{mn})$ can be estimated by computing the $d_{ij}d_{mn}$ value for each nucleotide site and then taking an average over the sites in the sequence, as in the computation of equation (29).

In the case of five taxa with taxa 4 and 5 as known outgroups, we have

$$x = [d_{13} + d_{2(45)}] - [d_{12} + d_{3(45)}], \tag{31a}$$

$$y = [d_{23} + d_{1(45)}] - [d_{12} + d_{3(45)}], \tag{31b}$$

where $d_{i(jk)} = (d_{ij} + d_{ik})/2$. The variance of $z = x + y$ can be obtained as above, though more tediously.

More Than Four Taxa

Sattath and Tversky (1977) proposed the following method of phylogenetic reconstruction for the case of more than four OTUs (see also Fitch, 1981). First, compute a distance matrix. For each OTU pair (ij), examine all other pairs (mn) and count the number of quadruples in which the four-point condition holds (i.e., $d_{ij} + d_{mn}$ is smaller than both $d_{im} + d_{jn}$ and $d_{in} + d_{jm}$). The pair with the highest score is selected and considered as a single (composite) OTU. Second, compute a new distance matrix and search for the next neighbor pair following the above procedure. The process is repeated until all OTUs are clustered. In this method, the distance between two OTU clusters is computed by the arithmetic mean of all the pairwise distances involved. For example, the distance between clusters (*ijk*) and (*mn*) is defined as

$$d_{(ijk)(mn)} = (d_{im} + d_{in} + d_{jm} + d_{jn} + d_{km} + d_{kn})/6.$$

After the topology is inferred by the above procedure, we propose to test its significance by testing whether all the subsets in the tree are significantly supported. Suppose that the inferred tree is that shown in Figure 12-4. We may first test the significance of the subset (12) by testing whether

$$z = d_{13} + d_{2(456)} + d_{23} + d_{1(456)} - 2[d_{12} + d_{3(456)}] \quad (32)$$

is significantly greater than 0. We can test the significance of the subset (56) in the same manner. Then there are only four OTUs left: (12), 3, 4, and (56). Since the hypothesis is that (12) and 3 form one pair of neighbors, and 4 and (56) form another pair, we test whether

$$z = [d_{(12)(56)} + d_{34}] + [d_{(12)4} + d_{3(56)}] - 2[d_{(12)3} + d_{4(56)}] \quad (33)$$

is significantly greater than 0.

Another way of testing the significance of a subset is as follows. For every possible pair of OTUs i and j in the subset, and for every possible pair of OTUs m and n not in the subset, we test whether

$$z = d_{im} + d_{jn} + d_{in} + d_{jm} - 2[d_{ij} + d_{mn}] \quad (34)$$

is significantly greater than 0. If every test is significant, then the subset is taken to be significant. This test is more stringent than the preceding test.

Some of the above formulas look complicated, but computer computation is straightforward. Computer programs are available from the authors.

LIKELIHOOD TESTS

The first application of the maximum likelihood (ML) method to tree reconstruction was made by Cavalli-Sforza and Edwards (1967), who used gene-frequency data. Later, Felsenstein (1973, 1981) developed ML algorithms for amino acid or nucleotide sequence data (see also Neyman, 1971; Kashyap and Subas, 1974). In the past, however, the ML method has not been used frequently, largely because of computational difficulties. Fortunately, computer computation is now much faster, and in recent years a number of authors have become interested in this method (Hasegawa et al., 1985, 1987; Felsenstein, 1987, 1988; Saitou, 1988, 1990; Smouse and Li, 1987; Kishino and Hasegawa, 1989, and references therein).

Likelihood Function

To use the ML method, one must first have a probabilistic model for the process of nucleotide substitution. That is, one must specify the transition

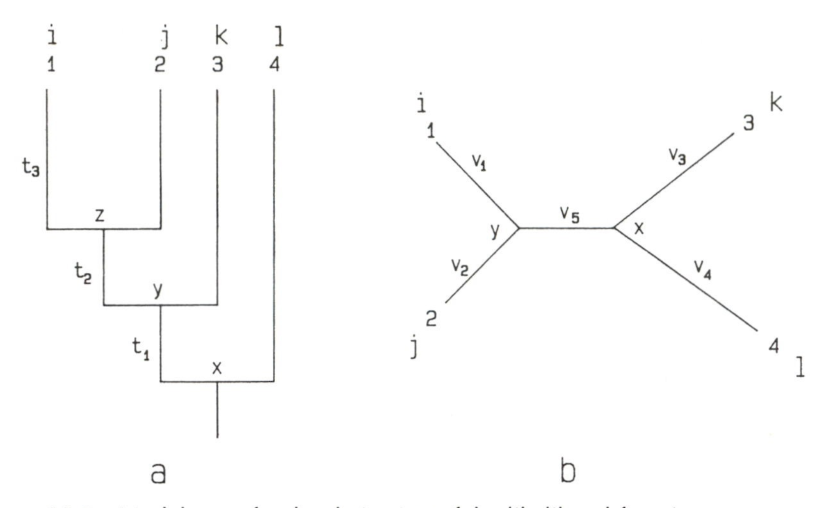

Figure 12-5 Model trees for the derivation of the likelihood function.

probability from one nucleotide state to another in a time interval. For example, in the one-parameter (or Jukes-Cantor) model, if the rate of substitution is λ per site per unit time, then the probability of changing from nucleotide i to nucleotide j per unit time is $\lambda/3$. Furthermore, given nucleotide i at time 0, the probability that the nucleotide at time t is also i is given by

$$P_{ii}(t) = \frac{1}{4} + \frac{3}{4}e^{-4\lambda t/3} \tag{35}$$

and the probability that the nucleotide at time t is j, $j \neq i$, is given by

$$P_{ij}(t) = \frac{1}{4} - \frac{1}{4}e^{-4\lambda t/3}. \tag{36}$$

In general, we can write

$$\mathrm{P(t)} = e^{t\mathrm{R}}, \tag{37}$$

where P(t) is the matrix of transition probabilities $P_{ij}(t)$ and the matrix R consists of the rate of change r_{ij} from nucleotide i to nucleotide j per unit time; the diagonal element r_{ii} is defined so that the sum of the elements in each row is equal to 0.

The next step is to set up the likelihood function. Let us use the case of four sequences (taxa) as an example. For the hypothetical tree given in Figure 12-5a, the likelihood function for a nucleotide site with nucleotides i, j, k, and l in sequences 1, 2, 3, and 4, respectively, can be computed as follows. Assume a constant rate of substitution. If the nucleotide at the ancestral node (the root) was x, the probability of having nucleotide l in sequence 4 is $P_{xl}(t_1 + t_2 + t_3)$, the probability of having nucleotide y at the common ancestral node of sequences 1, 2, and 3 is $P_{xy}(t_1)$, and so on.

Therefore, given x, y, and z at the ancestral node and the two other interior nodes, the probability of observing i, j, k, and l at the tips is equal to

$$P_{x\ell}(t_1 + t_2 + t_3)\, P_{xy}(t_1)\, P_{yk}(t_2 + t_3)\, P_{yz}(t_2)\, P_{zi}(t_3)\, P_{zj}(t_3).$$

All the above transition probabilities can be computed by using equations (35) and (36). Since in practice we do not know the ancestral nucleotide, we can only assign a probability g_x, which is usually assumed to be the frequency of nucleotide x in the sequence. Noting that x, y, and z can be any of the four nucleotides, we sum over all possibilities and obtain the following likelihood function.

$$h(i, j, k, l) = \sum_x g_x P_{x\ell}(t_1 + t_2 + t_3) \sum_y P_{xy}(t_1)\, P_{yk} (t_2 + t_3) \sum_z P_{yz}(t_2)\, P_{zi}(t_3)\, P_{zj}(t_3). \quad (38)$$

This equation is the same as equation (14) of Saitou (1988), except that the term $P_{yz}(t_2)$ is now placed inside rather than outside the summation over z (i.e., the last summation).

A formulation without the assumption of rate-constancy, though more tedious, can be done in the same manner. In this case, it is usually more convenient to consider the transition probability in terms of the branch length; for example, we consider $P_{ij}(v_\alpha)$ instead of $P_{ij}(t_\alpha)$, where $v_\alpha = \lambda_\alpha t_\alpha$ and λ_α and t_α are the values of λ and t for the α-th branch.

For an unrooted tree, we usually do not assume rate-constancy. For the unrooted tree given in Figure 12-5b, we obtain the following likelihood function:

$$h(i,j,k,\ell) = \sum_x g_x P_{x\ell}(v_4)\, P_{xk}(v_3) \sum_y P_{xy}(v_5)\, P_{yi}(v_1)\, P_{yj}(v_2), \quad (39)$$

if we assume that the interior node connecting taxa 3 and 4 is the ancestral node (Felsenstein, 1981; Saitou, 1988). If the process is time-reversible so that the transition probability from i to j in a time interval is the same as that from j to i (i.e., $P_{ij}(t) = P_{ji}(t)$, which is the case for the one-parameter model and many other current models of nucleotide substitution), then any node or point in the tree can be taken as the "ancestral" node. This is the so-called pulley-principle (Felsenstein, 1981). However, if the process is not time-reversible, then we must assume or infer the root of the tree.

In the above we considered a single site. The likelihood for all sites is the product of the likelihoods for individual sites if we assume that all the nucleotide sites involved are independent. Suppose that there are s homologous sequences each with N nucleotides. We can represent the data by a matrix $\mathbf{X}$, in which the element X_{ij} is the nucleotide at the j-th site in the i-th sequence. Let $\mathbf{X}_k$ be the vector $(X_{1k}, \ldots\ldots, X_{sk})$ (i.e., the transpose of the k-th column of the matrix $\mathbf{X}$). Then for a given phylogenetic tree, the likelihood for the entire sequence is

$$L(\theta|\mathbf{X}) = \prod_{k=1}^{N} f(\mathbf{X}_k|\theta), \quad (40)$$

where $f(\mathbf{x}|\theta) = f(x_1, x_2, \ldots\ldots, x_s|\theta)$ is the probability that at the nucleotide site under study, sequence i has the nucleotide x_i, and the vector θ denotes the unknown parameters such as the branching dates and the rates of nucleotide substitution along the branches of a tree (Kishino and Hasegawa, 1989).

Under the assumption that all sites are independent and equivalent, the likelihood function can also be derived by considering the nucleotide configurations among the sequences (Saitou, 1988). A nucleotide configuration at a site is defined as the pattern of nucleotide difference at that site among the sequences involved. Let us consider the case of four sequences. As shown in Table 12-5 there are 15 configuration patterns. The first pattern (i, i, i, i) means that the nucleotide is the same in all four sequences, the second pattern (i, i, i, j) means that the nucleotide is the same for the first three sequences, but is different for the fourth sequence, and so on. For the unrooted tree given in Figure 12-5b, and under the one-parameter substitution model, we can show that the probability of obtaining the first pattern is equal to

$$U_1 = 4\, h(i, i, i, i),$$

where $h(i, i, i, i)$ is given by equation (39), and the factor 4 arises because there are four possible nucleotides. The probability of obtaining the second pattern is equal to

$$U_2 = 12\, h(i, i, i, j),$$

in which the factor 12 arises because there are four possibilities for i, and for each i there are three possibilities for j. In the same manner, we can obtain the probabilities for the other patterns in Table 12-5 (Saitou, 1988).

Table 12-5 Nucleotide Configurations for Four Sequences

	Configuration*			
Number	A	B	C	D
1	*i*	*i*	*i*	*i*
2	*i*	*i*	*i*	*j*
3	*i*	*i*	*j*	*i*
4	*i*	*j*	*i*	*i*
5	*j*	*i*	*i*	*i*
6	*i*	*i*	*j*	*j*
7	*i*	*j*	*i*	*j*
8	*i*	*j*	*j*	*i*
9	*i*	*i*	*j*	*k*
10	*i*	*j*	*i*	*k*
11	*j*	*i*	*i*	*k*
12	*j*	*k*	*i*	*i*
13	*j*	*i*	*k*	*i*
14	*i*	*j*	*k*	*i*
15	*i*	*j*	*k*	*l*

*A, B, C, and D are sequences, and i, j, k, and l are different nucleotides.
(From Saitou, 1988.)

The likelihood function, L_j, for the j-th topology is

$$L_j = C \prod_{i=1}^{15} U_{ij}^{N_i}, \tag{41}$$

where U_{ij} is the probability of obtaining the i-th pattern for the j-th topology, N_i is the observed number of the i-th pattern, $C = N!/(N_1!\ N_2!\ \ldots\ldots N_{15}!)$ and N is the total number of nucleotide sites examined and is equal to the sum of N_i's.

Statistical Tests

After the likelihood function L is derived under a topology, one can search for the t_α or v_α values that maximize L. In the tree reconstruction procedure proposed by Felsenstein (1981), one computes the ML value for all or many topologies and chooses the one with the highest ML value as the most likely candidate for the true tree. It has been noted, however, that the likelihood function varies from topology to topology so that the ML values for different topologies are conditional and cannot be compared in the traditional statistical sense (Nei, 1987; Smouse and Li, 1987; Saitou, 1988). That is, the ML values for different topologies would be equivalent to the ML values computed under different hypotheses and thus cannot be compared in the traditional sense. However, Kishino and Hasegawa (1989) suggest that the comparison is meaningful in view of Akaike's (1974, 1985) theory. Akaike has shown that an information criterion (AIC) defined as

$$\text{AIC} = -2\ \ell n(\text{maximum likelihood}) + 2(\text{number of parameters}) \tag{42}$$

can be used as a measure of *badness* of a model. This theory is based on the Kullback-Leibler information, and two models can be compared even if they are non-nested; the one with a lower AIC is preferred (Akaike, 1974, 1985). Unfortunately, although Akaike's theory provides a logical basis for using the ML value as a criterion for choosing the best tree topology, in terms of statistical testing the theory does not remove all logical problems involved in non-nested models (see below).

In the case of four taxa, the situation is relatively simple. There are only three possible unrooted trees (Fig. 12-6A). Since each tree has only one internal branch, we may test its significance by testing the significance of the internal branch length. We use tree 1 as an example. We note that if the central branch is assumed to be 0, then tree 1 reduces to the starlike tree in Figure 12-6B. Thus, testing the significance of the central branch in tree 1 is equivalent to testing whether tree 1 is significantly better than the starlike tree. Since the latter is a submodel of the former, the test can be done in the traditional sense. The former model has one more parameter than does the latter, and so twice the logarithm of the ratio of the ML values for the two models is asymptotically distributed as a χ^2 with one degree of freedom (see Smouse and Li, 1987; Felsenstein, 1988). Strictly

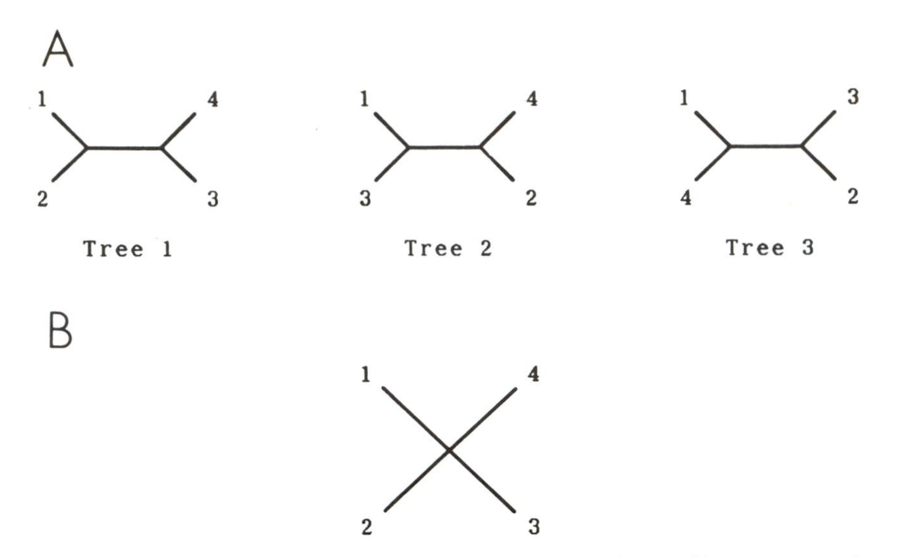

Figure 12-6 (A) Three possible topologies for an unrooted tree of four taxa. (B) The starlike tree for four taxa. (From Saitou, 1988.)

speaking, the starlike model is not completely inside tree 1, but is only on the boundary of the probability space. However, since the likelihood function is continuous at the boundary, this probably does not create a problem (E. Thompson, cited in Felsenstein, 1988). Usually, only one of the three trees in Figure 12-6A will have an ML value greater than the ML value for the starlike tree, and the other two trees assume the ML value when their internal branch is 0 (Saitou, 1988). However, in some cases more than one tree may have an ML value larger than that for the starlike tree. When such a situation occurs, we may follow Felsenstein's (1988) suggestion to test whether the difference in the ML values between the two trees is significant. Here a logical problem arises because the two ML values are estimated with the same number of parameters (and under different hypotheses), so there is no degree of freedom left for statistical testing. Felsenstein suggests to assume one degree of freedom and that twice the logarithm of the ratio of their ML values follows the χ^2 distribution. This test tends to be more stringent than the above test.

When more than four taxa are involved, the situation becomes complicated. In this case, testing of the significance of all internal branches tends to be more stringent than testing the significance of an inferred phylogeny, but is simpler and may be the only feasible way. Because of the large amount of computation involved, we consider only the ML tree (i.e., the tree with the highest ML value among all possible trees). Actually, even searching for this tree is extremely tedious. Therefore, Felsenstein (1981) and Saitou (1988) have proposed step-by-step search methods. Saitou's method for the case of five taxa is illustrated in Figure 12-7. At level I, one considers a starlike tree and computes the ML value. At level II, one

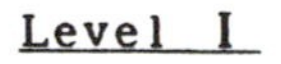

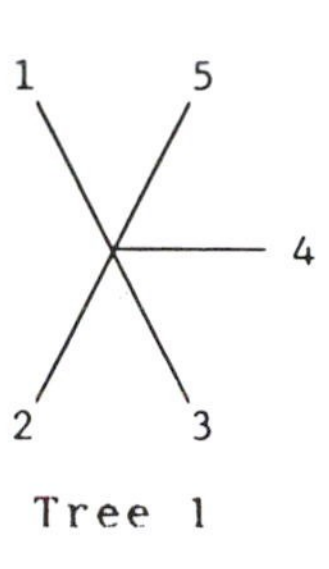

Level II

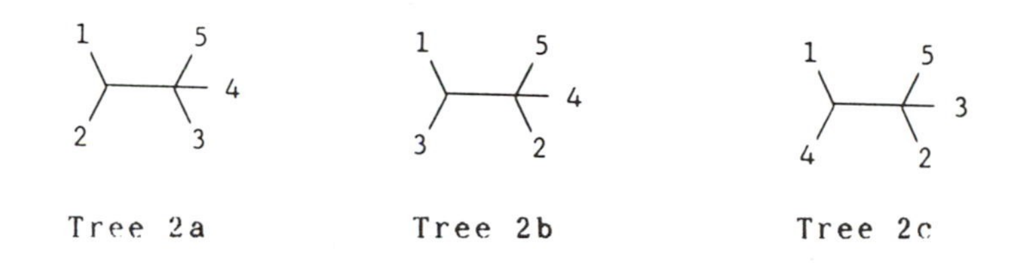

Level III

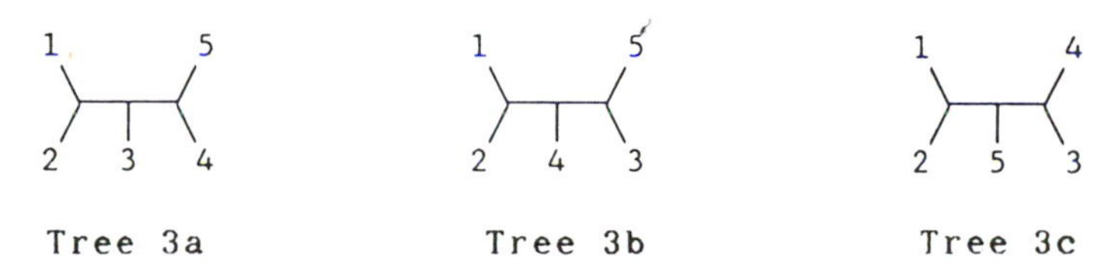

Figure 12-7 Three levels of unrooted trees for five taxa. (From Saitou, 1988.)

considers trees with a bifurcating node and a trifurcating node. There are ten such trees, three of which are shown in Figure 12-7. One computes the ML values for all ten trees and chooses the one with the highest ML value. Suppose it is tree 2a. This tree can be resolved into three bifurcating trees, trees 3a, 3b, and 3c (Fig. 12-7). One then computes the ML values for all three trees and chooses the tree with the highest ML value as the final tree. This search is not exhaustive, but is likely to lead to the ML tree. After the ML tree is inferred, we can test the significance of each internal branch. For example, to test the significance of the second internal branch of tree 3a, we note that when this branch becomes 0, tree 3a reduces to tree 2a. Since tree 2a is a submodel of tree 3a, we can test whether tree 3a is significantly better than tree 2a by the χ^2 test described in the preceding paragraph.

Instead of testing whether a particular tree is significant, we may test some plausible alternative trees against each other. Two approaches have been proposed. First, Felsenstein (1987, 1988) suggests in the case where the two topologies are adjacent (i.e., they each have a branch which, if its length is shrunk to 0, results in the same trifurcation) that we could test one topology against the other by pretending that there is one degree of freedom. For example, trees 3a and 3b in Figure 12-7 are two adjacent trees because they both reduce to tree 2a when the second internal branch is shrunk to 0. Three situations can occur in testing tree 3a against tree 3b. The first situation is that both trees assume an ML value when the length of the second internal branch is 0. In this case, both trees are not plausible. The second situation is that one tree, say tree 3a, assumes an ML value when the second internal branch is greater than 0, while tree 3b assumes an ML value when the second internal branch is 0. In this case, testing tree 3a against tree 3b is equivalent to testing the significance of the second internal branch of tree 3a or testing tree 3a against tree 2b. The third situation is that both trees 3a and 3b assume their ML value when the second internal branch is greater than 0. In this case, the testing is more stringent than testing the significance of the second internal branch in either tree 3a or 3b. The third situation should arise rarely, because at least one of the two trees is erroneous and would have a low probability of having a positive second internal branch. As noted above, Felsenstein's approach has a logical problem because the ML values for both trees are estimated with the same number of parameters and under different hypotheses. However, the test tends to be conservative.

Second, Kishino and Hasegawa (1989) propose to test one tree against another by computing the mean and variance of the difference in the ML value between two trees. Assume that all nucleotide sites in the sequence are independent and equivalent. Consider trees 1 and 2. Let $L_i(\theta_i|X)$ be the likelihood function of tree i, and $\ell(\hat{\theta}_i|X)$ be the logarithm of the ML value for tree i, where $\hat{\theta}$ is the maximum likelihood estimator of θ_i. Then the difference in mean is given by $\ell_1(\hat{\theta}_1|X) - \ell_2(\hat{\theta}_2|X)$. Kishino and Hasegawa (1989) propose to estimate the variance by the sample variance, that is,

$$V[\ell_1(\hat{\theta}_1|X) - \ell_2(\hat{\theta}_2|X)]$$
$$= \frac{N}{N-1}\sum_{k=1}^{N}\left(\ell n\frac{f_{(2)}(X_k|\hat{\theta}_2)}{f_{(1)}(X_k|\hat{\theta}_1)} - \frac{1}{N}\sum_{j=1}^{N}\ell n\frac{f_{(2)}(X_j|\hat{\theta}_2)}{f_{(1)}(X_j|\hat{\theta}_1)}\right)^2$$

where $f_i(X_j|\hat{\theta}_i)$ is given by equation (40). If the sample size N is large, then the standard normal test may be used. Note that like Felsenstein's approach, this one is also heuristic because the two likelihood functions are not constructed from the same topology (i.e., are not from the same probability space). However, an application of this method to the DNA sequence data from apes and man seems to give reasonable results (Kishino and Hasegawa, 1989).

Finally, it should be noted that all the above tests are based on the large-sample theory and so a large amount of data is required.

DISCUSSION

Problem of Statistical Inconsistency

In phylogenetic reconstruction, as well as in any other statistical inference problems, it is important for a method to be statistically consistent in the sense that the probability for the method to give the correct tree increases with the amount of data. When the degree of sequence divergence is small or moderate (e.g., 50% or lower), the maximum parsimony method is likely to be consistent. However, when the degree of divergence is high, and when the assumption of rate-constancy is violated, it may not be consistent (Felsenstein, 1978; Li et al., 1987b). It should be noted that a statistical test does not correct the inconsistency problem inherent in the tree-reconstruction method used, and so a strongly supported tree may in fact be an erroneous one. The evolutionary parsimony (EP) method was intended to overcome the effects of unequal rates of evolution. However, the EP method can also become statistically inconsistent if the transversional rates are unequal (Jin and Nei, 1990). Indeed, in an application of this method to the small-subunit rRNA sequences from human, *Drosophila*, rice, and *Physarum*, Gouy and Li (1989) obtained a χ^2 of 4.8 for the tree with human and rice in a clade, the χ^2's for the other two trees being 0.6 and 0.3. The favored tree is obviously erroneous as humans are much more closely related to *Drosophila* than to rice.

Heterogeneous Data

Phylogenetic studies often use sequence data from different DNA regions. If all the regions studied have similar rates of nucleotide substitution, then all the data can be combined together into a single set. However, if considerable variation in rates exist, regions with different rates should be treated separately, particularly when the distance matrix approach is used, because the internodal distances are dependent on the rate of evolution. The question then arises as to how to test the significance when the results from different data sets are combined. A simple test procedure is the inverse χ^2 method (Fisher, 1932). Suppose that there are k different data sets. Let p_i be the significance level (probability) estimated from the i-th data set. If the null hypothesis is true, then $-2\ell n(p_i)$ has a χ^2 distribution with two degrees of freedom and

$$P = -2 \sum_{i=1}^{k} \ell n\,(p_i)$$

has a χ^2 distribution with $2k$ degrees of freedom.

In the case of likelihood tests, it is easy to combine results from different regions because the log ML values from different regions can be added.

A more difficult problem arises when the substitution rate is uneven over a sequence, but it is unclear how to divide the sequence into separate regions. Such a situation would occur often in coding sequences. The neighbors relation tests proposed above would be robust against rate heterogeneity because the variances are estimated from sample variances, which should include variation due to heterogeneous rates. The tests based on the variance of internal branch lengths would be less robust because the variances are computed under the assumption of a homogeneous sequence and so would tend to be underestimates. In the likelihood test proposed by Kishino and Hasegawa (1989), the variances are estimated from sample variances and so the test may be robust, but it is not clear whether the other likelihood tests discussed above are robust. It has often been argued that parsimony methods are robust against rate heterogeneity. This may be true when the degree of sequence divergence is small or moderate, but may not be true for divergent sequences. Since highly divergent sites can lead to inconsistent results, they should be given less weight in both phylogenetic reconstruction and statistical tests.

Strengths and Weaknesses of Current Tests

To date, no study seems to have been made that compares the relative strengths and weaknesses of different tests. Therefore, our discussion below involves intuitive arguments and speculations. A rigorous comparison should be made in the future by using both computer simulation and empirical data.

The parsimony tests developed by Cavender (1978) and Felsenstein (1985) are not difficult to compute. However, since Cavender's test assumes extremely unfavorable conditions, it has low power. Felsenstein's tests assume an equal probability (i.e., ⅓) for an informative site to support each of the three possible unrooted trees. This assumption is close to reality if the branching order for the three taxa under study is close to a trichotomy, and if the rate of evolution is approximately the same among the three lineages. In this situation, the tests would perhaps perform better than the tests based on distance methods or likelihood methods because the confidence level can be calculated more precisely than in the other tests. However, if the central branch (c in Figure 12-3a) is substantially long, Felsenstein's assumption would not hold well, and his tests may not perform as well as the tests based on distance methods or likelihood methods. Parsimony tests are difficult to develop when the number of taxa involved is larger than four. The performance of the variance test by Kishino and Hasegawa (1989) has not been studied. This test is computationally difficult when the number of taxa is large.

The test based on the variance of internal branch lengths is rather easy to compute when the number of taxa under study is five or fewer. This test would tend to be more powerful than the neighbors relation test because it is formulated under more specific assumptions on the model of nucleotide substitution and because it takes better account of the correlation between sequences. However, since it is dependent on more assumptions, it is less robust than the neighbors relation test. When the number of taxa becomes more than five, the former test becomes computationally tedious. Moreover, the estimates of mean and variance of a branch length depends on the assumption of the tree topology. For example, the mean and variance of the first internal branch in Figure 12-4 would not be accurate if the true branching order of OTUs 4, 5, and 6 is different from that shown in the figure. These problems are less serious (or do not arise) if we use the neighbors relation test.

The maximum likelihood is a widely used statistical method. Its use in phylogenetic reconstruction can be justified, although there may be some logical problems as discussed above. The major problem with this approach is computational difficulties. When the number of taxa under study is five or fewer, the computation is not too tedious and one can even use a sophisticated model of nucleotide substitution. Unfortunately, as the number of taxa increases, the computation becomes prohibitively tedious. If the computational difficulties can be overcome, then the likelihood approach should enjoy more attention than its current status.

ACKNOWLEDGMENTS

We thank N. Saitou for sending us his figures. This study was supported by NIH grant GM30998.

REFERENCES

Akaike, H. (1974) A new look at the statistical model identification. *IEEE Trans. Autom. Contr. AC*–**19**:716–723.

Akaike, H. (1985) Prediction and entropy. Pp. 1–24 in *A Celebration of Statistics* (A.C. Atkinson and S.E. Fienberg, eds.). Springer-Verlag, New York.

Brown, W.M., E.M. Prager, A. Wang, and A.C. Wilson. (1982) Mitochondrial DNA sequences of primates: tempo and mode of evolution. *J. Mol. Evol.* **18**:225–239.

Buneman, P. (1971) The recovery of trees from measurements of dissimilarity. Pp. 387–395 in *Mathematics in the Archeological and Historical Sciences* (F.R. Hodson, D.G. Kendall, and P. Tautu, eds.). Edinburgh University Press, Edinburgh.

Cavalli-Sforza, L.L., and A.W.F. Edwards. (1967) Phylogenetic analysis models and estimation procedures. *Am. J. Hum. Genet.* **19**:233–257.

Cavender, J.A. (1978) Taxonomy with confidence. *Math. Biosci.* **40**:271–280.

Cavender, J.A. (1981) Tests of phylogenetic alternatives under generalized models. *Math. Biosci.* **54**:217–229.

Chakraborty, R. (1977) Estimation of time of divergence from phylogenetic studies. *Can. J. Genet. Cytol.* **19**:217–223.

Edwards, A.W.F., and L.L. Cavalli-Sforza. (1964) Reconstruction of evolutionary trees. Pp. 67–76 in *Phenetic and Phylogenetic Classification* (V.H. Heywood and J. McNeill, eds.). Syst. Assn. Publ. No. 6., Systematics Association, London.

Farris, J.S. (1977) On the phenetic approach to vertebrate classification. Pp. 823–850 in *Major Patterns in Vertebrate Evolution* (M.K. Hecht, P.C. Goody, and B.M. Hecht, eds.). Plenum Press, New York.

Felsenstein, J. (1973) Maximum-likelihood and minimum-steps methods for estimating evolutionary trees from data on discrete characters. *Syst. Zool.* **22**:240–249.

Felsenstein, J. (1978) Cases in which parsimony or compatibility methods will be positively misleading. *Syst. Zool.* **27**:401–410.

Felsenstein, J. (1981) Evolutionary trees from DNA sequences: a maximum likelihood approach. *J. Mol. Evol.* **17**:368–376.

Felsenstein, J. (1983) Inferring evolutionary trees from DNA sequences. Pp. 133–150 in *Statistical Analysis of DNA Sequence Data* (B.S. Weir, ed.). Dekker, New York.

Felsenstein, J. (1985) Confidence limits on phylogenies with a molecular clock. *Syst. Zool.* **34**:152–161.

Felsenstein, J. (1987) Estimation of hominoid phylogeny from a DNA hybridization data set. *J. Mol. Evol.* **26**:123–131.

Felsenstein, J. (1988) Phylogenies from molecular sequences: inference and reliability. *Ann. Rev. Genet.* **22**:521–565.

Fisher, R.A. (1932) *Statistical Methods for Research Workers*. Oliver and Boyd, London.

Fitch, W.M. (1981) A non-sequential method for constructing trees and hierarchical classifications. *J. Mol. Evol.* **18**:30–37.

Gouy, M., and W.-H. Li (1989) Phylogenetic analysis based on rRNA sequences supports the archaebacterial rather than the eocyte tree. *Nature* **339**:145–147.

Hasegawa, M., H. Kishino, and T. Yano. (1985) Dating of the human-ape splitting by a molecular clock of mitochondrial DNA. *J. Mol. Evol.* **22**:160–174.

Hasegawa, M., H. Kishino, and T. Yano. (1987) Man's place in Hominoidea as inferred from molecular clocks of DNA. *J. Mol. Evol.* **26**:132–147.

Jin, L., and M. Nei. (1990) Limitations of the evolutionary parsimony method of phylogenetic analysis. *Mol. Biol. Evol.* **7**:82–102.

Jukes, T.H., and C.R. Cantor. (1969) Evolution of protein molecules. Pp. 21–132 in *Mammalian Protein Metabolism* (H. N. Munro, ed.). Academic Press, New York.

Kashyap, R.L., and S. Subas. (1974) Statistical estimation of parameters in a phylogenetic tree using a dynamic model of the substitutional process. *J. Theor. Biol.* **47**:75–101.

Kimura, M. (1980) A simple method for estimating evolutionary rates of base substitutions through comparative studies of nucleotide sequences. *J. Mol. Evol.* **16**:111–120.

Kimura, M., and T. Ohta. (1972) On the stochastic model for estimation of mutational distance between homologous proteins. *J. Mol. Evol.* **2**:87–90.

Kishino, H., and M. Hasegawa. (1989) Evaluation of the maximum likelihood estimate of the evolutionary tree topologies from DNA sequence data, and the branching order in Hominoidea. *J. Mol. Evol.* **29**:170–179.

Klotz, L.C., N. Komar, R.L. Blanken, and R.M. Mitchell. (1979) Calculation of evolutionary trees from sequence data. *Proc. Natl. Acad. Sci. USA* **76**:4516–4520.

Li, W.-H. (1981) Simple method for constructing phylogenetic trees from distance matrices. *Proc. Natl. Acad. Sci. USA* **78**:1085–1089.

Li, W.-H. (1989) A statistical test of phylogenies estimated from sequence data. *Mol. Biol. Evol.* **6**:424–435.

Li, W.-H., and M. Gouy. (1990) Statistical tests of molecular phylogenies. Pp. 645–659 in *Molecular Evolution: Computer Analysis of Protein and Nucleic Acid Sequences* (R.F. Doolittle, ed.). Methods in Enzymology, vol. 183. Academic Press, New York.

Li, W.-H., M. Tanimura, and P.M. Sharp. (1987a) An evaluation of the molecular clock hypothesis using mammalian DNA sequences. *J. Mol. Evol.* **25**:330–342.

Li, W.-H., K.H. Wolfe, J. Sourdis, and P.M. Sharp. (1987b) Reconstruction of phylogenetic trees and estimation of divergence times under nonconstant rates of evolution. *Cold Spring Harbor Symp. Quant. Biol.* **52**:847–856.

Li, W.-H., C.-I. Wu, and C.-C. Luo. (1984) Nonrandomness of point mutation as reflected in nucleotide substitutions in pseudogenes and its evolutionary implications. *J. Mol. Evol.* **21**:58–71.

Mueller, L.D., and F.J. Ayala. (1982) Estimation and interpretation of genetic distance in empirical studies. *Genet. Res.* **40**:127–137.

Nei, M. (1987) *Molecular Evolutionary Genetics*. Columbia University Press, New York.

Nei, M., J. C. Stephens, and N. Saitou. (1985) Methods for computing the standard errors of branching points in an evolutionary tree and their application to molecular data from humans and apes. *Mol. Biol. Evol.* **2**:66–85.

Neyman, J. (1971) Molecular studies of evolution: a source of novel statistical problems. Pp. 1–27 in *Statistical Decision Theory and Related Topics* (S.S. Gupta and J. Yackel, eds.). Academic Press, New York.

Saitou, N. (1988) Property and efficiency of the maximum likelihood method for molecular phylogeny. *J. Mol. Evol.* **27**:261–273.

Saitou, N. (1990) Maximum likelihood methods. Pp. 584–598 in *Molecular Evolution: Computer Analysis of Protein and Nucleic Acid Sequences* (R.F. Doolittle, ed.). Methods in Enzymology, vol. 183. Academic Press, New York.

Saitou, N., and T. Imanishi. (1989) Relative efficiencies of the Fitch-Margoliash, maximum-parsimony, maximum-likelihood, minimum-evolution, and neighbor-joining methods of phylogenetic tree construction in obtaining the correct tree. *Mol. Biol. Evol.* **6**:514–525.

Saitou, N., and M. Nei. (1986) The number of nucleotides required to determine the branching order of three species with special reference to the human-chimpanzee-gorilla divergence. *J. Mol. Evol.* **24**:189–204.

Saitou, N. and M. Nei. (1987) The neighbor-joining method: a new method for reconstructing phylogenetic trees. *Mol. Biol. Evol.* **4**:406–425.

Sattath, S., and A. Tversky. (1977) Additive similarity trees. *Psychometrika* **42**:319–345.

Smouse, P.E., and W.-H. Li. (1987) Likelihood analysis of mitochondrial restriction-cleavage patterns for the human-chimpanzee-gorilla trichotomy. *Evolution* **41**:1162–1176.

Sneath, P.H.A., and R.R. Sokal. (1973) *Numerical Taxonomy*. W.H. Freeman, San Francisco.

Templeton, A.R. (1983) Phylogenetic inference from restriction endonuclease cleavage site maps with particular reference to the evolution of humans and the apes. *Evolution* **37**:221–244.

13

Discriminating Between Phylogenetic Signal and Random Noise in DNA Sequences

DAVID M. HILLIS

> Most parsimonious trees are really neat. I am all for them, but have you any idea about the distribution of all the trees in the universe that you might have sampled? Fitch (1984)

Nucleic acid sequences have become a major source of phylogenetic information. In order to use these data appropriately, it is critical to distinguish sequences that are saturated by change from those that are phylogenetically informative. In this chapter, I will call variation that is potentially informative about phylogenetic history "signal," and use "noise" to describe variation that is not informative. Distinguishing between signal and noise is not something that can be accomplished by use of alignment criteria, because two sequences may be identical at a large number of functionally constrained sites, and yet essentially randomized at sites where variation is tolerated (e.g., the third positions of codons). This chapter explores the use of the distributions of tree lengths as a guide to the detection of phylogenetic signal in comparative data sets.

In parsimony analysis, changes at nucleotide positions among aligned sequences are mapped onto a tree, and the number of evolutionary changes required to accommodate that tree with the data is calculated as the tree length. For any given data set, this procedure may be repeated for many thousands of trees; the tree that yields the shortest length is designated the optimal tree under the parsimony criterion. The optimal tree is thus the one that requires the fewest number of evolutionary changes. In this chapter, I argue that the shape of the distribution of tree lengths contains information useful in deciding whether or not the data set contains phylogenetic signal.

BACKGROUND

Fitch (1979) and Goodman et al. (1979a) were among the first authors to explore tree-length distributions in a phylogenetic context. These authors compared the relative phylogenetic information in two protein sequence data sets that had been used to infer relationships among orders of mammals. As can be seen in Figure 13-1, the distribution of tree lengths from the α-hemoglobin sequences is nearly perfectly symmetrical, whereas that from the α-crystallin sequences is strongly skewed with a long left tail. Therefore, there are many fewer solutions near the optimal (most-parsimonious) solution for the α-crystallin data set than for the α-hemoglobin data set. The shape of these tree-length distributions suggests that the α-

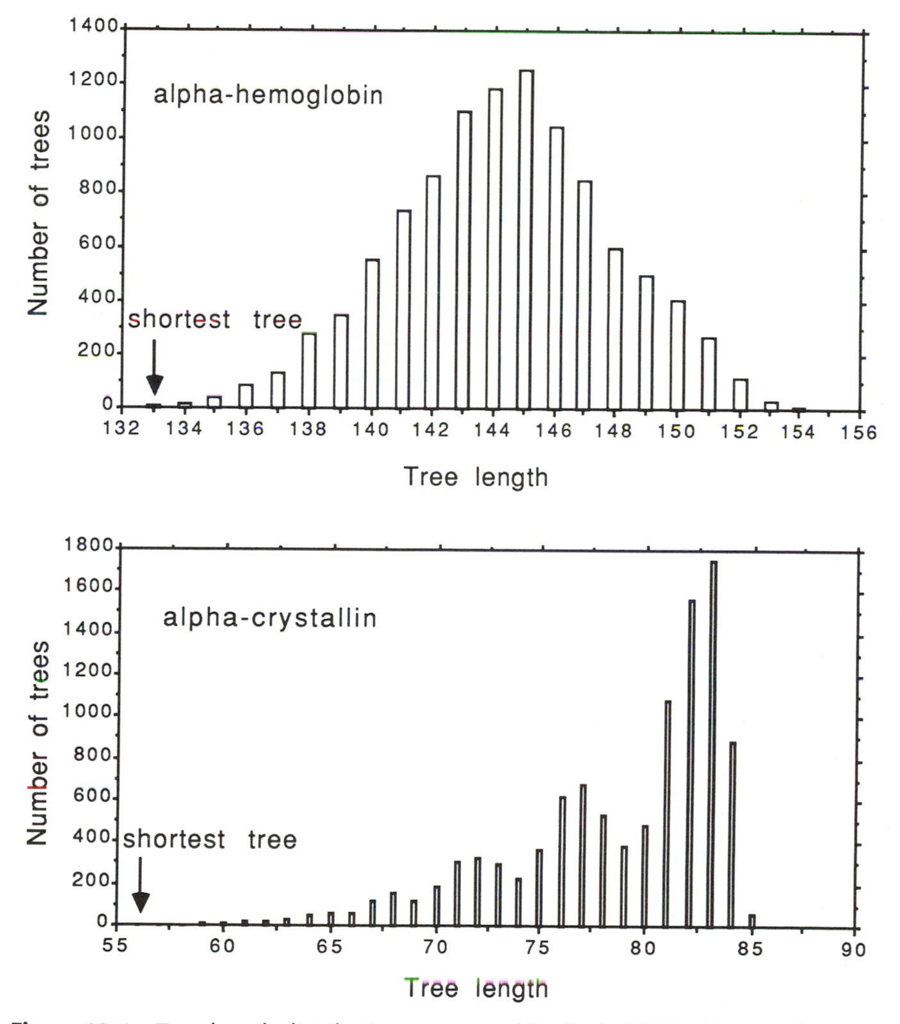

Figure 13-1 Tree-length distributions compared by Fitch (1979, 1984) and Goodman et al. (1979a).

crystallin data set is more informative (because it allows better discrimination among near-optimal solutions) than the α-hemoglobin data set.

Fitch (1984) suggested that symmetric distributions such as that for α-hemoglobin are "most likely for bushy trees where all lines emanate from a single point," whereas asymmetric distributions like that for α-crystallin are "most likely for the 'stringy' trees, those which are maximally asymmetric in their splitting and have long intervals between splits." Several authors have presented strongly skewed tree-length distributions and have suggested that such distributions are evidence for strong phylogenetic signal (e.g., Fitch, 1984; Goodman et al., 1979a; Hillis, 1985; Werman, 1986; Hillis and de Sá, 1988; Hillis and Dixon, 1989). I will argue that skewness of tree-length distributions can be a useful indicator of phylogenetic signal and is largely (but not completely) independent of tree topology.

Having looked at tree-length distributions for many data sets, I developed the view illustrated in Figure 13-2. If all tree topologies for a given set of sequences are equally optimal (i.e., of equal length), then there is neither signal nor noise in the data set. Such a data set would result in the completely unresolved bush shown in Figure 13-2a. If the real phylogeny is in the form of a bush, however, then an equal length for all possible trees is an unlikely outcome; random variation would result in some trees being shorter than others by chance alone. The distribution of trees would include some topologies that were shorter than average, and a similar number that were longer than average, so the distribution of these lengths would produce a nearly symmetrical distribution, such as those shown in Figure 13-2b and 13-2b′. The optimal topology (or topologies) in this case might be a completely symmetrical tree (Fig. 13-2b), a completely asymmetrical tree (Fig. 13-2b′), or something in between. However, if the true phylogeny was not a single polytomy and the sequences were constrained by history (i.e., contain phylogenetic information), then the tree-length distributions shown in Figures 13-2b and 13-2b′ are highly unlikely. Sequences constrained by history would produce distributions of tree lengths similar to those in Figure 13-2c and 13-2c′: highly asymmetrical distributions with few trees near the optimal solution. These asymmetric distributions are a consequence of congruence among characters as a result of a common phylogenetic history. Thus, many characters support the topology that reflects this history, and conflict with a large number of alternative trees.

If the view described above is correct, then it suggests a means of escaping a common problem in phylogenetic analysis. Any comparative data set can be subjected to phylogenetic analysis, even if the data contain no historical information (i.e., they are too noisy for meaningful phylogenetic analysis). Nonetheless, random variation will usually result in one or a relatively few optimal topologies (at least compared to the much larger number of possibilities). What is needed is a means of distinguishing such data sets from those that are informative about phylogenetic history. This chapter explores the causes of skewed tree-length distributions, examines the proposition

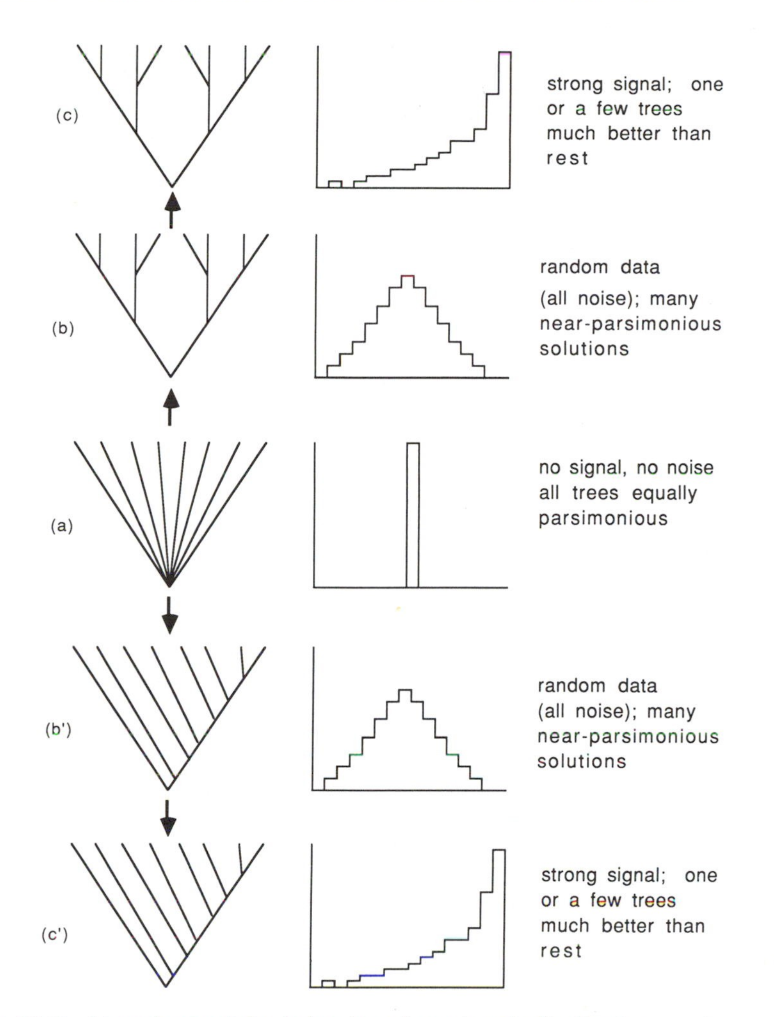

Figure 13-2 Hypothesis of the behavior of tree-length distributions under several model phylogenies with differing amounts of signal and noise. See text for explanation.

that random sequence data produce symmetrical tree-length distributions, explores possible tests for skewness, evaluates the amount of signal necessary to produce significant skewness, compares real data sets to random simulations, suggests a possible "stopping algorithm" to prevent the over-resolution of phylogenies, and points to areas in need of additional research.

THE CAUSES OF SKEWED TREE-LENGTH DISTRIBUTIONS

To understand why data sets with strong phylogenetic signal produce skewed tree-length distributions, consider a data set on eight taxa (Figs. 13-3–13-

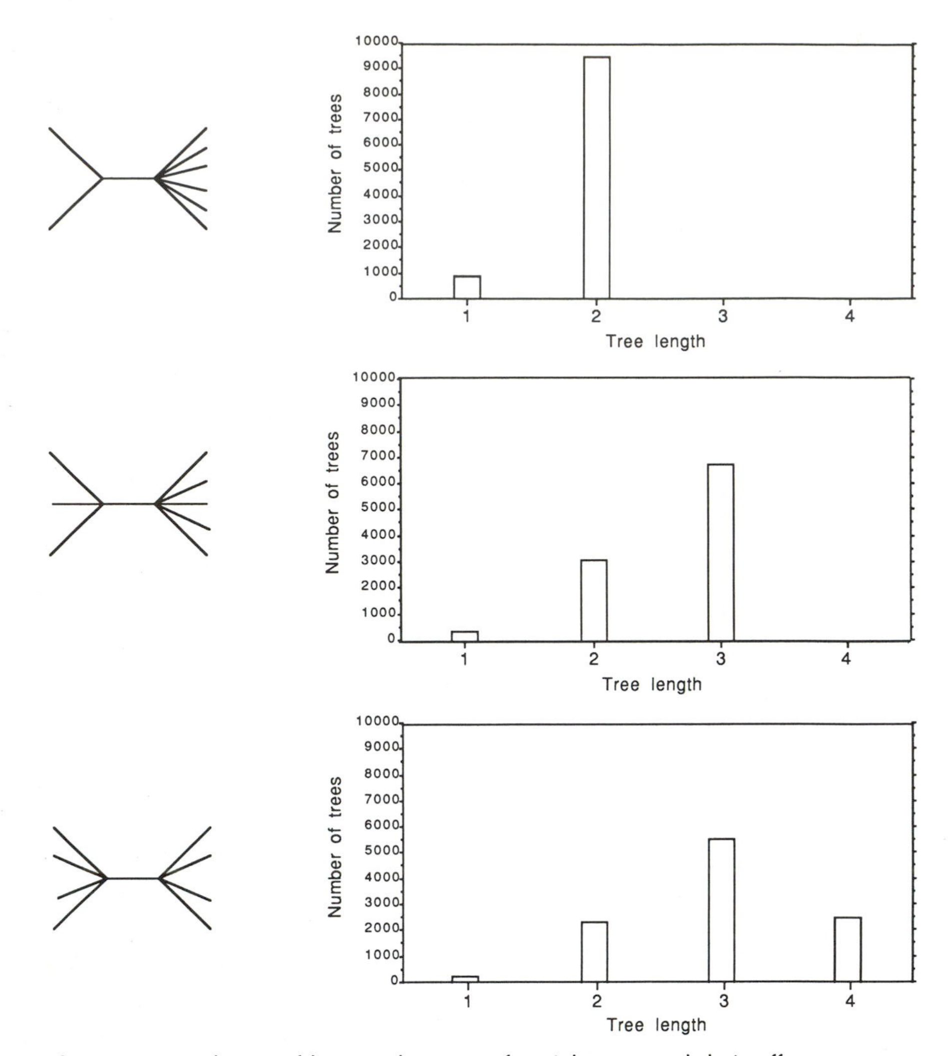

Figure 13-3 Classes of binary characters for eight taxa and their effects on tree-length distributions.

5).[1] For eight taxa, there are only three types of binary characters that affect tree length (Fig. 13-3): those for which two taxa have one state and the others have the alternative state ("two/six characters"); those for which the alternative states are distributed between three and five taxa, respectively ("three/five characters"); and those for which the alternative states are evenly divided between two subsets of four taxa each ("four/four char-

[1]Because tree lengths are unaffected by the rooting point of the tree, all examples in this discussion will concern unrooted trees. Also, for simplicity, examples in this section will involve ditypic (binary) characters, although the concepts can be extended easily to multistate characters.

acters"). Characters with states limited to a single taxon and characters that are uniform among all species are compatible with all possible tree topologies and thus do not affect the shape of the tree-length distribution.

Figure 13-3 shows the distributions of tree lengths for data sets limited to each of these types of characters. All three types of characters produce a tree-length distribution with a left-hand skew. Two/six characters produce distributions with the strongest skew but the smallest range; this is because only a very few trees connect the pair of species with the unique state, whereas all other trees require the character-state to evolve twice. Three/five characters also produce a skewed distribution, but more of the possible trees are compatible or partially compatible with this type of character. Although few trees unite the three species with the same state, many more unite at least two of the three species on the tree, and most trees split all three species. This produces a skewed distribution with a range of two steps (compared to a range of one step for a two/six character; see Fig. 13-3). The four/four characters produce the distribution with the greatest span but the least skewness, because there are fewer trees that unite none of the similar taxa than unite only two of the four similar taxa (Fig. 13-3). Therefore, any informative character can add to the skewness of a tree-length distribution, although different types of characters do not contribute to skewness equally.

The effects of tree topology on tree-length distributions can be assessed by considering the kinds of characters that would be produced by the different topologies. For eight taxa, there are only four unrooted, unlabeled topologies (Fig. 13-4). Each internal branch in the topology is supported by either a two/six, a three/five, or a four/four character; the numbers of each of these types of internal branches are shown in Figure 13-4. If one assigns two characters to each internal branch, the distributions for each topology would be as shown in Figure 13-5. Note that all of these distributions are strongly skewed, although not all to the same degree. A common measure of skewness is the g_1 statistic—the third central moment divided by the cube of the standard deviation (Sokal and Rohlf, 1981). For n trees of length T, g_1 is calculated as

$$g_1 = \frac{\sum_{i=1}^{n} (T_i - \overline{T})^3}{n\, s^3}$$

where s is the standard deviation of the tree lengths. This statistic is negative for distributions with left-skew, 0 for symmetric distributions, and positive for distributions with right-skew. The skewness of the distributions in Figure 13-5 varies from $g_1 = -1.096$ to $g_1 = -0.6637$. Therefore, although tree topology has a quantitative influence on skewness, the qualitative result (strong left-skew) is the same for all topologies. It is interesting to note that symmetry of the tree topology appears to have little influence on skewness.

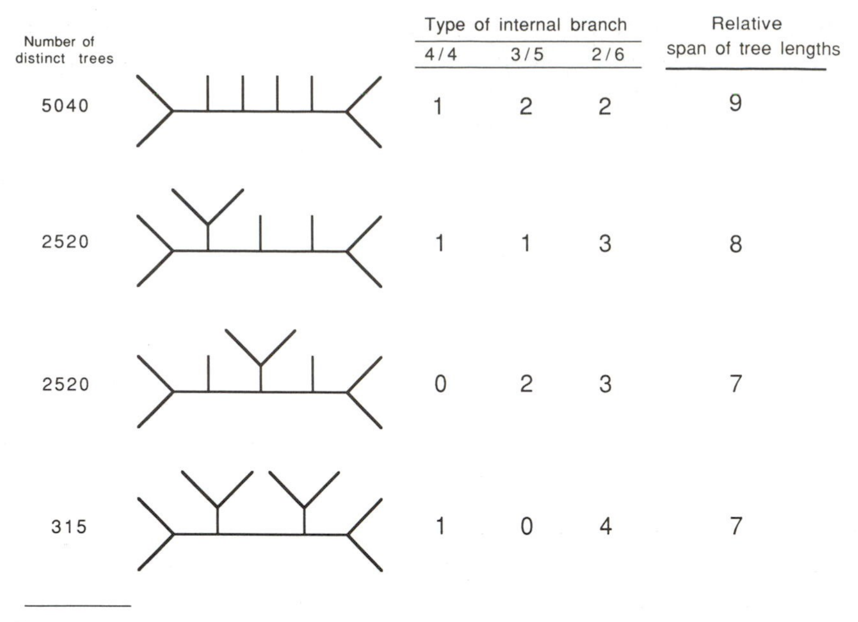

Number of distinct trees
Type of internal branch
4/4
3/5
2/6
Relative span of tree lengths
5040
1
2
2
9
2520
1
1
3
8
2520
0
2
3
7
315
1
0
4
7
Total: 10,395

Figure 13-4 Distinct unrooted tree topologies for eight taxa, numbers of internal branches supported by each class of binary character, and the relative spans of the tree-length distributions.

EXPERIMENTS WITH RANDOM SEQUENCES

To address the behavior of tree-length distributions in the absence of phylogenetic signal, random data matrices were constructed in the following manner (Fig. 13-6). One hundred matrices each with six, seven, or eight sequences, each 100 nucleotides long, were created using the random data generator in the computer program MacClade (version 2.97; Maddison and Maddison, 1991). This version of the program uses the random number generator in the Macintosh tool box to generate random data (D. Maddison, personal communication). The four nucleotides were given equal probabilities of occurring, so the frequency of each nucleotide was approximately 0.25 in each sequence. All possible tree topologies were analyzed for all 300 random matrices using Swofford's (1990) Phylogenetic Analysis Using Parsimony computer software package (PAUP version 3.0).

To address the effects of the length of sequences on skewness, two additional sets of 100 matrices each of random data were generated for eight taxa. These data sets consisted of sequences 30 positions long and 200 positions long, respectively.

Two examples of tree-length distributions from the eight-taxa data sets are shown in Figure 13-7; figured are the distributions with the strongest left-hand and right-hand skews, respectively. The distribution of g_1 scores for the tree-length distributions from all the random data sets are shown

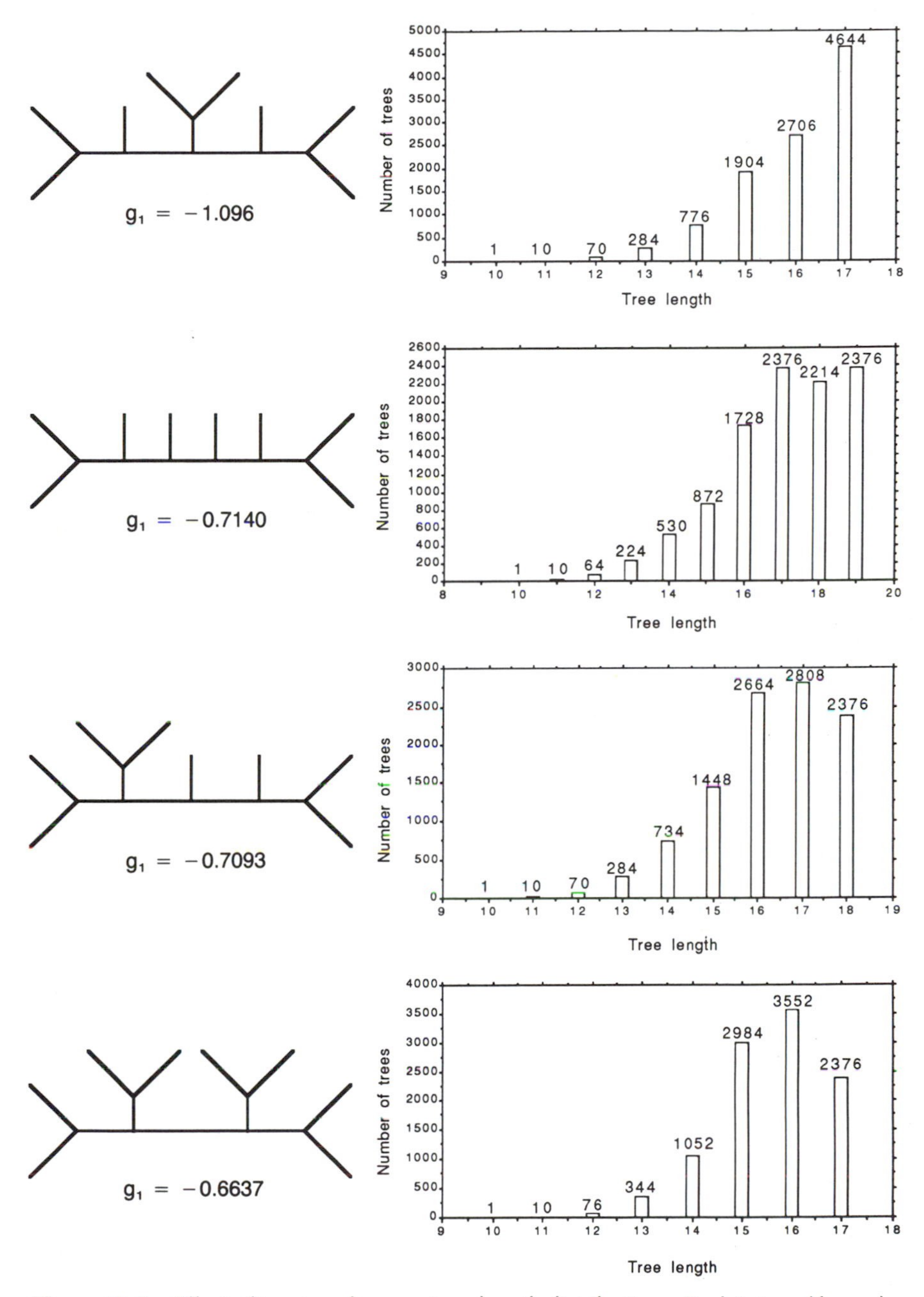

Figure 13-5 Effect of tree topology on tree-length distributions. Each internal branch is assigned a length of two character changes.

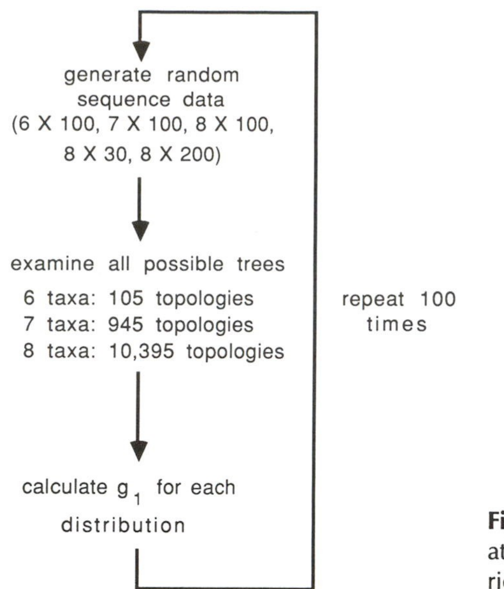
generate random sequence data
(6 X 100, 7 X 100, 8 X 100, 8 X 30, 8 X 200)
examine all possible trees
6 taxa: 105 topologies
7 taxa: 945 topologies
8 taxa: 10,395 topologies
calculate g 1 for each distribution
repeat 100 times

Figure 13-6 Protocol for the generation of random sequence-data matrices and analysis of skewness.

in Figure 13-8. Two trends are clear from these scores. First, the mean and modal values of g_1 across data sets are slightly negative, as predicted for binary data by Le Quesne (1989). Even with random data, approximately twice as many tree-length distributions are skewed slightly to the left as to the right. Second, the average value of the g_1 scores approaches 0 with increasing number of taxa: the tree-length distributions become closer to symmetrical.

Although the only published tests for skewness of tree-length distributions have used the normal distribution as a null model (e.g., Werman, 1986), the random data simulations indicate that this is an inappropriate test (Fig. 13-9). Only a small percentage of tree-length distributions based on random sequences fall into the 95% confidence limits of a normal distribution. The percentage of distributions that fall within the 95% confidence limits for skewness of a normal distribution drops with increasing number of taxa, despite the fact that the distributions become more symmetrical (Fig. 13-9). This is because the degree of expected departures from symmetry for a normal distribution decreases with increasing sample size, and the sample size (number of tree topologies) increases very rapidly with an increase in the number of taxa (Felsenstein, 1978). Thus, less than 20% of the tree-length distributions produced from random sequences fall within the expectations of a normal distribution with just eight taxa (Fig. 13-9).

The random data simulations provide a simple means for testing tree-length distributions for significantly greater left-handed skewness than would be expected from random data, without making any *a priori* assumptions about the shape of the distribution of expected g_1 values. The lower 5% and 1% of the g_1 distributions, shown in Figure 13-8, can be used to obtain

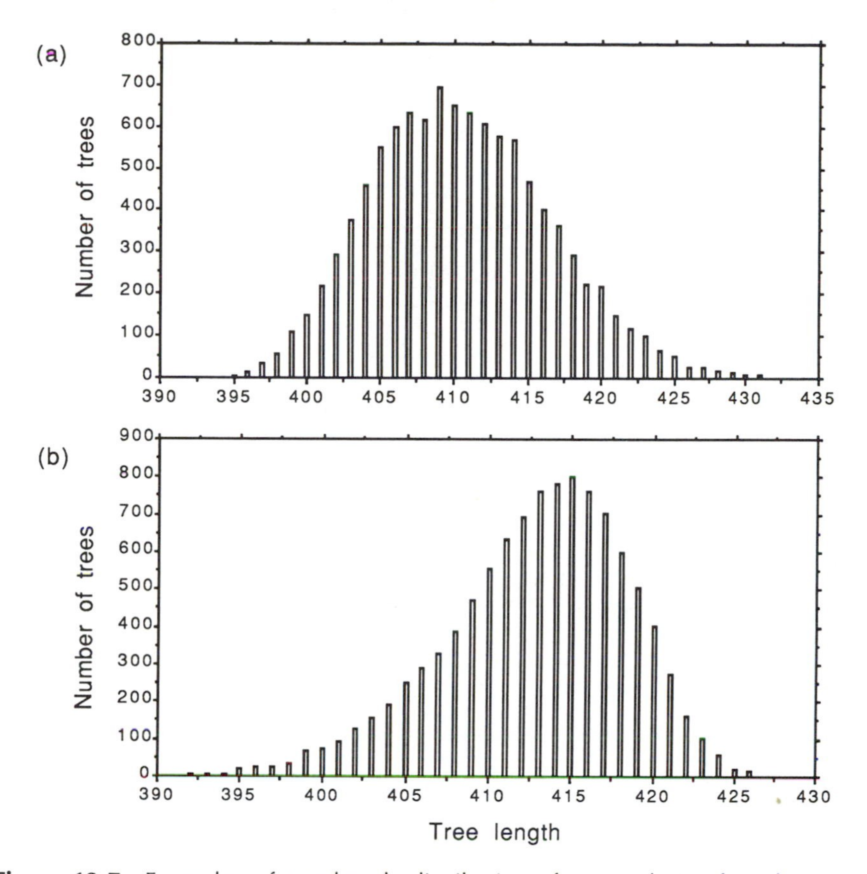

Figure 13-7 Examples of tree-length distributions from analysis of random sequence-data matrices for eight taxa. Shown are the most right-skewed (a; g_1 = 0.3055) and the most left-skewed (b; g_1 = −0.5278) distributions produced from 100 random matrices.

critical values of g_1 for a given number of taxa. A test employing these values (Table 13-1) requires a minimum of 100 variable sequence positions. For fewer than 100 positions, the critical values are slightly lower (more negative); there is very little change in the critical value for data sets with more than 100 sequence positions (Fig. 13-10).

How much phylogenetic signal is needed to produce significant skewness? To examine this question, I added characters consistent with the branches of a single tree (signal) to the eight-taxa random matrices (noise). For most data sets, 10% signal (e.g., 11 characters out of 110) was sufficient to produce tree-length distributions that were significantly skewed, as indicated by the test discussed above (see Fig. 13-11). The most right-skewed distribution examined (g_1 = 0.3055) required 20% signal to become significantly left-skewed (g_1 = −0.4037). Therefore, the test appears to be fairly sensitive; even small percentages of phylogenetic signal can be detected in otherwise random data sets.

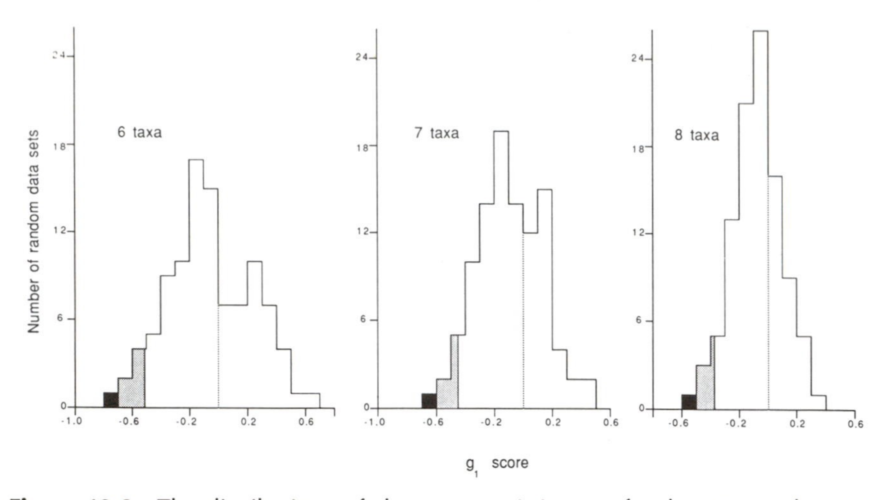

Figure 13-8 The distributions of skewness statistics (g_1) for the 300 random sequence matrices (six, seven, and eight taxa). The dashed vertical lines indicate the g_1 score for symmetrical distributions. The lower 5% (shaded) and 1% (black) of each distribution is indicated.

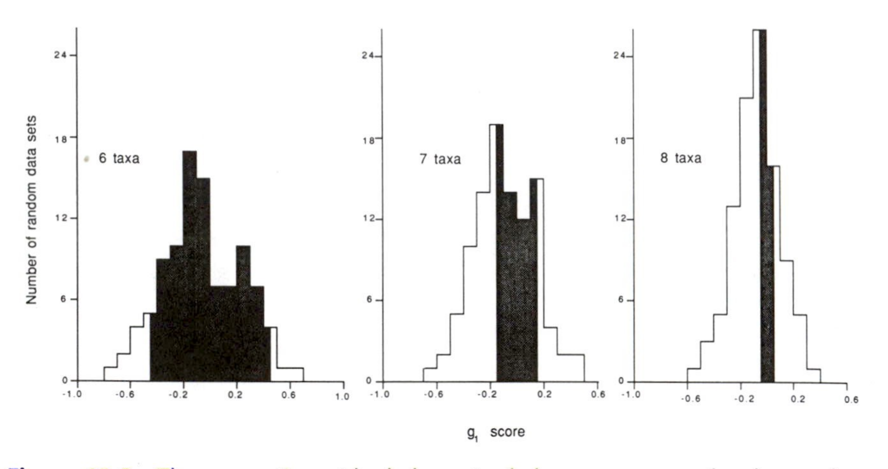

Figure 13-9 The proportions (shaded area) of skewness scores for the random sequence-data matrices that are not significantly different from scores for normal distributions (six, seven, and eight taxa).

Table 13-1 Critical Values for g_1 Statistics of Tree-Length Distributions for Six, Seven, and Eight Taxa*

	Number of Taxa		
p	6	7	8
0.05	−0.51	−0.45	−0.34
0.01	−0.67	−0.60	−0.47

*Scores lower than those shown are outside of the 95% or 99% confidence limits for distributions derived from random data.

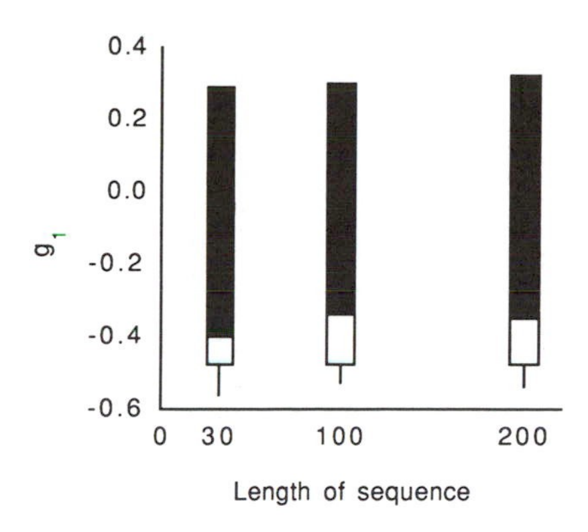

Figure 13-10 The effect of length of sequence on g_1 statistics. The three samples each contain 100 matrices of eight taxa; the lengths of the random sequences are 30, 100, and 200 nucleotides. The lowest 1% of the distribution is indicated by a single line, the lowest 5% of the distribution is indicated by an open bar plus the line, and the remaining portion of the distribution is indicated by a solid bar.

ANALYSIS OF REAL SEQUENCES

I have examined skewness of tree-length distributions from several real sequence data sets, all with at least eight taxa. Except for the α-hemoglobin data discussed by Fitch (1984) and above, all of these were significantly more skewed than the distributions from random data matrices ($p < 0.05$; Table 13-1 and Fig. 13-12). Only one of these distributions (that for mitochondrial cytochrome *b* genes) was even within the observed bounds of the g_1 values for the random matrices. The mitochondrial cytochrome *b* data set consists of sequences on eight species of vertebrates, including an actinopterygian fish, two amphibians, a bird, and four mammals. If one of the two most closely related taxa (sheep and cow) are removed from

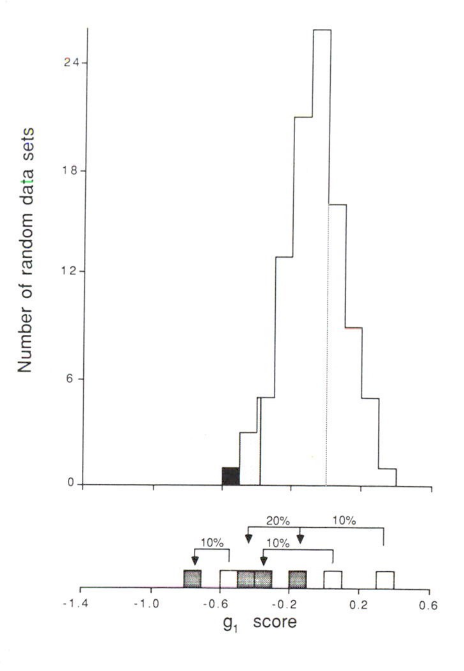

Figure 13-11 The effect on skewness of adding signal to otherwise random sequence matrices (see text). The g_1 scores represented by open boxes in the lower portion of the figure correspond to distributions produced from random data (the most left-skewed, symmetrical, and right-skewed distributions are used as examples). The shaded boxes indicate the g_1 scores for the distributions after the addition of 10% or 20% signal, as indicated.

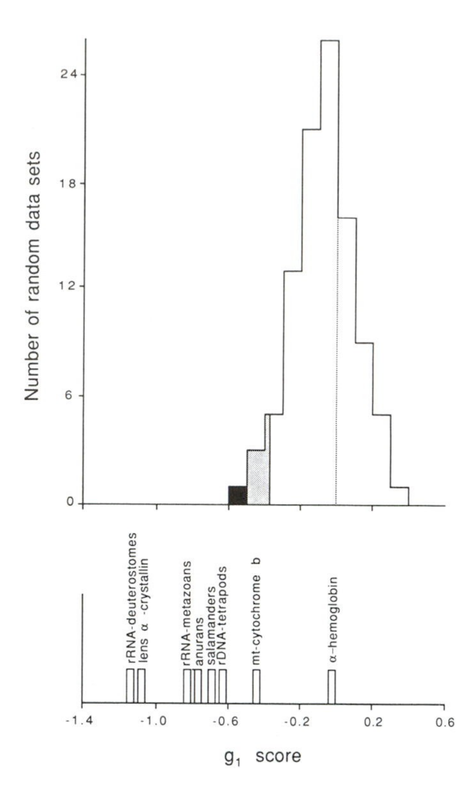

Figure 13-12 Skewness (g_1) scores for tree-length distributions produced from real data matrices (lower) compared to the distributions of g_1 scores produced from analysis of random data matrices (upper) for eight taxa. The rRNA data on deuterostomes (taxa shown in Fig. 13-13) and metazoans (representatives of eight phyla) are from Field et al. (1988); the lens α-crystallin data on eight orders of mammals are from De Jong et al. (1977); the α-hemoglobin data on eight orders of mammals are from Goodman et al. (1979b); the anuran and salamander data are from Hillis (1991); the ribosomal DNA data on tetrapods are from Hillis et al. (1991); and the mitochondrial cytochrome *b* sequences are eight species of vertebrates from Kocher and White (1989) and GenBank (IntelliGenetics, Mountain View, CA).

this data matrix, the tree-length distribution becomes much less skewed, indicating that it is the inclusion of these relatively closely related species that accounts for most of the phylogenetic signal in these sequences.

The observations from the mitochondrial cytochrome *b* data set suggest an algorithm that could be used to prevent over-resolution of phylogenies (reading beyond the signal) in sequence data sets. The execution of this algorithm is shown in Figure 13-13 for the ribosomal RNA (rRNA) sequences presented by Field et al. (1988) for eight species of deuterostomes.

1. Determine the skewness of the tree-length distribution for all possible trees (or a random sample thereof—MacClade and PAUP each have options for this purpose).
2. If the distribution is significantly skewed, then find the best-supported branch that unites two or more of the taxa (e.g., by bootstrapping or counting synapomorphies) and continue to step 3. If the distribution is no more skewed than would be expected from random data, then stop.
3. Repeat step 1 for all remaining tree topologies, given the branch determined in step 2.
4. Repeat step 2, saving any branches previously determined (continue repeating steps 1 and 2 until the distribution of all remaining trees is no more skewed than expected from random data).

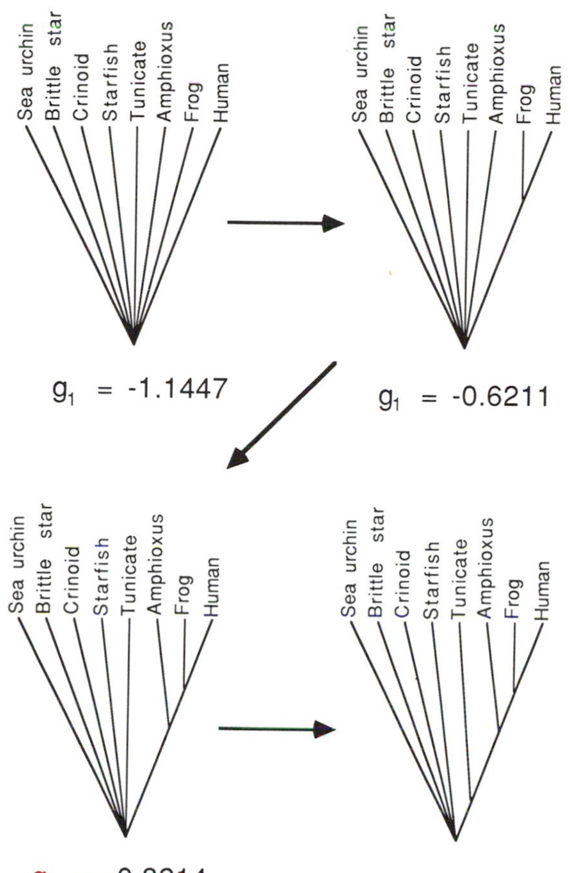

Figure 13-13 Resolution of relationships among eight species of deuterostomes (Field et al., 1988) using the "stopping algorithm" described in the text.

One objection to this algorithm is that tree resolution cannot proceed beyond a tree of five unresolved lineages (Fig. 13-13), because the number of possible tree topologies (15) is too small to examine skewness (see Table 13-2). However, I have found that for most sequence data sets, the tree-length distribution is no longer skewed well before this point; therefore, this limitation may not be too severe. Another objection is that the overall probability of making a type-I error increases as the number of resolved branches increases, so the initial level of α must be set quite low in order to keep the overall α below 0.05. This reduces the power of the test. There also may be objections that a single branch is determined at each step, whereas the phylogenetic information may be distributed across more than one branch. Nonetheless, I present this algorithm as an example of the possible uses of tree-length distributions, although I do not advocate its use until tree-length distributions have been studied in greater detail.

Table 13-2 Number of Distinct Unrooted Trees for Three or More Taxa, and the Recommended Methods of Analyzing Tree-Length Distributions

Number of Taxa	Number of Unrooted Trees	Method of Analysis
3	1	Too few trees
4	3	Too few trees
5	15	Too few trees
6	105	Exact enumeration
7	945	Exact enumeration
8	10,395	Exact enumeration
9	135,135	Exact enumeration or random sampling
10	2,027,025	Exact enumeration or random sampling
11	34,459,425	Exact enumeration or random sampling
≥12	≥654,729,075	Random sampling

IMPLEMENTATION AND DIRECTIONS FOR FUTURE RESEARCH

This study suggests that tree-length distributions can provide an accurate and sensitive indication of the presence of phylogenetic signal in comparative sequence data sets. Skewness statistics provide a simple means of evaluating these distributions, which are often determined as a by-product of phylogenetic analysis. In response to these findings, David L. Swofford and David R. Madison have recently modified PAUP and MacClade, respectively, so that these programs now provide g_1 statistics for tree-length distributions.

Although I have used the conventional measure of skewness (g_1) as a guide to the detection of phylogenetic signal, other measures (e.g., the other odd central moments) may prove to be even better indicators of signal. However, considerable additional research needs to be conducted before any measures of skewness can be used meaningfully with regularity. Simulations with random data (or exact solutions) need to be conducted for greater numbers of taxa (Table 13-2), the effects of base composition and length of sequence need to be studied in greater detail, and additional studies need to be conducted on the effects of different levels of signal and noise (or the effects of two sets of conflicting signal). Despite the need for these additional studies, this study indicates the general usefulness of tree-length distributions in phylogenetic analysis.

ACKNOWLEDGMENTS

I thank J. Coddington, B. I. Crother, M. T. Dixon, J. S. Farris, J. Felsenstein, K. Halanych, A. I. Hillis, W. M. Fitch, and M. White for helpful discussion, comments on the manuscript, and other assistance. D. R. Maddison and D. L. Swofford provided test versions of their respective programs, without which this study would not have been possible. J. J. Bull has provided constant input and advice on this work, and I am especially grateful for his efforts. This study was supported by National Science Foundation grants BSR-8657640 and BSR-8796293.

REFERENCES

De Jong, W. W., J. T. Gleaves, and D. Boulter. (1977) Evolutionary changes of α-crystallin and the phylogeny of mammalian orders. *J. Mol. Evol.* **10**:123–135.

Felsenstein, J. (1978) The number of evolutionary trees. *Syst. Zool.* **27**:27–33.

Field, K. G., G. J. Olsen, D. J. Lane, S. J. Giovannoni, M. T. Ghiselin, E. C. Raff, N. R. Pace, and R. A. Raff. (1988) Molecular phylogeny of the animal kingdom. *Science* **239**:748–753.

Fitch, W. M. (1979) Cautionary remarks on using gene expression events in parsimony procedures. *Syst. Zool.* **28**:375–379.

Fitch W. M. (1984) Cladistic and other methods: problems, pitfalls, and potentials. Pp. 221–252 in *Cladistics: Perspectives on the Reconstruction of Evolutionary History* (T. Duncan and T. F. Stuessy, eds.). Columbia University Press, New York.

Goodman, M., J. Czelusniak, and G. W. Moore. (1979a) Further remarks on the parameter of gene duplication and expression events in parsimony reconstructions. *Syst. Zool.* **28**:379–385.

Goodman, M., J. Czelusniak, G. W. Moore, A. Romero-Herrera, and G. Matsuda. (1979b) Fitting the gene lineage into its species lineage, a parsimony strategy illustrated by cladograms constructed from globin sequences. *Syst. Zool.* **28**:132–163.

Hillis, D. M. (1985) Evolutionary genetics of the Andean lizard genus *Pholidobolus* (Sauria: Gymnophthalmidae): phylogeny, biogeography, and a comparison of tree construction techniques. *Syst. Zool.* **34**:109–126.

Hillis, D. M. (1991) The phylogeny of amphibians: current knowledge and the role of cytogenetics. Pp. 7–31 in *Amphibian Cytogenetics and Evolution* (D. M. Green and S. K. Sessions, eds.). Academic Press, New York.

Hillis, D. M., and R. de Sá. (1988) Phylogeny and taxonomy of the *Rana palmipes* group (Salientia: Ranidae). *Herpetol. Monographs* **2**:1–26.

Hillis, D. M., and M. T. Dixon. (1989) Vertebrate phylogeny: evidence from 28S ribosomal DNA sequences. Pp. 355–367 in *The Hierarchy of Life. Molecules and Morphology in Phylogenetic Analysis* (B. Fernholm, K. Bremer, and H. Jörnvall, eds.). Elsevier Science Publishers B. V., Amsterdam.

Hillis, D. M., M. T. Dixon, and L. K. Ammerman. (1991) The relationships of coelacanths: evidence from sequences of vertebrate 28S ribosomal RNA genes. In press in *Coelacanth Biology and Evolution* (J. A. Musick, ed.). Special volume of *Environ. Biol. Fishes* **32**.

Kocher, T. D., and T. J. White. (1989) Evolutionary analysis via PCR. Pp. 137–147 in *PCR Technology. Principles and Applications for DNA Amplification* (H. A. Erlich, ed.). Stockton Press, New York.

Le Quesne, W. J. (1989) Frequency distributions of lengths of possible networks from a data matrix. *Cladistics* **5**:395–407.

Maddison, W. P., and D. R. Maddison. (1991) *MacClade: Interactive Analysis of Phylogeny and Character Evolution*. Sinauer Associates, Sunderland.

Sokal, R. R., and F. J. Rohlf. (1981) *Biometry*. 2nd ed. W. H. Freeman, San Francisco.

Swofford, D. L. (1990) *PAUP: Phylogenetic Analysis Using Parsimony*, Version 3.0. Illinois Natural History Survey, Champaign.

Werman, S. D. (1986) Phylogenetic relationships of the true viper, *Eristocophis mcmahoni*, based on parsimony analysis of allozyme characters. *Copeia* **1986**:1014–1020.

14

When are Phylogeny Estimates From Molecular and Morphological Data Incongruent?

DAVID L. SWOFFORD

Many sources of information may be available for use in inferring phylogenetic relationships among a group of taxa. Because different character sets share a common evolutionary history, we expect reliable methods of phylogenetic analysis to recover the correct evolutionary tree regardless of the kind of data employed. To the extent that character sets "tell the truth" about their past, phylogenies inferred from different character sets should be congruent with the true tree and therefore with each other. This line of reasoning suggests that congruence among data sets might provide the strongest achievable evidence that a proposed phylogeny is accurate (e.g., Penny and Hendy, 1986). In practice, however, the ideal of perfect congruence is frequently not achieved. Phylogenies estimated from additional character sets typically disagree in minor details and sometimes contain major discrepancies.

When phylogeny estimates from two character sets disagree in nontrivial ways, we are confronted with a dilemma: both of the estimates cannot be correct, so how do we reconcile the differences? One explanation is that one of the data sets is simply unreliable and that no method of phylogenetic reconstruction could be expected to recover the correct tree given the poor quality of the data. At one time or another, most of us have witnessed or participated in vigorous debates about the relative strengths of alternative sources of information or about the power versus futility of particular techniques. The increase in prominence of molecular approaches to systematics has increased the frequency of such exchanges, especially when molecular evidence controverts traditional notions of phylogenetic relationship within a set of taxa. A fundamental question that must be addressed when apparent conflicts arise is whether the incongruence is "real" (i.e., unavoidable) or merely "spurious" (explicable due to sampling error or to inappropriate assumptions or analytical methods) (Hillis, 1987).

This chapter has two purposes. First, in keeping with the general theme of this volume, I will review several methods currently being used to assess levels of congruence, presenting my own views on the strengths and limitations of each. Second, I will suggest some additional procedures, facilitated by recent improvements in computer software, that allow a more comprehensive examination of the question posed in the title.

Terminology

I will deliberately avoid excessive formalism. In the remainder of the chapter, collections of terminal taxa are referred to simply as *groups*. *Nontrivial* or *informative* groups contain at least two terminal taxa, but not all of the terminal taxa under study (the *universal* group). (In general, the term "group" as used herein can be considered synonymous with the terms *subset* and *component* used by some authors.) Attention is restricted to rooted trees with *labeled* terminal nodes, representing actual terminal taxa, and *unlabeled* internal nodes, representing ancestral taxa that are purely hypothetical. Under these restrictions, the terms "tree" and "cladogram" can safely be used interchangeably. ("Terminal taxa" are equivalent to the *tips, leaves, terms*, or *OTUs* of some authors; hypothetical ancestral taxa are sometimes referred to as *HTUs*.) If the word "taxon" is used without a modifier, it is assumed to mean "terminal taxon."

The term *monophyletic* is used in the sense of Hennig (1966): a group is monophyletic on a tree if, and only if, it contains all of the descendants of its most recent common ancestor. A statement that a group "appears" or "occurs" on a tree, or that a tree "contains" a group, implies that the group is monophyletic on the tree. A *clade* is a monophyletic section (subtree) of a tree, and is identified by the names of its terminal taxa; these terminal taxa constitute a *cluster*.

"TAXONOMIC" CONGRUENCE: CONSENSUS TREES, CONSENSUS INDICES, AND TREE COMPARISON METRICS

Taxonomic congruence refers to the extent to which independent classifications for the same set of taxa support the same groupings. The most commonly used approach for assessing taxonomic congruence, pioneered by Mickevich (1978), involves the calculation of *consensus trees* that summarize areas of agreement among classifications. The amount of structure retained by a consensus tree provides a measure of the congruence among the original classifications. The general approach is illustrated in Figure 14-1. First, two (or more) *rival trees* are constructed using different char-

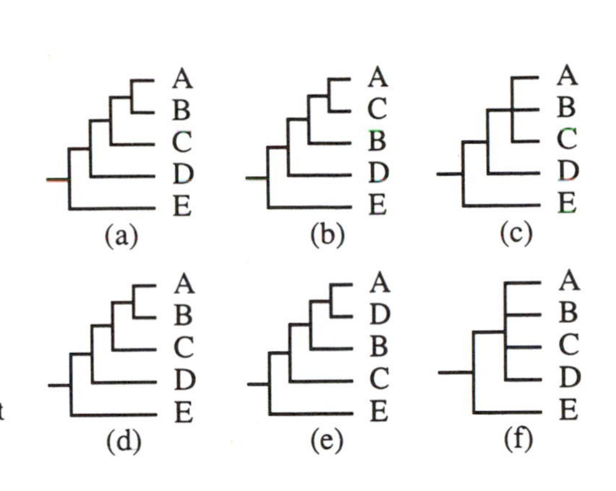

Figure 14-1 Illustration of basic strategy used by Mickevich to compare methods with respect to taxonomic congruence. Two trees (a,b) calculated using method one and a consensus (c). Two trees (d,e) calculated using method two and a consensus (f). Because the consensus tree from method two is less resolved, one can conclude that for this data set, method one generates more congruent classifications.

acters sets (Fig. 14-1a,b). Second, a consensus method summarizes the groupings shared by the rival trees, producing a consensus tree (Fig. 14-1c). A highly resolved consensus tree reflects substantial congruence among the rival trees, whereas a poorly resolved consensus tree may indicate areas of significant disagreement. If another method of tree estimation is applied to the same pair of data sets, different rival trees may result (e.g., Fig. 14-1d,e), perhaps yielding a more poorly resolved consensus (Fig. 14-1f). In this example, the first method generated more congruent classifications than the second.

Mickevich's (1978) study compared various phenetic and numerical cladistic methods with respect to their ability to construct congruent classifications from different data sets. She used the only consensus method available at the time, that of Adams (1972). Since then, a number of additional consensus methods have been proposed (Nelson, 1979; Margush and McMorris, 1981; Sokal and Rohlf, 1981; McMorris and Neumann, 1983; Neumann, 1983; Shao, 1983; Stinebrickner, 1984; Barthélemy and McMorris, 1986; Bremer, 1990). Some of the more commonly used methods are described below.

Strict and Majority-Rule Consensus

Of the existing consensus tree methods, the *strict consensus* is conceptually the simplest. The strict consensus tree, apparently used first by Schuh and Polhemus (1981), was defined by Sokal and Rohlf (1981) as the unique tree that contains only those groups that appear on all of the rival trees. Unfortunately, strict consensus trees were called "Nelson trees" (after Nelson, 1979) by Schuh and Farris (1981). Page (1989a) clearly demonstrates that "strict" and "Nelson" trees are not equivalent, but the improper synonymy has become entrenched in some circles.

Examples of strict consensus trees are shown in Figure 14-2. The greatest advantage of the strict consensus is simplicity of interpretation. If a group

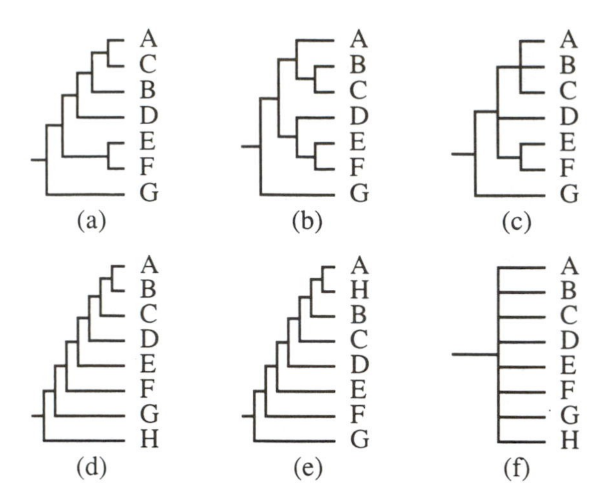

Figure 14-2 Examples of strict consensus. The top row shows two trees (a,b) and their strict consensus tree (c). The bottom row shows two trees (d,e) that are identical except for the placement of taxon H. Their consensus tree is completely unresolved (f), indicating that no groups are shared by the two trees.

appears on the strict consensus tree, we can immediately be certain that it appears on all of the rival trees. However, strict consensus also has a serious liability. Two trees that support identical relationships for all but one of the taxa (e.g., taxon H in Fig. 14-2d,e) may share no groupings when this last taxon is included, so that the consensus tree is a completely unresolved bush (Fig. 14-2f). Adams (1986) and Funk and Brooks (1990) show similar examples.[1] This property has caused some workers to conclude that strict consensus trees are "too strict."

When several trees are being compared, a *majority-rule consensus* (Margush and McMorris, 1981) may be preferable to a strict consensus. Instead of including in the consensus tree only those groups that occur on all of the rival trees, we retain all groups that appear on more than some prespecified proportion of the rival trees. Typically, a proportion of 0.5 is used, so that all groups occurring on more than half of the rival trees are retained in the "50% majority-rule" consensus. Note that a group must be present on *more* than half of the trees to be retained in the consensus, because two groups occurring in exactly half of the trees might not be able to coexist on the same tree. Of course, if only two trees are being compared, the strict and majority-rule methods are equivalent.

Adams Consensus

To my knowledge, the consensus method of Adams (1972) predates all others. Adams actually described two variants on his consensus method,

[1] I am not sure who first called attention to this property. I first heard it described by Joseph Felsenstein at the October 1981 meeting of the Numerical Taxonomy Conference in Ann Arbor, Michigan.

the first applying when the rival trees are "fully labeled" (i.e., internal nodes represent actual ancestral taxa rather than purely hypothetical ones), and the second when all internal nodes are "unlabeled" as is usually the case in phylogenetic analysis. Consequently, Adams consensus trees are sometimes more precisely called *Adams-2 trees*, eliminating any uncertainty about which of Adams' (1972) methods is being used.

Adams consensus trees often preserve more of the structure found in the rival trees than do strict consensuses. For example, Figure 14-3c shows the Adams consensus of the trees in Figure 14-3a and b. It reflects the following statements about relationship made by both rival trees: (1) taxa A and C are more closely related to each other than either is to taxa D, E, or F; (2) taxa E and F are more closely related to each other than either is to taxa A, C, or D; and (3) taxon D is more closely related both to taxa E and F than it is to either taxa A or C. However, in no case does the same group of taxa appear on both of the rival trees; consequently, the strict consensus is a completely unresolved bush. Funk and Brooks (1990) present a similar example (their figs. 14–16). Regrettably, the same authors also present another tree (their fig. 9) that is incorrectly called an "Adams consensus" (of the trees in their figs. 7 and 8), stating that "the Adams tree is better resolved than either of the [rival] cladograms because neither . . . conflicts with the other." Actually, the preservation of all uncontradicted groups by a consensus technique is a property of the combinable component (Bremer, 1990) and Nelson (1979) consensus methods (see below). Although this example was intended to illustrate the differences between Adams and strict consensus methods, when performed correctly the two methods yield the same consensus trees in this case. Furthermore, because they consider strict and Nelson trees to be equivalent, the tree referred to as a Nelson consensus tree (NCT:6) is actually an Adams consensus tree, and vice versa. Thus, Funk and Brooks' (1990) conclusion that "an Adams tree answers those questions about which the clades are not in conflict" is less appropriate than the characterization of Mickevich and Platnick (1989): "In an Adams consensus . . . any taxonomic statements shared by the classifications being compared are included in the consensus, regardless of whether they constitute completely uncontradicted components." That is, the statement that taxa E and F are more closely related to each other than to any of the remaining taxa, made by Funk

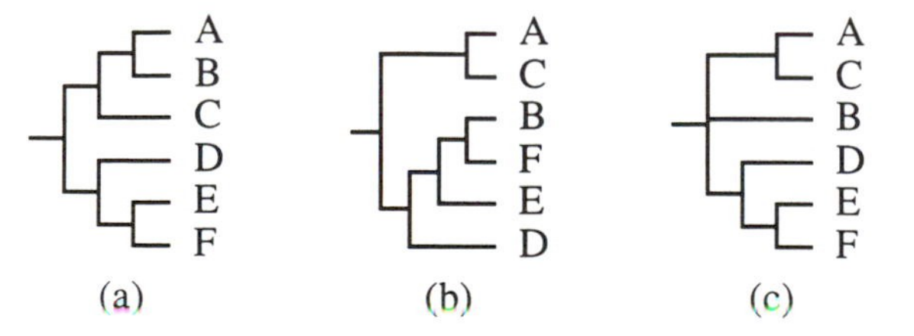

Figure 14-3 An example of Adams (1972) consensus. Tree (c) is the Adams consensus of trees (a) and (b).

and Brooks' cladogram 8, is not *shared by* their cladogram 7, with the result that an EF cluster is not represented in the consensus.

McMorris et al. (1983) suggested that Adams' method lacks a compelling justification and has achieved popularity primarily owing to historical precedence. However, Adams (1986) addressed that criticism by characterizing the Adams consensus as the "best tree that represents all the information shared among a set of trees." In this second paper, Adams suggests that a tree be thought of as a set of "leaf subset nestings" rather than as a "set of clusters." In biological terms, one group of taxa *nests* within a larger one if the more recent common ancestor of the smaller group is a descendant of the most recent common ancestor of the larger group. (Note that this definition does not require monophyly of either group.) He then proves that the Adams consensus tree, as defined in his earlier (1972) paper, is the unique tree that satisfies the following two conditions: (1) any nesting found in all of the rival trees must also occur in the consensus tree; and (2) any nesting that reflects clusters of the consensus tree (i.e., a nesting involving the inclusion of one monophyletic group within a larger monophyletic group) must be found in all of the rival trees.

Adams' (1972) method has also been criticized (Sokal and Rohlf, 1981; Rohlf, 1982; Rohlf et al., 1983) because it may produce consensus trees containing clusters not found on any of the rival trees (see Fig. 14-4). This property mildly complicates the interpretation of Adams consensus trees, but it must be accepted if one agrees that the structure of a tree encompasses more information than a simple listing of its clusters (Adams, 1986). Under Adams' interpretation, this additional information comes in the form of nestings, which are loosely analogous to the "taxonomic statements" of Mickevich and Platnick (1989).

Combinable-Component and Nelson Consensus

When one or more of the rival trees is not fully dichotomous, groups that are never contradicted may occur on some, but not all, of the trees. For example, suppose that three trees contain a clade (A,B) and a fourth tree exhibits an unresolved (A,B,C) trichotomy. In this case, the AB group would not be retained in a strict consensus of the four trees, even though it occurred in three-fourths of the rival trees and was not contradicted by any of them. In this case, we might choose to include the group AB in the

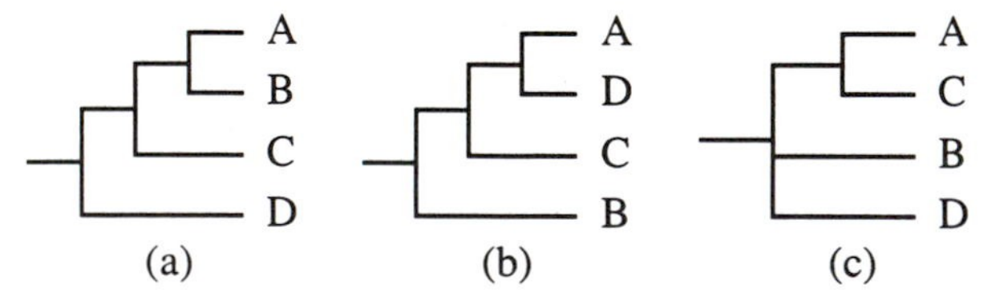

Figure 14-4 An example illustrating that an Adams consensus tree (c) may contain groups not found on any of the rival trees (a, b).

consensus. Bremer (1990) formally described this approach as "combinable-component consensus;" essentially the same idea was put forth by Hillis (1987). Two groups are "combinable" (Nelson, 1979) if either (1) they have no taxa in common ("exclusion"); (2) they are identical ("replication"); or (3) one group is a proper subset (further resolution) of the other ("inclusion"). The combinable-component consensus tree is defined by the set of all combinable groups (i.e., each group retained in the consensus is equal to or combinable with all groups of every rival tree). When all rival trees are fully dichotomous, the strict and combinable-component consensus methods are equivalent.

Considerable confusion exists as to exactly what is meant by "Nelson consensus." Bremer (1990) argues (correctly, in my view) that when exactly two trees are compared, the "Nelson tree" becomes equivalent to the combinable-component consensus tree. Likewise, Mickevich and Platnick (1989:40) state: "In a Nelson consensus, only those components [groups] that are uncontradicted in the classifications being compared are included in the consensus tree." This definition would include combinable, but not necessarily universally occurring, groups in the consensus. But two pages later (Mickevich and Platnick, 1989:42) they assert that "a Nelson consensus includes only those subtaxa *exactly* replicated in the cladograms being compared" (emphasis mine), which is the definition of a strict consensus tree.

Nelson (1979) described his method for obtaining a "general cladogram" as a two-stage procedure. First, a tree containing only those groups found on at least two trees ("replicated components") is constructed. Then, if there are any other unreplicated groups that are combinable with all of the replicated groups, they are included in the consensus as well. If all unreplicated groups are combinable with all replicated groups, as will always be the case when exactly two trees are compared, Nelson's approach becomes equivalent to Bremer's (1990) combinable-component consensus. Thus, Page's (1989a) contention that Nelson's method is equivalent to strict consensus for the two-tree case is true only if the two trees are both fully dichotomous, in which case any unreplicated group will be noncombinable with at least one group appearing on the other tree. For more than two trees, however, Nelson's method can produce results that differ strikingly from both strict and combinable-component consensus trees. Figure 14-5 shows three rival trees in which the replicated groups are AB (trees 5a and 5c), ABC (trees 5a and 5b), and ABCD (all three trees). The two unreplicated groups (AC and ABD) are both noncombinable with at least one of the replicated groups, therefore the Nelson consensus tree includes only the three replicated groups (Fig. 14-5d). On the other hand, the only group retained by the strict (and combinable-component) consensus is ABCD (Fig. 14-5e). In this example, the Nelson and majority-rule consensus trees are identical, but this equivalence does not hold in general.

Unfortunately, Nelson (1979) did not consider the possibility of replicated, but noncombinable, groups, leading McMorris et al. (1983) to conclude that his method need not even specify a tree. Page (1989a) addressed this oversight, providing a formal, unambiguous definition of "Nelson consensus" that is equivalent to Nelson's original procedure when all replicated groups are combinable, but also treats the case of noncombinable replicated groups. Page's method uses standard graph theoretical techniques (e.g., Bron and Kerbosch, 1973) to find all *cliques* of combinable groups: sets of groups such that every group in the set is combinable with every other group in the set. Each clique is assigned a score equal to the sum, over groups included in the clique, of the number of times each group is replicated in the rival trees. (The replication count for a group is simply one less than the number of times the group occurs on the rival trees, so that unreplicated groups do not contribute to the score.) If there is a single clique with the highest replication score, the groups included in this clique define the Nelson consensus tree. If more than one such clique exists, the Nelson consensus contains only those groups found in all of the maximal-replication cliques. In this case, groups found in some, but not all, of the maximal-replication cliques are classified as "ambiguous."

Although no one seems to have suggested it, an alternative procedure would be to include unreplicated as well as replicated groups in the score for each clique, directly analogous to character compatibility analysis (Le Quesne, 1969; Estabrook et al., 1976). Nelson's (1979) emphasis on replicated groups was motivated by his calculations of the small probability of replicating particular groups by chance alone. However, it is not obvious that this preoccupation with replication is justified. Suppose that in a study of 20 taxa, two completely asymmetrical (pectinate) trees contain the group AB. According to Nelson, the probability that the second tree would replicate the AB group found on the first tree equals one divided by the number of distinct ways of choosing a pair of objects (without replacement) from a pool of 20:[2]

$$P(\text{replication}) = \binom{20}{2}^{-1} = 1/190 \approx 0.0053.$$

But if taxa A and B were known to be part of a larger "true" group ABC, then the probability that two random resolutions of a trichotomy involving these three taxa would generate an AB group in both cases is 1/9, greater by a factor of 10. Nelson uses similar logic to establish that combinability of unreplicated groups with replicated groups is not particularly surprising; it is unclear why this line of reasoning should not be extended to some replicated groups as well.

[2] I do not agree with his procedure for calculating these probabilities, but that is a separate issue. A more appropriate procedure (Simberloff, 1987) is to generate a large number of pairs of random trees under a Markovian model and count the number of times a given component is replicated due to chance alone.

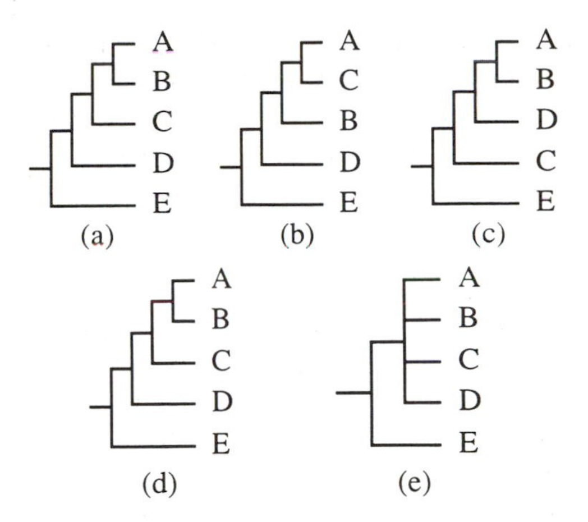

Figure 14-5 An example of Nelson (1979) consensus. Tree (d) is the Nelson consensus of trees (a), (b), and (c). The strict tree (e) is much less resolved than the Nelson consensus, demonstrating that the two methods are not equivalent.

Median Consensus

The median consensus procedure (Barthélemy and Monjardet, 1981; Barthélemy and McMorris, 1986) is closely related to majority-rule consensus. In this method, a tree comparison metric that reflects the topological distance (i.e., disagreement) between any pair of trees (see below) is defined. If we represent by $d(T_i,T_j)$ the distance between a pair of trees T_i and T_j, then the total distance of any tree T to a set of k rival trees is given by

$$d_T = \sum_{i=1}^{k} d(T,T_i).$$

A tree T_m is a *median* if its total distance d_m to the rival trees is less than that for any other tree. If the tree comparison metric is the symmetric-difference distance (Robinson and Foulds, 1981; Hendy et al., 1984; see below), the 50% majority-rule consensus tree is a median tree (Barthélemy and McMorris, 1986). If the number of rival trees (k) is odd or if there are no groups that appear in exactly half of the rivals, the majority-rule tree is the only median tree. If k is even, any tree representing a combination of the majority-rule consensus tree with one or more (combinable) groups appearing in exactly half of the rival trees is also a median tree.

Barthélemy and McMorris (1986) stated that the consensus tree published by Penny et al. (1982) was an example of a median consensus tree. Although Penny et al. (1982) did not use the term "median," D. Penny (personal communication) has also recently suggested that it be used for the consensus method described in their paper. However, the procedures of Penny et al. (1982) and of Barthélemy and McMorris (1986) are different. Penny et al.'s (1982) method chooses the *binary* (fully dichotomous) tree that minimizes the total distance to the rival (binary) trees. The "consensus tree" of Penny et al. (1982) and the majority-rule consensus tree have total symmetric-difference distances of 216 and 192, respectively, to the 39 rival trees, demonstrating that the former is not a median tree under the criterion of Barthélemy and McMorris (1986). To avoid further confusion, I suggest that "median tree" be restricted to its original usage and that "median binary tree" be used for the Penny et al. (1982) procedure. The latter seems particularly useful as a means for selecting a "representative" tree from a set of rival binary trees.

Largest Common Pruned Trees

In all of the consensus methods discussed above, the consensus tree contains all of the taxa included in the rival trees. A rather different approach (Gordon, 1980; Finden and Gordon, 1985) is to find the largest subset (or subsets) of terminal taxa for which the rival trees specify identical relationships. In this method, taxa and their associated branches are removed ("pruned") from the rival trees until the reduced trees become topologically equivalent, yielding a *common pruned tree*. By removing the smallest possible number of terminal taxa, one obtains a *largest* common pruned tree (there may be more than one such tree).

This approach is particularly useful when only a few taxa are responsible for the incongruence among trees, thereby providing a means of identifying "unstable" taxa. For example, the strict consensus of the two trees shown in Figure 14-6a and b is poorly resolved (Fig. 14-6c). However, these two trees are equivalent in topology if taxon B is pruned from both trees (Fig. 14-6d). Terminal taxa not included in a common pruned tree can subsequently be "regrafted" to the tree. An obvious way to perform the regrafting (Gordon, 1980) is to reconnect removed taxa such that any group appearing on all rival trees is preserved in the regrafted tree. The resulting "common pruned-and-regrafted tree" is identical to or a resolution of the strict consensus tree (e.g., Fig. 14-6e). In this example, the pruned-and-regrafted tree is also equivalent to an Adams consensus tree, but the two methods do not always give the same results.

Like other consensus methods, the largest-common-pruned-tree approach has some undesirable properties that limit its utility. One drawback

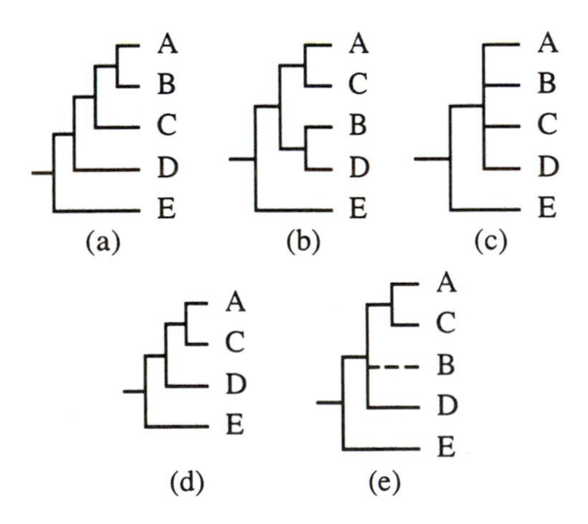

Figure 14-6 An example of largest common pruned trees. Two rival trees (a, b). Strict consensus (c) of these two trees. Largest common pruned tree (d). Common pruned-and-regrafted tree (e) obtained by reattaching taxon B.

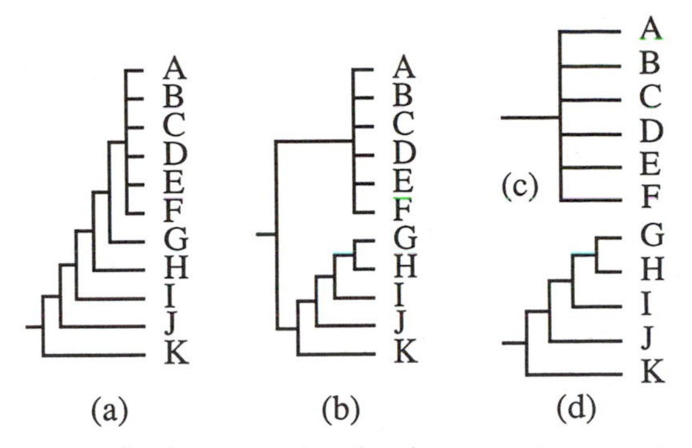

Figure 14-7 An example demonstrating that *largest* common pruned trees are not always preferable. Tree (c), the largest common pruned tree for rival trees (a) and (b), conveys much less information than does a smaller common pruned tree (d).

of the common-pruned-tree method is the difficulty of identifying *largest* common pruned trees. An obvious "brute-force" approach is to examine all ways of pruning k taxa and checking the resulting pruned trees for equivalence, increasing k until a common pruned tree is found. As noted by Finden and Gordon (1985), this method is computationally infeasible for all but small data sets. These authors describe heuristic algorithms that appear to be reasonably effective, but an exact algorithm that avoids the computational disadvantages of the brute-force approach would be more satisfying.

Even if finding them were easy, *largest* common pruned trees are not always preferable, as demonstrated by the example of Figure 14-7. The largest common pruned tree for trees 7a and 7b (Fig. 14-7c), which retains six taxa, is clearly less useful than is the smaller common pruned tree of Figure 14-7d, which retains only five taxa. This example suggests that consideration of all common pruned trees of nontrivial size may provide useful information that is lost if attention is restricted to the largest common pruned tree(s).

Common pruned trees have received little attention from systematists, probably due to their unavailability in widely distributed computer packages such as COMPONENT (Page, 1989b), Hennig86 (Farris, 1988), PAUP (Swofford, 1991), and PHYLIP (Felsenstein, 1990). I hope to include a facility for computing them in a future version of PAUP.

Consensus Indices

Examination of a consensus tree can provide a rough idea of the extent to which a collection of rival trees agrees upon relationships; a *consensus index* provides a quantitative measure of this agreement. Consensus indices typically vary between 0 (implying no agreement among the rival classifications) and 1 (implying the maximum agreement possible). A variety of consensus indices have been proposed, reflecting different perceptions among authors as to what aspects of "consensus" should be measured and how these aspects should be quantified. In this regard, Day and McMorris (1985) argue that the term "consensus index" should be reserved for measures that strictly measure agreement among rival trees. Other measures that consider additional attributes such as information content (including all of the "consensus indices" commonly used by systematists) are called *consensus object invariants*. This distinction highlights an interesting contrast in the perspectives of mathematicians versus biologists. To Day and McMorris (1985), a group of identical trees exhibits "perfect" agreement even if the trees are totally unresolved. These authors consider it axiomatic that a consensus index should take the value of 1 in the presence of such agreement, regardless of the nature of the "vote." On the other hand, a biologist views unanimous agreement on a "bush" as agreement on nothing at all (e.g., Mickevich and Platnick, 1989), and therefore wants the index to take the value 0. In the following discussion, I will use the term "consensus index" in the less restrictive sense. Rohlf (1982) and Mickevich and Platnick (1989) have presented thorough and readable discussions of consensus indices, including computational details. Consequently, I will confine my remarks to brief characterizations of some of the more commonly used measures.

The simplest consensus index simply quantifies the amount of resolution of a consensus tree and is equal to the number of nontrivial clusters that the tree contains (= component information of Nelson, 1979) divided by the maximum possible such number ($t - 2$, where t is the number of terminal taxa). Colless (1980, 1981) has referred to this quantity as the "consensus fork" index. However, the degree of resolution of a classification does not adequately reflect the amount of information it contains. For example, both trees in Figure 14-8 contain two nontrivial groups, but the tree of Figure 14-8a clearly allows more statements about relationship to be made, and therefore conveys more information (see Mickevich and Platnick, 1989 for an eloquent discussion of this issue).

Several measures that combine information and agreement into a single index have been proposed. These indices all contain a numerator (score) representing information of some sort divided by a limiting value (nor-

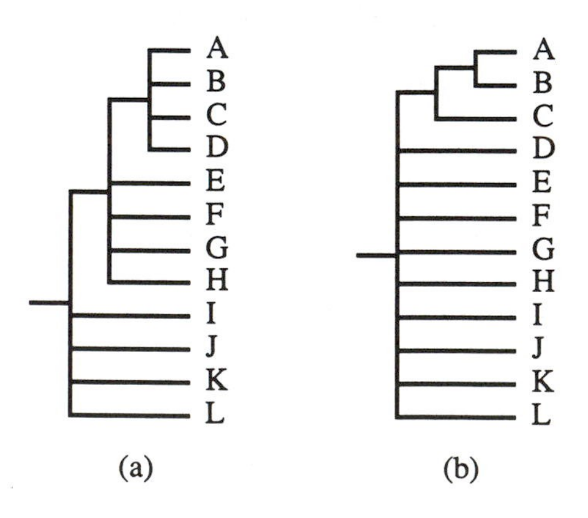

Figure 14-8 Demonstration that consensus trees with the same degree of resolution (number of nontrivial groups) convey different amounts of information. Tree (a) allows more statements of relationship to be made than does tree (b).

malizing factor) representing the highest achievable score for any tree. The earliest measure, Mickevich's (1978) *CI*, was defined in terms of the number of "extra steps" required for a set of artificial cluster variables (describing the nontrivial clusters on the consensus tree) to evolve on a bush under the Wagner parsimony criterion (Farris, 1970). Rohlf (1982) showed that the numerator of Mickevich's index could be expressed equivalently as a sum, over all nontrivial clusters contained in the consensus, of cluster "weights:" the weight for the *i*-th cluster equals $\min(n_i - 1, t - n_i)$, where n_i is the number of terminal taxa in cluster i and t is the total number of terminal taxa. Colless (1980) suggested simply using n_i as the weight for each cluster, which yields a *weighted consensus fork* index proportional to Nelson's (1979) *total information*.[3] Mickevich and Platnick (1989) apparently rediscovered the same index, calling it P_m, the "proportion of the maximal total information." They noted that total information could be defined using artificial variables, analogously to Mickevich's original formulation: total information is equivalent to "WISS total steps," the total number of steps required for the artificial variables to evolve on a bush under the Camin-Sokal parsimony criterion (Camin and Sokal, 1965; Farris et al., 1970). Yet another related measure is the "levels sum" (Schuh and Farris, 1981), which counts the number of times distinct pairs of terminal taxa occur together in the same informative cluster. It is computed by summing, over all pairs of taxa, the number of informative clusters that contain both members of each pair. Rohlf (1982) demonstrated that the procedure was equivalent to using a cluster weight of $n_i(n_i - 1)/2$ and determined the limiting value to be $t(t - 1)(t - 2)/6$ (note correction of typographical error in his formula).

A property of all of the indices described in the above paragraph is that they are sensitive to the symmetry or balance of the tree (Colless, 1980; Rohlf, 1982; Shao and Rohlf, 1983; Shao and Sokal, 1990). The normalized total-information measure and the levels sum can achieve their maximum value only when the consensus tree is fully resolved and maximally asymmetrical (Colless, 1980; Rohlf, 1982). Rohlf (1982) makes the same claim for Mickevich's (1978) *CI*. However, at least as originally defined by Mickevich, *CI* can in fact achieve its maximum value on a tree that is maximally symmetrical (Fig. 14-9).

An important implication of asymmetry bias in a consensus index is that two different asymmetrical trees can achieve a higher score on the index than two identical symmetrical trees. Whether this property is viewed as an asset or a liability depends upon one's point of view. If we want a consensus index to reflect only agreement among trees, we clearly do not want to use a measure that is also sensitive to asymmetry. If we instead want an index to reflect, in some sense, the amount of information shared by a set of trees, then we must accept the possibility that different, asym-

[3] The formula given by Colless for calculating the limiting value of his index contained a typographical error and should instead be $(n + 1)(n - 2)/2$. It is clear from his calculations that he used the correct formula.

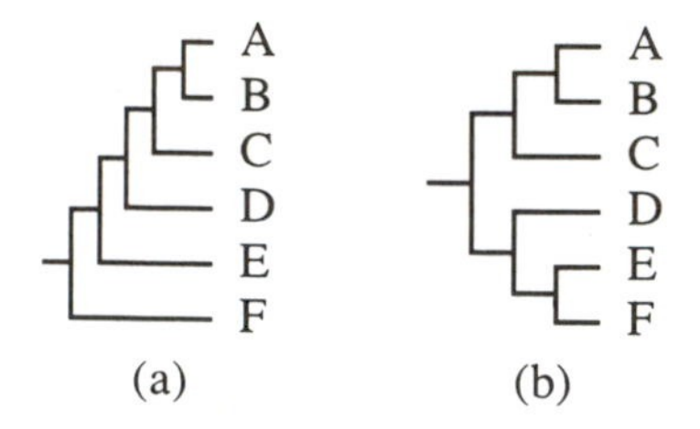

Figure 14-9 Demonstration that Mickevich's consensus index can achieve its maximum value on a tree that is not maximally asymmetrical. $CI = 1.0$ for both trees.

metrical trees may allow more shared statements about relationship than trees that are identical but symmetrical, and we should therefore design our index accordingly. If we then use this index to evaluate the extent to which alternative methodological approaches generate congruent classifications (e.g., Mickevich, 1978), we must be careful to avoid overstating our conclusions. Specifically, the finding that two different data sets produce the same estimate of phylogenetic relationships using method "X" says nothing about the truth of those relationships (i.e., the "false confidence" problem cited by Faith, 1988). For example, two data sets will always produce identical estimates of a phylogeny if the "estimation" procedure consists of placing the terminal taxa in alphabetical order on a completely pectinate tree. Conversely, if we are willing to assume that a method is reliable in general, and we want to use taxonomic congruence as a means of assessing confidence in a particular result, we would not want to use a consensus index that penalized the method for getting the correct result if that result happened to correspond to a symmetrical tree.

It is possible to design consensus indices that preserve the desirable group-size properties without suffering from excessive tree asymmetry bias. For example, Rohlf's (1982) CI_1 always takes the value 1 for a fully resolved consensus tree (i.e., identical, fully resolved rival trees), regardless of the degree of asymmetry. Mickevich and Platnick's (1989) statement that Rohlf's measure amounted to "mutating the consensus information index (Mickevich, 1978) to exclude considerations beyond resolution alone" represents a misunderstanding of Rohlf's index. It is clear that Rohlf wants the index to reflect more than resolution alone, but he does not want it to take a value less than 1 if two or more fully resolved rival trees agree completely. As such, CI_1 does not deserve to be tossed onto the scrap heap along with Colless' consensus fork index. Indeed, for the trees shown in Figure 14-10a and b, which are equally well resolved and therefore have equal consensus fork indices, both Rohlf's CI_1 and Mickevich's CI are exactly three times greater for the "more informative" tree 10a than for tree 10b. However, CI_1 is greater for tree 10c, which implies complete agreement of the rival trees, than for tree 10d, which might represent the consensus of two trees that specified different relationships for taxa A, B, and C. In contrast, both Mickevich's CI and the weighted consensus fork (P_m) are higher for tree 10d than for tree 10c.

Rohlf (1982) also suggested another intriguing index: the proportion of all possible binary trees that contain the clusters found in the consensus

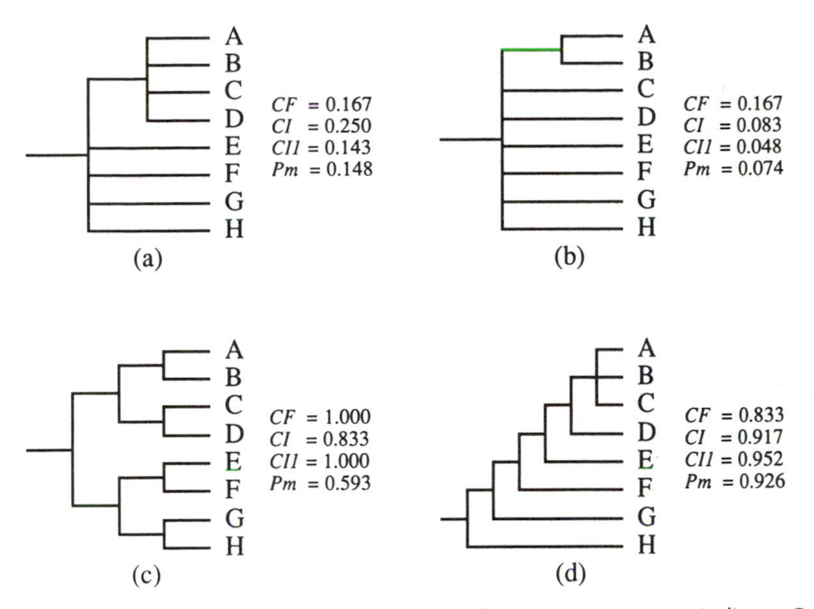

Figure 14-10 Effect of tree symmetry and resolution on consensus indices. Consensus trees (a) and (b) both contain a single nontrivial group, but measures such as Rohlf's (1982) CI_1, Mickevich's (1978) CI, and Mickevich and Platnick's (1989) P_m (= Colless' weighted consensus fork) take higher values for tree (a) than for tree (b), reflecting the higher information content of (a). Although consensus tree (c) is fully resolved, it takes lower values for CI and P_m than the partially unresolved, but more asymmetrical consensus tree (d). (CF = Colless' unweighted consensus fork.)

tree. The number of binary trees that contain the clusters of the consensus tree is equivalent to the "number of fully resolved cladograms allowed by a [strict] consensus" (Mickevich and Platnick, 1989; see also Mickevich and Farris, 1981). For example, suppose that a strict consensus tree with eight terminal taxa contains two dichotomies, one trichotomy, and one tetrachotomy. We can fully resolve the trichotomy in three different ways and the tetrachotomy in 15 different ways, so there are $3 \times 15 = 45$ binary trees allowed by the consensus. (The number of ways to "fully resolve" a k-tomy is equivalent to the number of rooted binary trees for k taxa; e.g., Felsenstein, 1978.) Since there are 135,135 possible rooted binary trees for eight taxa, Rohlf's (1982) CI_2 is then $45/135{,}135 = 3.33 \times 10^{-4}$. Thus, the number of trees allowed by the consensus is a tiny fraction of the number of possible trees. Mickevich and Platnick (1989) use the complement of CI_2, corresponding to the "number of prohibited resolutions," which follows the usual convention of letting the value 1 indicate maximum resolution. Mickevich and Platnick also tackle the more difficult problem of computing the number of resolutions prohibited by an Adams (1972) consensus tree. They recommend using the product of the proportion of prohibited Adams resolutions and the normalized total-information index as a "general information index."

A legitimate criticism of all of the consensus indices described above is that they depend only on the consensus tree; once this tree has been calculated, all contact with the original rival trees is abandoned (Day and McMorris, 1985). Thus, for example, it is not possible to determine whether a consensus index is low because there was substantial disagreement among a set of well-resolved trees or because there was general agreement among a set of poorly resolved trees. Day and McMorris (1985) categorize all of these indices as "type 1 paradigms." In paradigms of "type 2" the consensus index is computed after calculation of a consensus tree, but the index depends on both the form of the consensus tree and on the original trees. An example is the α index of Day (1983), which does not appear to have an exact algorithm for its calculation. It has been ignored by systematists, perhaps because it satisfies the Day-McMorris axiom specifying that it always take the value 1 for identical rival trees, even if they are unresolved. In "type 3 paradigms," the consensus tree and index are specified simultaneously by an incremental process in which addition of groups to the consensus tree determines corresponding changes to the consensus index (e.g., *s*-consensus index of Stinebrickner, 1984, 1986; consensus index ci_n of Adams, 1986). Unfortunately, none of these indices provides what we really need: an index that is sensitive to both agreement among and information content of the original trees, but that also allows us to quantify the relative contributions of each of these aspects to lack of resolution in the consensus. We cannot hope to achieve such precision and versatility in a one-dimensional index, but further work in this area may prove fruitful.

What Does the Magnitude of a Consensus Index Really Mean?

What does it mean to say that a consensus index equals, say, 0.7? Technically, for many indices, this value implies that the rival trees share 70% as much information on relationships as does a set of identical, maximally asymmetrical trees. But does a value of 70% indicate substantial agreement and/or shared information, or should we be dejected over the failure to find the congruence we anticipated? This problem of interpretation presents a serious obstacle to the use of consensus methods in quantifying levels of congruence. It may be useful to test, using Monte Carlo methods, whether a given consensus index is greater than that expected for a set of randomly chosen trees (Shao and Rohlf, 1983; Shao and Sokal, 1986). For the case of two rival trees, Shao and Sokal (1986) provide critical-value tables for eight different consensus indices over a range of taxon numbers. A similar approach has been used by Simberloff (1987). One difficulty with these methods is that the null hypothesis is relatively uninteresting; it would be extremely unlikely for the amount of congruence to be so low that we would be unable to reject the hypothesis that our trees were no more congruent than a random sample of trees. Unfortunately, it is not obvious how a "more interesting" null hypothesis would be formulated.

General Criticisms of Consensus Methods

Miyamoto (1985) and Carpenter (1988) have criticized the use of consensus trees as a basis for classification because, in general, they provide a less parsimonious explanation of the original character data—and therefore less "explanatory power"—than do the original trees from which the consensus is derived. Technically, this criticism does not pertain to the use of consensus methods for the study of congruence, because there is no necessary connection between the calculation of a consensus tree as the basis of a consensus index and the establishment of a classification based on that consensus tree. However, the fitting of character data to consensus trees in this way is inappropriate in any case. Consensus trees are simply statements about areas of agreement among trees; they should not be interpreted as phylogenies. Polytomous nodes on a consensus tree do not indicate simultaneous cladogenetic events ("hard" polytomies; Maddison, 1989), as assumed by the character-state reconstruction methods used by Miyamoto (1985) and Carpenter (1988). Rather, polytomies on the consensus tree indicate areas of uncertain resolution ("soft" polytomies). Maddison (1989) has described algorithms for reconstructing character changes under the uncertain-resolution (soft polytomy) interpretation that effectively allow a polytomy to be resolved in the way that is most favorable for each character considered independently. These algorithms thus provide a more appropriate mechanism for fitting character data to a consensus tree, and the criticisms of Miyamoto (1985) and Carpenter (1988) no longer apply if they are used.

More to the point is Miyamoto's (1985) argument that because the original character data are no longer relevant once the "fundamental" trees have been constructed, all information about the relative amounts of character support favoring alternative resolutions is discarded (see also Cracraft, 1983). For example, suppose that the groupings AB, AC, and BC are supported by three, two, and one nucleotide position characters, respectively. Further suppose that the AC group is supported by 20 characters in a morphological data set, with no characters supporting either the AB or BC groupings. A consensus method would then see only the conflicting AB and AC groupings from the molecular and morphological trees, respectively, and would be forced to accept an (A,B,C) trichotomy, despite the weak, ambiguous support for AB in the molecular data, and the strong unambiguous support for AC in the morphological data. Because of these considerations, Miyamoto (1985) and others (e.g., Kluge, 1989) have recommended combining all of the available data sets into a single pooled data set rather than computing trees separately for each. Again, while this strategy may be desirable for classificatory purposes, it is less useful if we wish to examine the extent to which different data sets support congruent taxonomic relationships.

Yet another difficulty with the use of consensus approaches to evaluate congruence concerns the selection of trees for comparison. If we compare only the optimal (i.e., most-parsimonious, or best-fitting according to some other criterion) trees for each data set, we effectively ignore the uncertainty associated with the estimate of the tree. For example, the best tree for a molecular data set may be highly incongruent with the best tree for a morphological data set. Yet, there may be another tree that explains the molecular data *nearly* as well as the "best" one that is virtually identical to the tree derived from the morphological data. Evidence is accumulating that this possibility is a very real one. Maddison (1991) shows several examples of data sets that generate two or more "islands" of equally parsimonious trees: trees within an island are (by definition) relatively similar in topology to other trees in the same island, but trees in different islands can be strikingly different, as shown in Figure 14-11. Hendy et al. (1988) have also discussed this phenomenon. Clearly, before reaching a conclusion that phylogeny estimates from two data sets are incongruent, we need to examine the possibility that consideration of near-optimal trees for one or both data sets may reduce the apparent amount of incongruence. The hypothetical example of Figure 14-12 illustrates this point. The most-parsimonious trees for a morphological (Fig. 14-12b) and a molecular (Fig. 14-12c) data set are relatively incongruent. However, a third tree (Fig. 14-12d), while not the best tree for either data set, is the *second* best tree for both, and is only one step longer than the most-parsimonious tree for the combined data set.

Tree Comparison Metrics

Another approach to assessing taxonomic congruence is through the use of *tree comparison metrics* that quantify differences in the shapes of pairs of trees. These methods have the advantage of not requiring preliminary calculation of a consensus tree. On the other hand, the distance between two trees—just a number—is more difficult to interpret than a consensus tree that immediately indicates areas of agreement and a consensus index that varies between predictable bounds. Penny and Hendy (1985) outline a number of interesting applications of tree comparison metrics. For instance, if the probability distribution of a particular metric is known, we can ask whether any pair of trees is more similar than would be expected by chance alone (e.g., Penny et al., 1982). We can also test whether the positioning of one taxon (or a small set of taxa) is particularly unstable by watching the decrease in tree-to-tree distances as we prune one (or a few) taxa from the trees.

The most widely used tree comparison metric has been the symmetric-difference distance, or partition metric (Robinson and Foulds, 1979; Robinson and Foulds, 1981), defined as the number of groups that appear on one tree or the other but not on both. (It can be defined equivalently as the minimum number of branch contractions and decontractions, in any order, required to transform one tree into the other, see Robinson and Foulds, 1981.) Its chief advantages are that it is easy to calculate and it

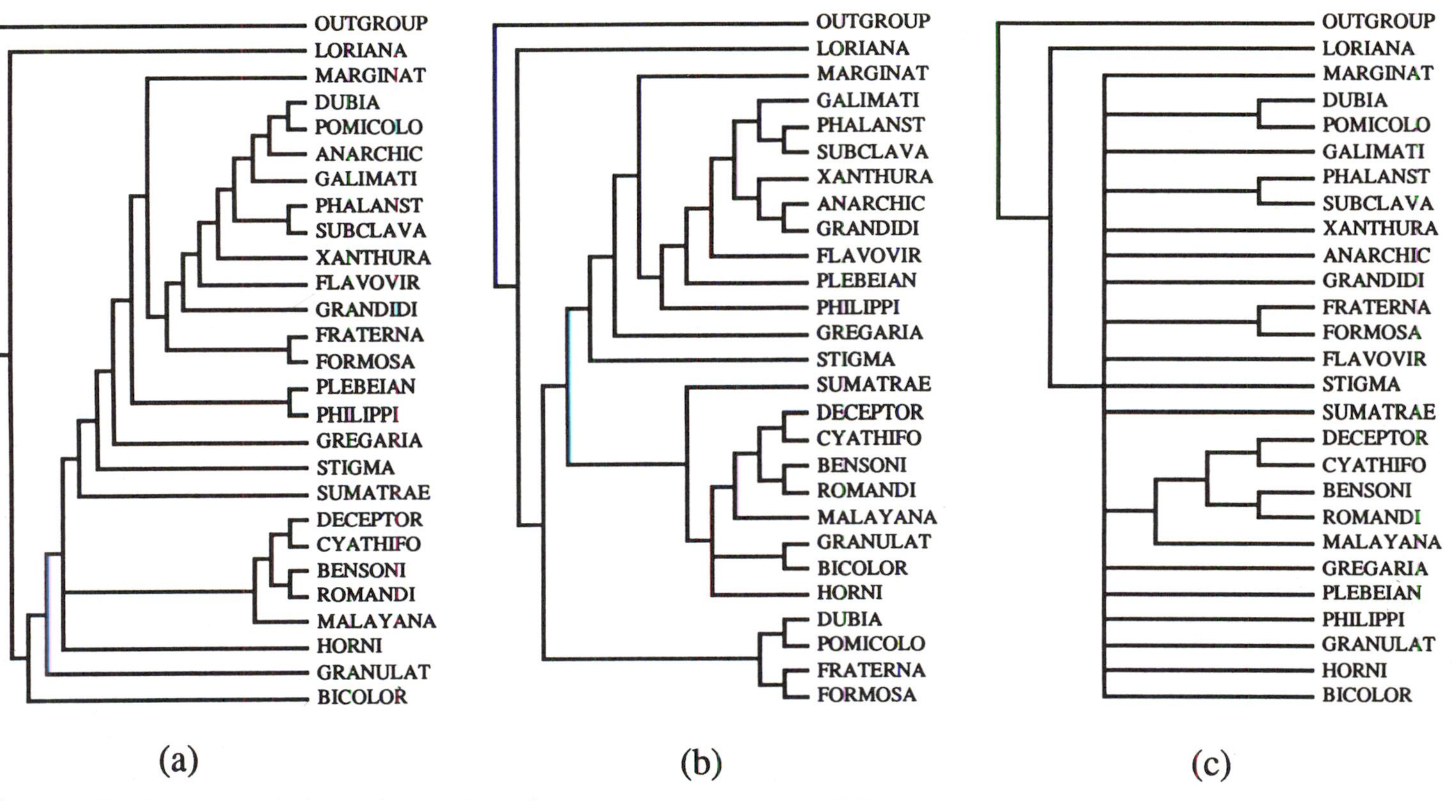

Figure 14-11 Two trees (a,b) from different "islands" for James Carpenter's "ROPA10" data set (see Maddison, 1991). Their strict consensus is shown in (c); see text.

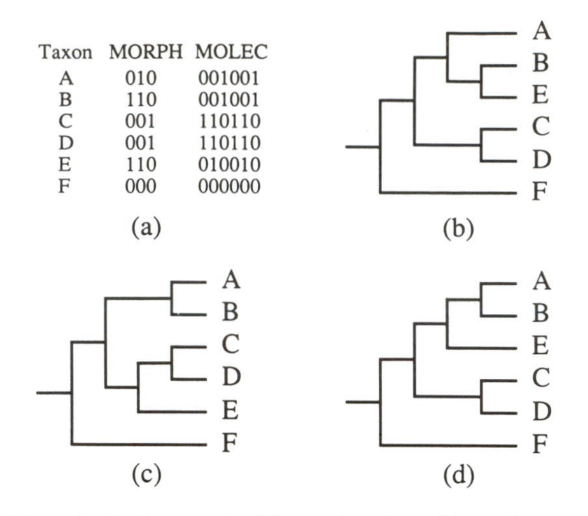

Taxon	MORPH	MOLEC
A	010	001001
B	110	001001
C	001	110110
D	001	110110
E	110	010010
F	000	000000

Figure 14-12 Hypothetical example illustrating that although a tree may not be optimal for either of two data sets, it may explain both data sets reasonably well. A morphological data set (MORPH) containing three characters and a molecular data set (MOLEC) containing six characters (a). Most-parsimonious tree (b) for the morphological data set (length for MORPH = 3; length for MOLEC = 10; total length = 13). Most-parsimonious tree (c) for the molecular data (length for MORPH = 5; length for MOLEC = 6; total length = 11). Second most-parsimonious tree (d) for both data sets (length for MORPH = 4; length for MOLEC = 8; total length = 12).

works equally well for binary and nonbinary trees. Because of its computational tractability, Hendy et al. (1984) have been able to calculate its probability distribution for up to 16 taxa on binary trees and 12 taxa allowing nonbinary trees. It does possess a drawback similar to that of strict consensus trees, however, in that two trees that are identical except for the placement of one terminal taxon can take the maximum possible distance value. Penny and Hendy (1985) suggest strategies for coping with this problem.

A well-studied, but relatively little used tree comparison metric is the "crossover" or "nearest-neighbor interchange" metric (e.g., Robinson, 1971; Waterman and Smith, 1978). This distance measure lacks an efficient exact algorithm, but useful approximations have been developed (Brown and Day, 1984). Its probability distribution is known for up to eight taxa on binary trees (Jarvis et al., 1983). Other measures of tree dissimilarity have been suggested by Farris (1973), Estabrook et al. (1985), and Crozier et al. (1986).

"CHARACTER" CONGRUENCE

Character congruence (Kluge, 1989) refers to the extent to which all available characters make a unified, internally consistent statement about the relationships of the taxa under study. For the purpose of studying congruence between data sets, placing the emphasis squarely on the primary data (characters) rather than on an abstraction of the primary data (trees derived from those characters) has tremendous appeal. In particular, such an approach permits a more direct assessment of the relative strength of support for incongruent character distributions.

Measures of Character Fit

The *consistency index* (Kluge and Farris, 1969) has traditionally been used to evaluate the fit of character data on phylogenetic hypotheses. Following the notation and terminology of Farris (1989), the consistency index for a single character, c, equals m/s, where m is the minimum amount of change that the character may show on any tree and s is the amount of change required by the character on the tree being evaluated. The quantity $(1 - c)$ then represents the proportion of change which can only be explained as homoplasy. An "ensemble" consistency index C for a suite of characters (e.g., those included in a data matrix) can be calculated as M/S, where M and S are the sums over all characters in the suite of the individual-character m and s values, respectively. Ideally, one could use the consistency index as a direct measure of character congruence, but several complications arise. For instance, as more taxa are added to a study, our ability to detect instances of homoplasy improves. The natural prediction that the consistency index would be negatively correlated with the number of taxa in a data set has been confirmed empirically (Archie, 1989; Sanderson and Donoghue, 1989), demonstrating that C cannot stand alone as a measure of character incongruence.

Another problem with the consistency index is that it is strongly affected by the distribution of character-state frequencies. For example, C is always equal to 1 for a character for which a single taxon possesses state 1 and the rest state 0. Consequently, it is often suggested that autapomorphies be excluded from the data set before calculating the ensemble consistency index. But that strategy just shifts the problem slightly: binary characters with a "$(2, t - 2)$" distribution of states cannot possibly have a consistency index less than 1/2, and so on. Farris (1989) proposed two new indices, the *retention index* and the *rescaled consistency index*. The *retention index*, r, is defined as $(g - s)/(g - m)$, where s and m are as defined above, and g is the maximum possible amount of change that a character could possibly require on any tree (equal to the length of the character on a completely unresolved bush). Thus when a character fits the tree as poorly as possible, its retention index will be 0, so that it can be more meaningfully interpreted as a congruence statistic. Note that for uninformative (e.g., autapomorphic) characters, $m = g$ so that r is undefined. The ensemble retention index R is defined analogously to the ensemble consistency index; that is, $(G - S)/(G - M) = (\Sigma g - \Sigma s)/(\Sigma g - \Sigma m)$. Farris (1989) recommends using r as a factor for scaling c between 0 and 1, hence rc and RC are called the "rescaled consistency index" and the "ensemble rescaled consistency index," respectively.

Archie (1989) has adopted an alternative approach to measuring character incongruence based on random permutations of rows (taxa) within columns (characters) of a data matrix. He estimates a statistic called the *homoplasy excess ratio (HER)* which can be thought of as a modified ensemble retention index in which G, the length of the longest possible tree, is replaced by *expected* length of a tree from the randomized data (his "observed homoplasy excess"). He suggests that the *HER-Maximum*, equivalent to Farris's retention index, can be used to approximate *HER*.

Character Incongruence Indices

Mickevich and Farris (1981) outlined a method for measuring character incongruence that partitions the total incongruence i_T (extra steps or homoplasies) into a within-data-set component i_W and a between-data-set component i_B. The partitioning is accomplished by computing most-parsimonious (minimal) trees for each data set separately and after pooling the data into a single combined data set. Let $E(X)$ and $E(Y)$ represent the number of extra steps required by minimal trees for two data sets X and Y, respectively, and let i_T represent the total incongruence (number of extra steps required by a minimal tree for the combined data set). The within-data-set incongruence, i_W, is then equal to $E(X) + E(Y)$ and the extra length i_B required due to incongruence between data sets is given by $i_T - i_W$. In other words, any excess homoplasy required by the combined data beyond that needed to explain each data set separately is attributed to incongruence between the two data sets. The Mickevich-Farris incongruence index I_{MF} can be expressed as the proportion of the total incongruence i_T that is due to between-data-set incongruence:

$$I_{MF} = i_B/i_T.$$

While this approach has intuitive appeal, it is not without problems. One difficulty is the need to combine the data sets into a single matrix. If, for example, one data set were cranial morphology and the other ribosomal RNA (rRNA) sequences, the morphological data set could easily be overwhelmed by the large number of molecular characters. Miyamoto (1985) suggested that when the numbers of characters in each data set become too unbalanced, character weights could be assigned so that each data set would exert the same total influence. A potentially more serious limitation is illustrated in the following example. Suppose that a given data set for taxa A, B, C, and an outgroup may contain only two types of characters: "type I" characters that support the tree ((AB)C) and "type II" characters that support the tree (A(BC)). Further, suppose that the distribution of character types is 10:0 in data set X (i.e., 10 characters of type I, and 0 characters of type II) and 5:5 in data set Y. The most-parsimonious tree for the combined data set, ((AB)C), requires five extra steps, and I_{MF}

equals [5 − (0 + 5)]/5 = 0. We would therefore conclude that there is no between-data-set incongruence, even though half of the characters in data set Y contradict the tree unanimously supported by the characters in data set X. This conclusion seems unreasonable.

M. Miyamoto (in personal communication to Kluge, 1989) suggested an alternative procedure for partitioning between- from within-data-set incongruence. As for Farris and Mickevich's method, minimal trees are first constructed for each data set treated separately. However, rather than combining the two data sets, each data set is mapped onto the tree computed for the *other* data set (the possibility of equally parsimonious trees is ignored for the moment). Let $F(A^*b)$ denote the number of extra steps required to explain the evolution of data set A on tree b, a minimal tree for data set B. The within-data-set incongruence is calculated as before:

$$i_W = F(X^*x) + F(Y^*y),$$

but the total incongruence is defined as

$$i_T^* = F(X^*y) + F(Y^*x).$$

The extra length due to incongruence between two data sets X and Y, expressed as a proportion of the total incongruence, is

$$I_M = (i_T^* - i_W)/i_T^*$$

(Kluge, 1989). Thus, if tree y explains data set X nearly as well as does tree x, and if tree x explains data set Y nearly as well as does tree y, then the incongruence value will be small, *regardless of how dissimilar trees x and y are*. Miyamoto's technique therefore survives one of the objections to consensus tree approaches raised above. A slight complication arises if there is more than one minimal tree for one or both data sets. In this case, a data set may require a different length depending on which tree from the opposite data set is chosen. One could use either the best-fitting tree, or average the lengths over all of the trees; the first alternative seems more in keeping with the spirit of parsimony.

Unfortunately, I_M shares some of the undesirable properties of I_{MF}. Modifying slightly the example given above (M. Miyamoto, personal communication), suppose that the distribution of character types I:II is 9:0 in data set X and 5:4 in data set Y. Tree ((AB)C) is then the minimal tree for both data sets, with I_M equal to [(0 + 4) − (0 + 4)]/4 = 0. As for I_{MF}, it does not seem reasonable to conclude that there is no between-data-set incongruence when nearly half of the characters in one data set contradict a tree supported by all of the characters in the other data set. Furthermore, if the distribution of character types were instead 4:5 in data set Y, the most-parsimonious tree for Y would be (A(BC)), with which all of the characters in X would be incongruent. In this case, I_M = [(9 + 5) − (0 + 4)]/(9 + 5) = 0.71. The relatively minor change from a 4:5 to a 5:4 distribution of character types I:II changes the between-data-set incongruence measure from none to over 70%! Because of these considerations, it is probably unreasonable to expect the complex notion of between-data-set incongruence to be adequately captured by a simple index.

A MULTIFACETED APPROACH TO THE STUDY OF CONGRUENCE: AN EXAMPLE USING KLUGE'S *EPICRATES* DATA

Kluge (1989) examined congruence in two data sets for snakes of the genus *Epicrates*: a biochemical data set composed of 24 skin and scent gland lipid characters (Tolson, 1987), and a morphological data set composed of 53 skeletal and external phenotypic characters. Although the biochemical data are not, strictly speaking, "molecular," this example is particularly useful for illustrating many of the concepts discussed above. Kluge calculated minimal trees separately for the morphological and biochemical data (first two trees in Fig. 14-13 and first 10 trees in Fig. 14-14, respectively; see also consensus trees in Figs. 14-15a–c) and for the combined data set (see Kluge, 1989). The morphological, biochemical, and combined data sets required 35, 4, and 44 extra steps, respectively. The Mickevich-Farris incongruence index (I_{MF}) is thus equal to $[44 - (35 + 4)]/44 = 0.114$. On this basis, Kluge (1989:16–17) concluded:

> Only 11.4% (5/44) of the total character incongruence is due to the disparity between data sets . . . Thus, disagreement between the biochemical and morphological characters is small relative to incongruence among the characters within each set. The two qualitatively independent sources of information are not strikingly contradictory in this example.

Although this conclusion seems reasonable on the surface, it deserves further scrutiny. For example, the group (*E. chrysogaster, E. exsul*) is supported by six unique and unreversed synapomorphies on all of the shortest trees for the biochemical data. However, this group is well separated on both of the shortest trees for the morphological data, with *E. exsul* more closely related to five other taxa than it is to *E. chrysogaster*. Also, the group (*E. inornatus, E. subflavus*) is supported by three unique and unreversed synapomorphies on both minimal trees for the morphological data but does not appear in half of the minimal trees for the biochemical data. In order to examine the question of congruence in *Epicrates* in more detail, I have conducted a number of additional analyses that extend those of Kluge (1989). Many of these analyses were facilitated by features new to version 3.0 of PAUP (Swofford, 1991).

The Incongruence Index I_M

The incongruence index I_M can be calculated using PAUP by first "excluding" the biochemical character set from consideration and finding minimal trees for the remaining data in order to obtain the within-data-set incongruence for the morphological characters. The incongruence of the biochemical characters with the trees obtained from the morphological data can be calculated by reincluding the biochemical character set, then excluding the morphological character set and requesting the length required by the biochemical characters on the same set of trees. An analogous procedure is used to quantify incongruence within the biochemical data set and incongruence of morphological characters on the trees computed for the biochemical data. For the *Epicrates* data, the biochemical characters require 34 and 36 steps, respectively, on the two minimal trees for the morphological data (Fig. 14-13) and the morphological data require from 105 to 116 steps on the 10 minimal trees for the biochemical data (Fig. 14-14). Subtracting the minimum possible number of steps for each data set, we find that at least $34 - 24 = 10$ extra steps are required by the biochemical data on the "morphological" trees and at least $105 - 65 = 40$ extra steps are required by the morphological characters on the "biochemical" trees. Thus, I_M equals $[(10 + 40) - (35 + 4)]/(10 + 40) = 0.220$, which suggests about twice as much between-data-set incongruence as does I_{MF}.

Consensus Trees and Indices

A strict consensus tree for the two shortest morphological and 10 shortest biochemical trees is shown in Figure 14-15d. Relatively little structure is preserved, as reflected by the low consensus index values in Table 14-1. The Adams consensus tree (Fig. 14-15e) is somewhat more resolved, however. The presence of a (*E. chrysogaster, E. exsul*) group in the Adams consensus may seem surprising given the marked separation of these two taxa on both morphological trees. However, this pair of taxa consistently nests within a larger group containing all taxa exclusive of *E. cenchria* and *E. angulifer*, and is therefore retained.

One of several largest common pruned trees for the same 12 trees is shown in Figure 14-15f. (Others can be obtained by substituting *E. inornatus* for *E. subflavus* and/or *E. fordii* for either *E. gracilis* or *E. monensis*.) While it is useful to know that all 12 of the trees agree on the relationships for these subsets of taxa, it is also disheartening that obtaining a common pruned tree requires the removal of nearly half of the taxa.

Suboptimal Trees

As discussed above, incongruence between two data sets may be more apparent than real if a tree can be found that is at most slightly suboptimal for both data sets. In the case of parsimony analysis, it is relatively simple to determine how many near-minimal trees exist by examining tree-length frequency distributions. Because the number of rooted binary trees for 10 taxa is so large (= 34,459,525), I approximated the shape of the frequency distribution of tree lengths by computing the lengths of 10,000 randomly chosen trees for each data set (Fig. 14-16). The distributions are strongly left-skewed for both, indicating that both contain strong phylogenetic signal

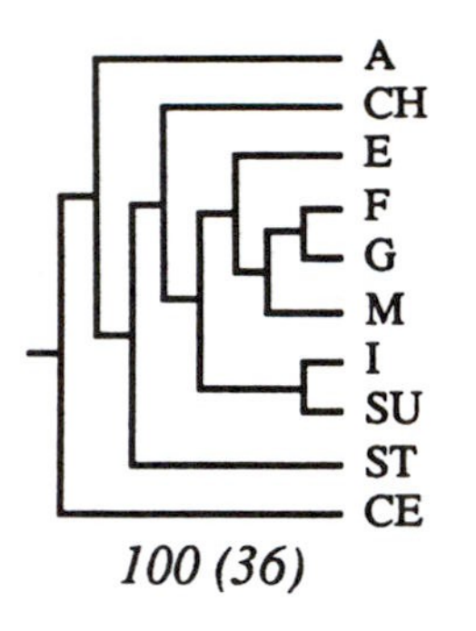

100 (36)

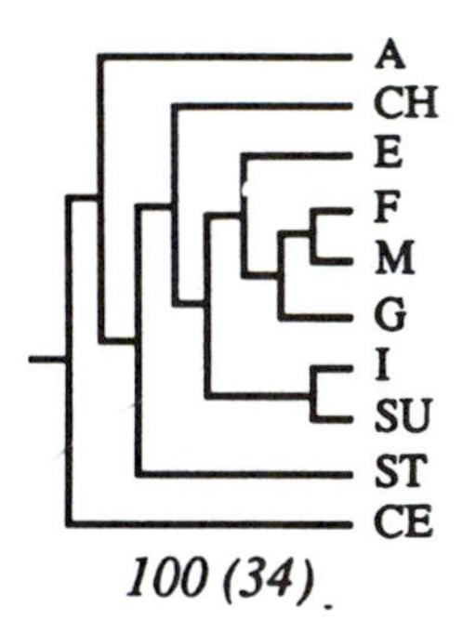

100 (34)

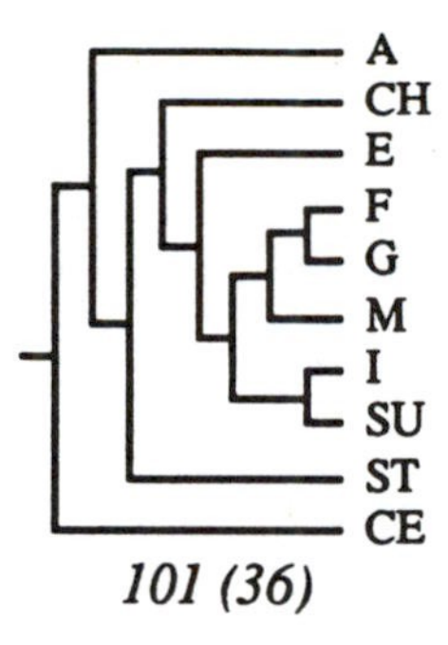

101 (36)

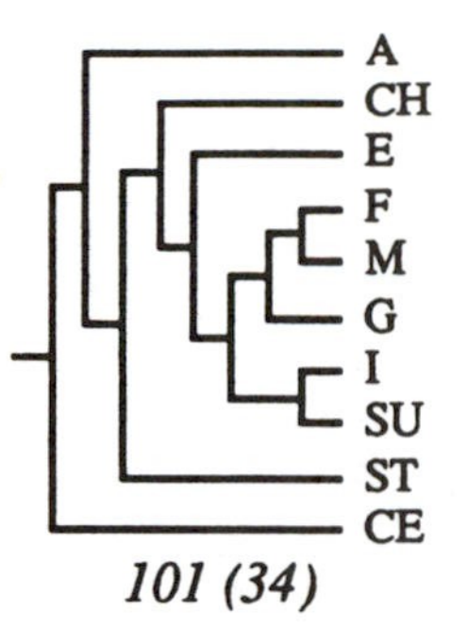

101 (34)

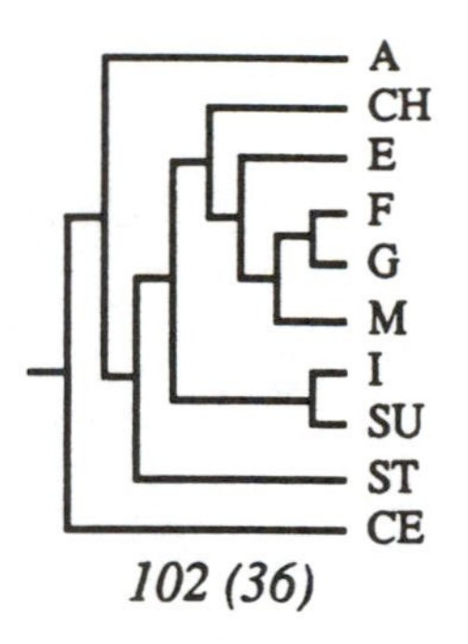

102 (36)

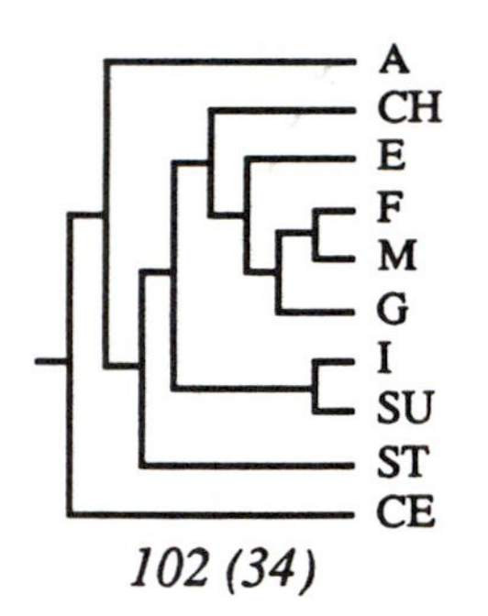

102 (34)

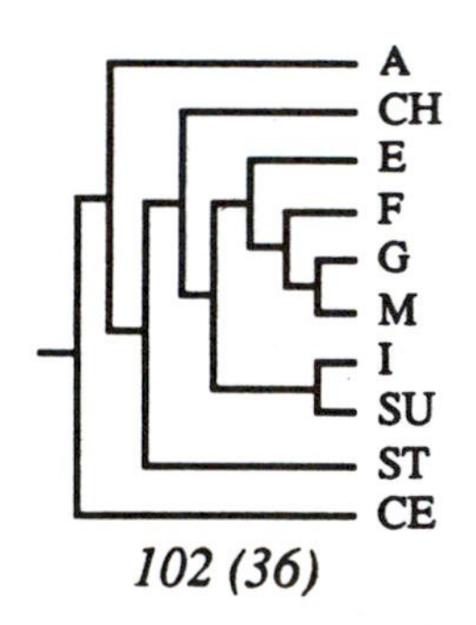

102 (36)

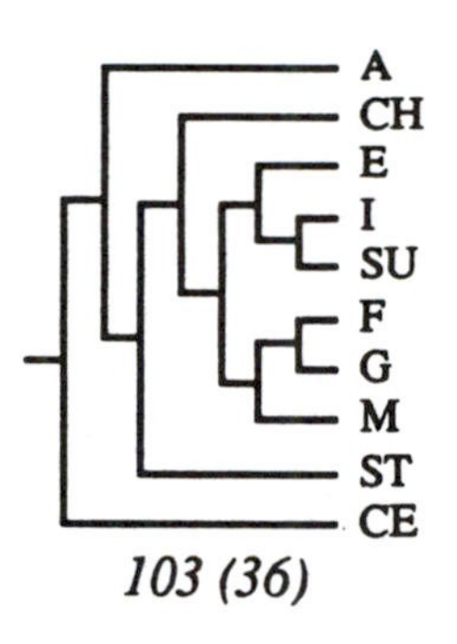

103 (36)

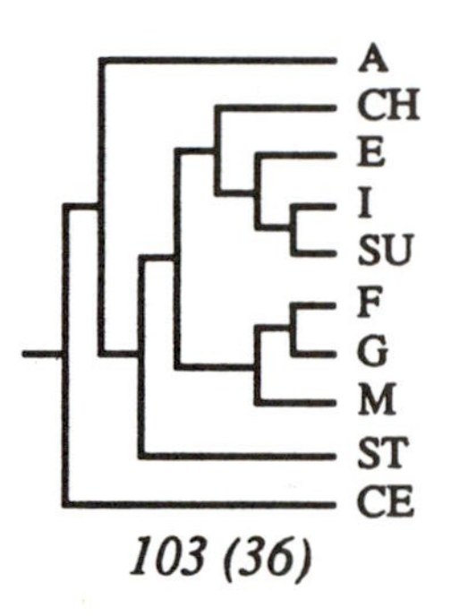

103 (36)

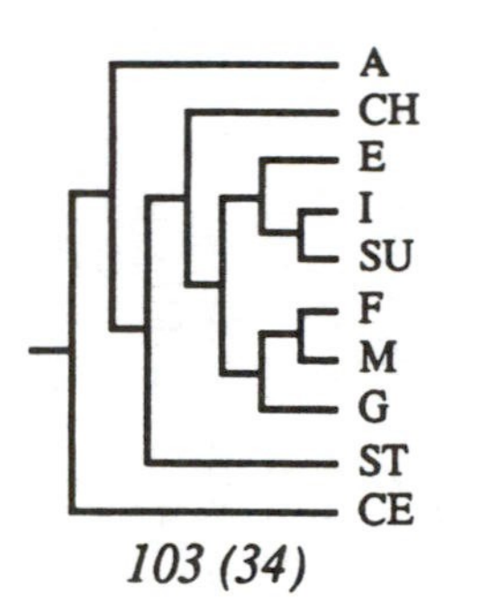

103 (34)

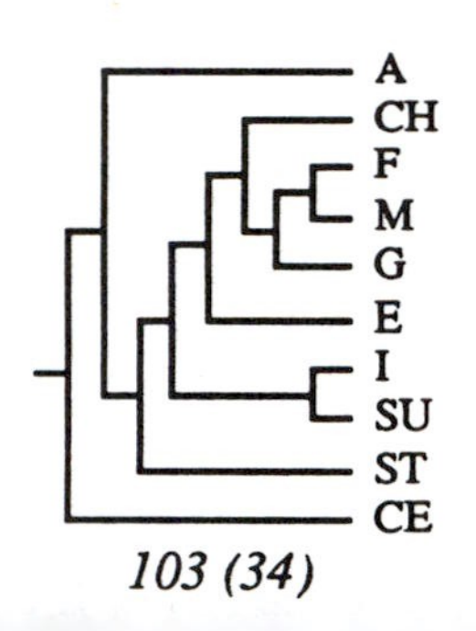

103 (34)

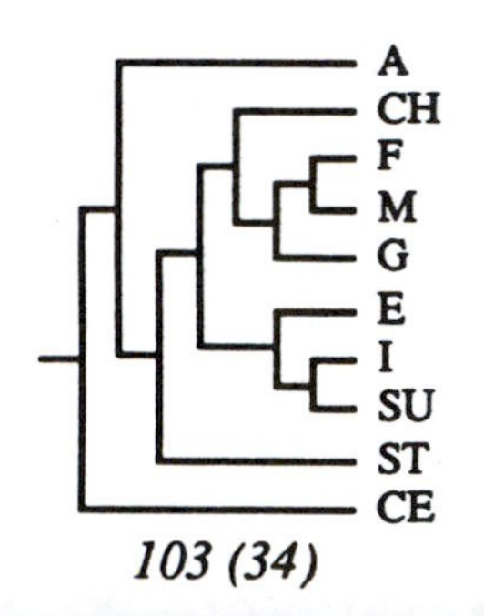

103 (34)

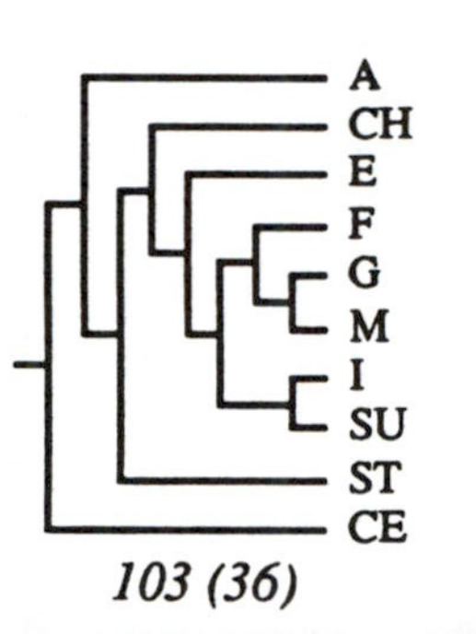

103 (36)

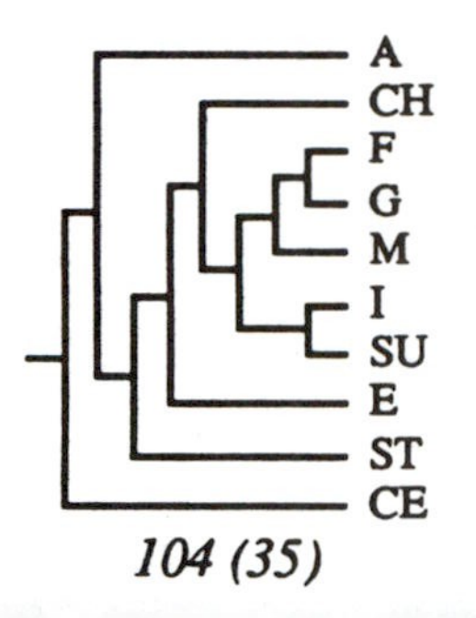

104 (35)

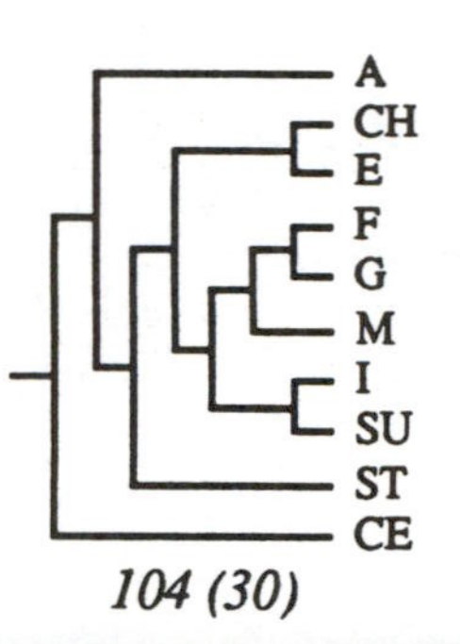

104 (30)

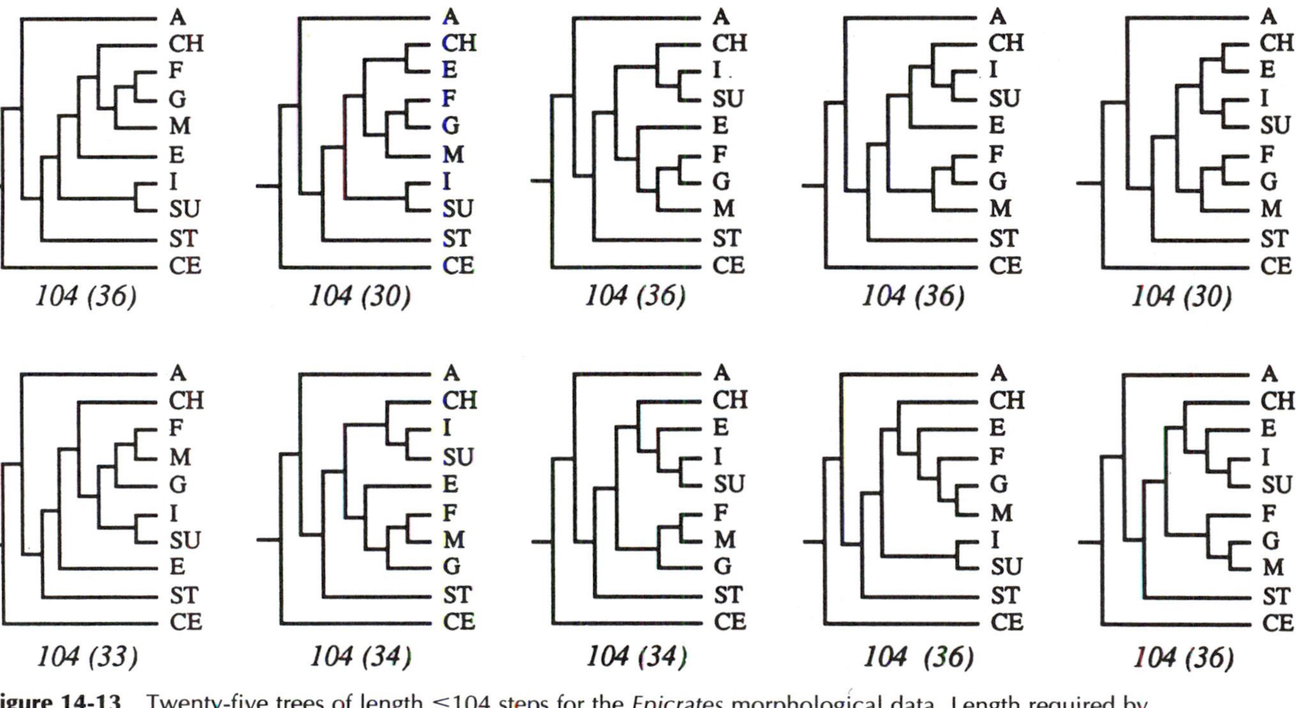

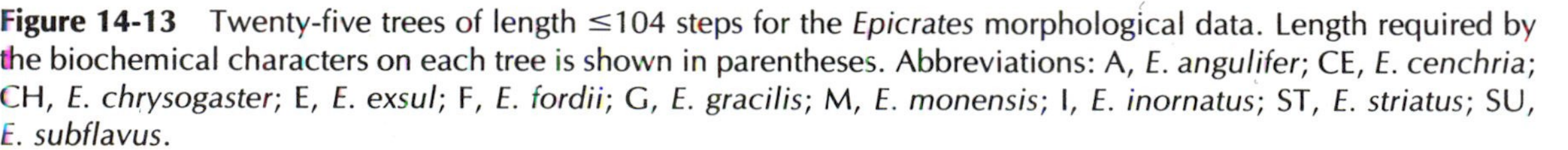

Figure 14-13 Twenty-five trees of length ≤104 steps for the *Epicrates* morphological data. Length required by the biochemical characters on each tree is shown in parentheses. Abbreviations: A, *E. angulifer*; CE, *E. cenchria*; CH, *E. chrysogaster*; E, *E. exsul*; F, *E. fordii*; G, *E. gracilis*; M, *E. monensis*; I, *E. inornatus*; ST, *E. striatus*; SU, *E. subflavus*.

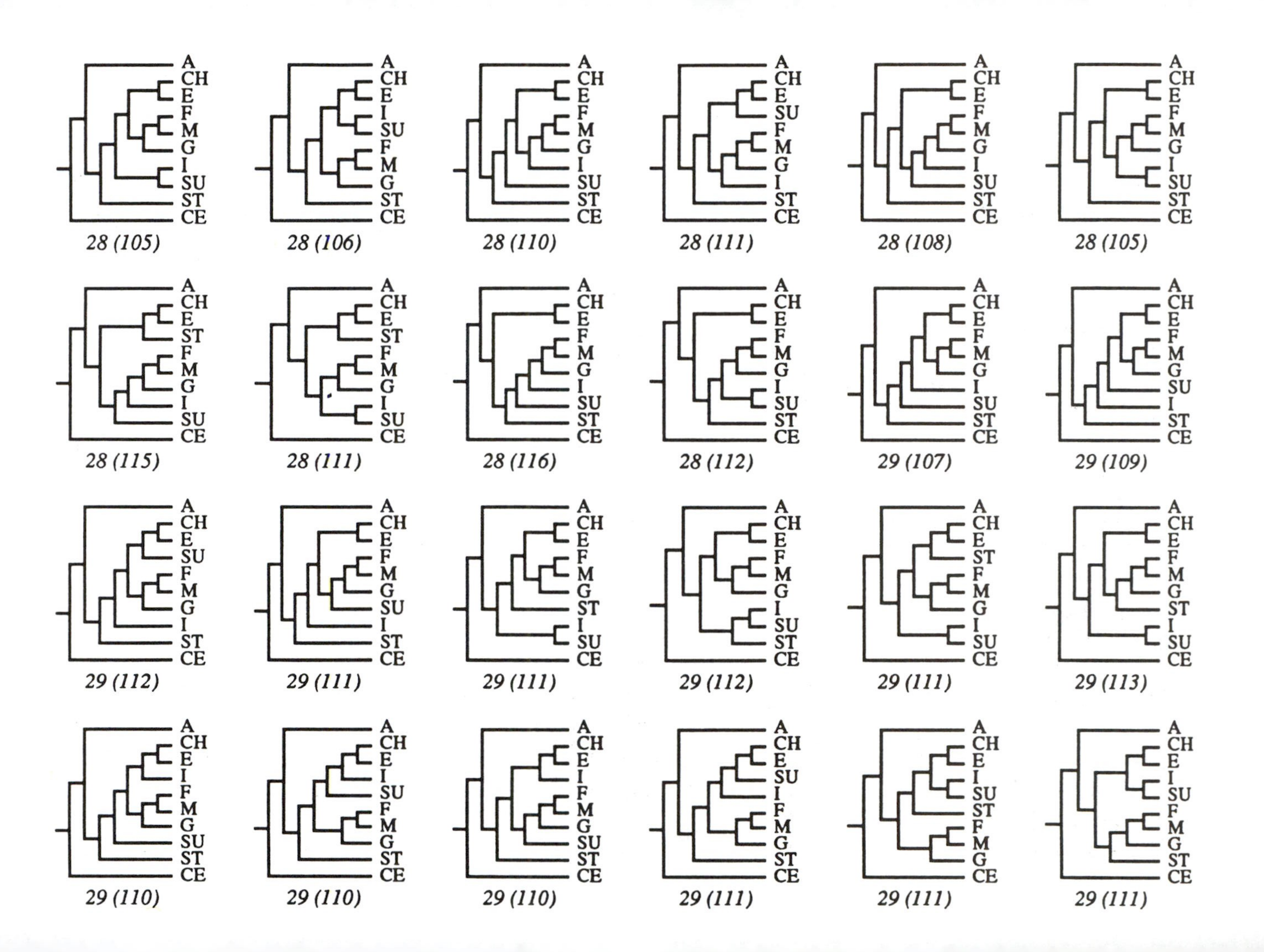
A CH E F M G I SU ST CE
28 (105)
A CH E I SU F M G ST CE
28 (106)
A CH E F M G I SU ST CE
28 (110)
A CH E SU F M G I ST CE
28 (111)
A CH E F M G I SU ST CE
28 (108)
A CH E F M G I SU ST CE
28 (105)
A CH E ST F M G I SU CE
28 (115)
A CH E ST F M G I SU CE
28 (111)
A CH E F M G I SU ST CE
28 (116)
A CH E F M G I SU ST CE
28 (112)
A CH E F M G I SU ST CE
29 (107)
A CH E F M G SU I ST CE
29 (109)
A CH E SU F M G I ST CE
29 (112)
A CH E F M G SU I ST CE
29 (111)
A CH E F M G ST I SU CE
29 (111)
A CH E F M G I SU ST CE
29 (112)
A CH E ST F M G I SU CE
29 (111)
A CH E F M G ST I SU CE
29 (113)
A CH E I F M G SU ST CE
29 (110)
A CH E I SU F M G ST CE
29 (110)
A CH E I F M G SU ST CE
29 (110)
A CH E SU I F M G ST CE
29 (111)
A CH E I SU ST F M G CE
29 (111)
A CH E I SU F M G ST CE
29 (111)

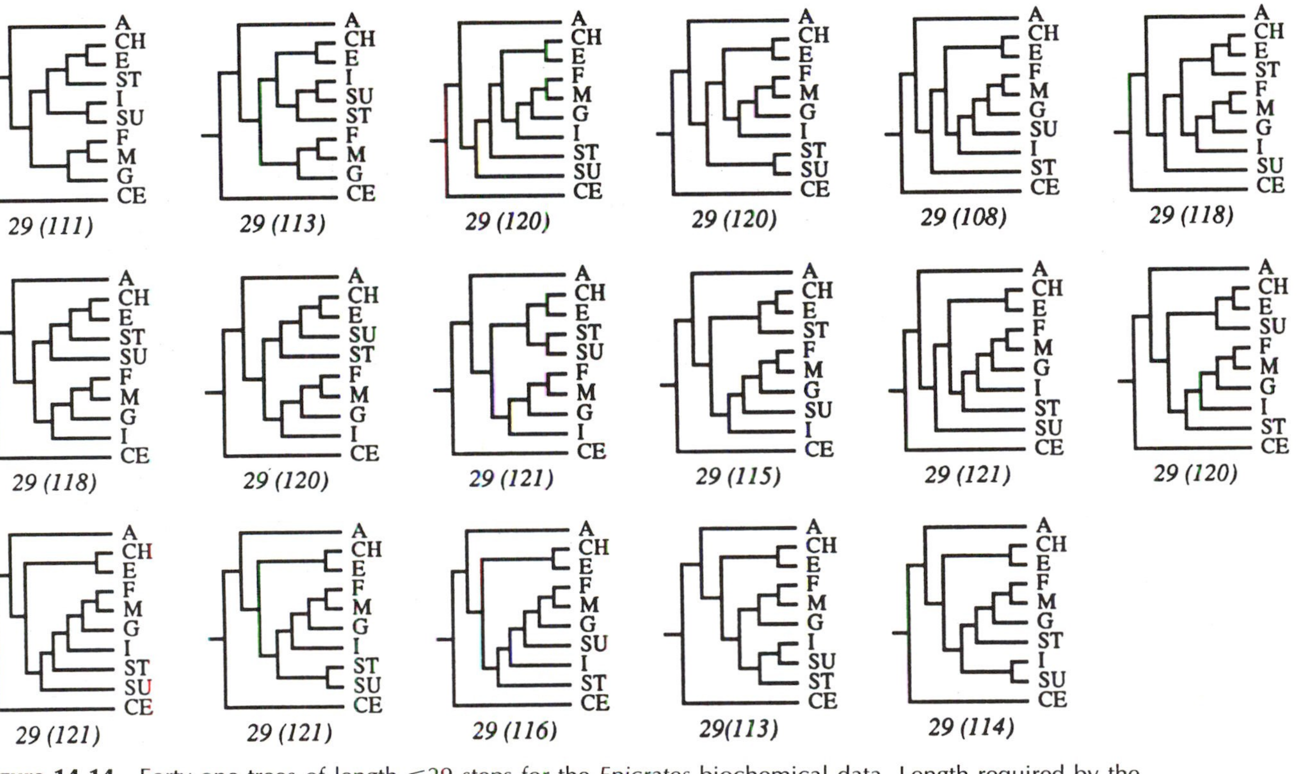

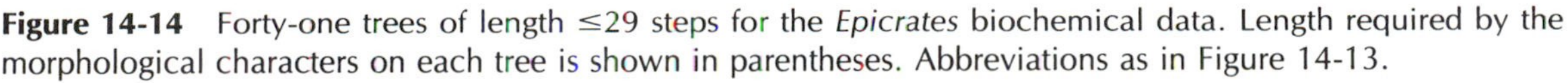

Figure 14-14 Forty-one trees of length ≤29 steps for the *Epicrates* biochemical data. Length required by the morphological characters on each tree is shown in parentheses. Abbreviations as in Figure 14-13.

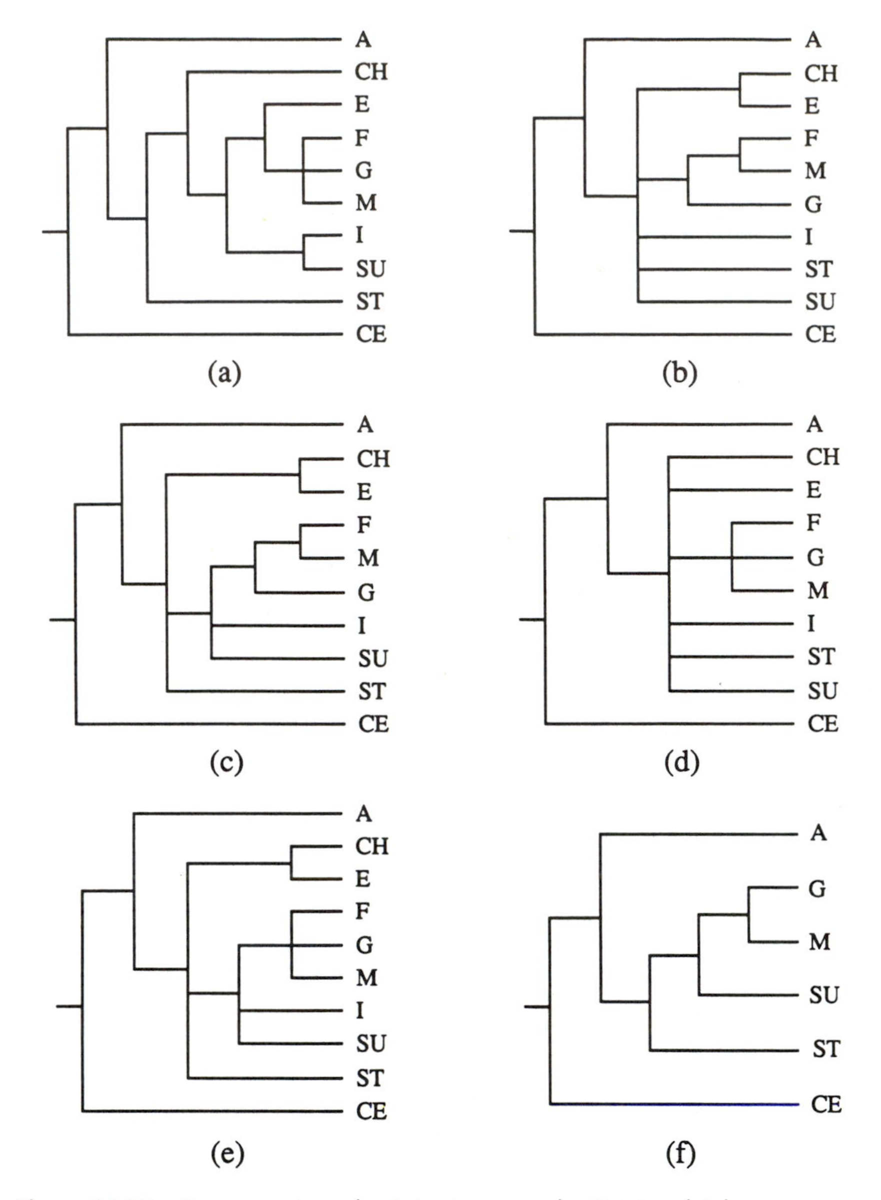

Figure 14-15 Consensus trees for *Epicrates* example. Strict and Adams consensus (a) of two minimal trees for morphological data. Strict (b) and Adams (c) consensus of 10 minimal trees for biochemical data. Strict (d) and Adams (e) consensus of two minimal trees for morphological data plus 10 minimal trees for biochemical data. A largest common pruned subtree (f) for the set of trees used for (d) and (e). Abbreviations as in Figure 14-13.

Table 14-1 Consensus Indices for Consensus Trees in Figure 14-15

Consensus Index	Tree				
	a	b	c	d	e
Colless CF	0.875	0.625	0.750	0.375	0.625
Mickevich-Platnick P_m (= Colless' weighted *CF*)	0.886	0.545	0.659	0.455	0.614
Mickevich *CI*	0.800	0.350	0.550	0.250	0.500
Rohlf CI_1	0.970	0.559	0.719	0.472	0.688
Rohlf CI_2	8.71×10^{-8}	3.05×10^{-6}	2.61×10^{-7}	8.23×10^{-5}	7.84×10^{-7}

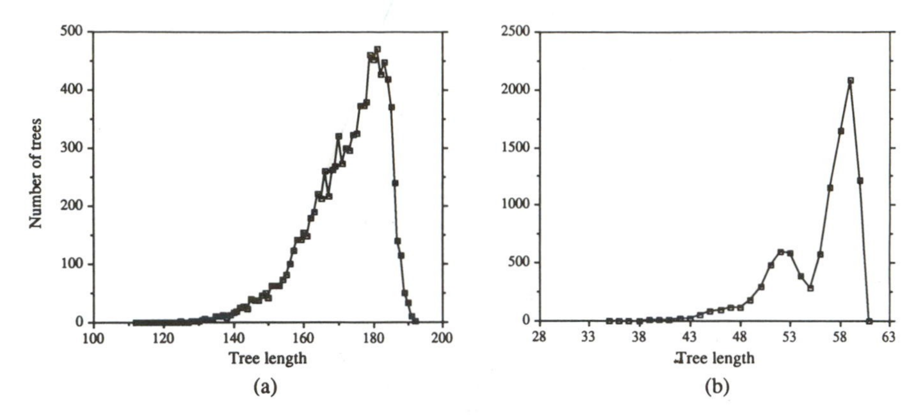

Figure 14-16 Approximate frequency distributions of tree lengths for *Epicrates* morphological (a) and biochemical (b) data. Each distribution was obtained by random sampling of 10,000 trees under a model in which every possible tree is equally likely.

(see Chapter 13). However, it is apparent from Figure 14-17, which shows the exact frequencies of tree lengths in the left tails of the distributions, that signal strength is a relative concept. For example, the vast majority of trees are longer than 120 steps for the morphological data (Fig. 14-16a), but there are 4,961 trees of length 120 or less (Fig. 14-17a). Obviously, there is no set of trees that is clearly better than the rest for either data set.

In order to determine whether trees exist that explain both data sets reasonably well, I initially decided to consider all trees that were less than 5% longer than the shortest trees for each data set. There are 25 trees of length less than or equal to 104 steps for the morphological data (Fig. 14-13) and 41 trees of length less than or equal to 29 steps for the biochemical data (Fig. 14-14). Surprisingly, these two sets of trees contain no tree in common! Additional analysis reveals that intersecting trees are not found until 105-step trees are included for the morphological data (two trees of length 28 steps for the biochemical data), or 30-step trees are included for the biochemical data (three trees of length 104 steps for the morphological data); these trees can be found in Figures 14-13 and 14-14.

Are phylogenies estimated for *Epicrates* from the two data sets incongruent? I will leave the final answer to this question to the reader, but I would merely suggest that the issue is much more complex than Kluge's (1989) assessment. I find it mildly distressing that the best 25 trees for one data set and the best 41 trees for another data set do not share even one tree in common. On the other hand, the number of trees in the union of all trees less than or equal to 105 steps for the morphological data, and all trees less than or equal to 30 steps for the biochemical data, is still a tiny fraction (188/34,459,525 = 0.00000546) of the total number of trees for 11 taxa. This observation suggests that there may indeed be substantial congruence between the two data sets, but that "congruence" is not quite what we had hoped it would be.

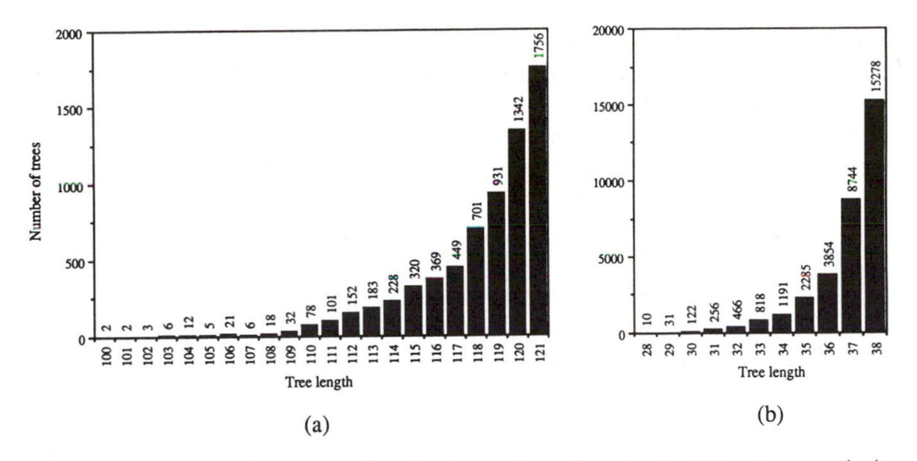

Figure 14-17 Exact frequency distributions of tree lengths for *Epicrates* morphological (a) and biochemical (b) data for lengths in the left tail of each distribution.

DISCUSSION AND CONCLUSIONS

That I have failed to provide definitive answers to the question posed in the title of this paper should not be surprising given the complexity of the problem. When phylogenies estimated from multiple data sets disagree, I believe that we should use a multiplicity of approaches to evaluate the incongruence, marshaling evidence for groups that are well supported and identifying those which should be subjected to further scrutiny. In addition to the techniques discussed in this chapter, additional methods for assessing the reliability of phylogenetic hypotheses (Penny and Hendy, 1986; Felsenstein, 1988; West and Faith, 1990) must be incorporated. A large number of researchers is just beginning to acquire molecular data for the purpose of "evaluating congruence of molecular phylogenies with morphologically based classifications;" it is imperative that these studies entail more than a simple comparison of the optimal trees for each data set.

To Combine or Not to Combine

One of the most controversial issues involving the handling of multiple data sets is whether or not they should be combined into a single data set for analysis. Kluge (1989) takes a strong position that classifications should be based on "total evidence." In his view, the preferred classification corresponds to the most-parsimonious tree for a data set containing every independent character from every available source of information. Miyamoto (1985) has also argued that most-parsimonious trees for combined data sets have maximum explanatory power and are therefore to be preferred.

Although there are situations in which combining data sets can be useful, I find compelling reasons to analyze logically independent data sets separately as well. Virtually all methods of phylogenetic inference assume independence of characters. This assumption is difficult if not impossible to verify, but it is usually much more plausible for characters in different data sets than for characters within the same data set. For example, the discovery that trees constructed separately from sequences for four different genes agree in most respects would inspire a reasonably high level of confidence in the conclusions. The corroboration provided by the independent estimates would not be available if the data sets were combined. On the other hand, the observation that three of the data sets agree completely with each other but disagree strikingly with the fourth would suggest an interesting problem for subsequent molecular evolutionary studies that could easily be missed if analyses were restricted to a pooled data set.

The problem of potential nonindependence can clearly be seen in the *Epicrates* example. Six lipid-compound characters are present in *E. chrysogaster* and *E. exsul* that are absent in the remaining taxa. Only one biochemical character (7) is incompatible with these six characters. A (*E. chrysogaster, E. exsul*) clade is therefore strongly supported by the biochemical data set: the shortest tree on which this group does not appear is 33 steps long, five steps longer than a minimal tree for the biochemical data. In contrast, this group is rather strongly rejected by the morphological data set: the shortest tree containing a (*E. chrysogaster, E. exsul*) clade requires 104 steps, four steps longer than a minimal tree. When the two data sets are combined, the cost of rejecting a (*E. chrysogaster, E. exsul*) group (six parallelisms involving independent acquisition of the same lipid compound in different lineages) outweighs the cost of including it (the four extra morphological steps cited above). Thus, the (*E. chrysogaster, E. exsul*) clade is included in the minimal trees for the combined data set. Furthermore, as Kluge (1989) emphasizes, it is "confirmed" by six "unique and unreversed diagnostic synapomorphies." However, as just demonstrated, this diagnosis is costly; four additional homoplasies must be postulated for the morphological data in order to preserve the (*E. chrysogaster, E. exsul*) clade in the trees for the combined data.

Although it is impossible to prove without a complete characterization of the genetic basis of the lipid variation, the incongruence between the biochemical characters supporting a (*E. chrysogaster, E. exsul*) grouping, and the morphological characters at least suggests the possibility that the biochemical characters are not truly independent. This prospect might have been overlooked entirely had the data sets not been analyzed separately.

Do "Classes" of Data Even Exist?

Another reason given for restricting analyses to combined data sets is that there are no truly recognizable "classes" of characters. Proponents of this argument suggest that "data are data," that nothing is gained by classifying data into categories such as molecular versus morphological, mitochondrial versus nuclear, or even small-subunit versus large-subunit rRNA. This is a philosophical position that I find unreasonable but cannot entirely refute. Presumably, even the most extreme advocate of this position would agree that the genes responsible for differentiation of skeletal elements are independent of the gene coding for lactate dehydrogenase. For the reasons discussed above, I believe that the advantages of comparing independent estimates of phylogeny outweigh the disadvantages.

Should We Ever Consider Suboptimal Trees?

Ardent defenders of the principle of maximum parsimony sometimes object to any consideration of nonminimal trees, suggesting that in doing so, one begins to "slide down a slippery slope" that does not end until every possible tree has been accepted as a candidate. Although I support the use of parsimony as a criterion for evaluating trees, I believe that much can be learned from approaches that examine nonminimal trees as in the *Epicrates* example above. Obviously, if one of the minimal trees for the morphological data set were the "true" tree and only the biochemical data were available, we would reject the true tree if we failed to consider nonminimal trees for the biochemical data.

I take the admittedly non-Popperian position that an ambiguous solution that contains the truth is, in many situations, preferable to an unambiguous solution that is wrong. If the ultimate goal is only to erect a classification for a group of taxa, then I accept the argument that a fully resolved, most-parsimonious tree is useful as a bold, testable hypothesis. It is becoming more and more common, however, for a systematist's *hypothesis* to become a comparative biologist's *assumption* (see, e.g., Donoghue, 1989 and references cited therein). In this context, systematists do a disservice to their fellow biologists if they do not provide any clues as to how well alternative, slightly suboptimal trees might explain a given set of data. For example, knowledge that a conclusion regarding adaptation receives some support from all trees within three steps of the shortest may be much more satisfying than the discovery that the same conclusion is strongly supported by the most-parsimonious tree, but decisively rejected by a tree one step longer.

ACKNOWLEDGMENTS

I am especially grateful to Mike Miyamoto and David Maddison for discussion and constructive criticism of the ideas presented herein. I also thank Joel Cracraft, Wayne Maddison, and participants in the University of Illinois Molecular Evolution Journal Club for their comments, and Jim Carpenter for permission to publish results based on his "Ropa10" data set. Finally, I thank Arnold Kluge and Peter Tolson for their efforts in obtaining the data on *Epicrates* used in the example.

REFERENCES

Adams, E. N. III. (1972) Consensus techniques and the comparison of taxonomic trees. *Syst. Zool.* **21:**390–397.

Adams, E. N. III. (1986) N-trees as nestings: complexity, similarity, and consensus. *J. Classif.* **3:**299–317.

Archie, J. W. (1989) Homoplasy excess ratios: new indices for measuring levels of homoplasy in phylogenetic systematics and a critique of the consistency index. *Syst. Zool.* **38:**253–269.

Barthélemy, J.-P., and F. R. McMorris. (1986) The median procedure for n-trees. *J. Classif.* **3:**329–334.

Barthélemy, J.-P., and B. Monjardet. (1981) The median procedure in cluster analysis and social choice theory. *Math. Soc. Sci.* **1:**235–267.

Bremer, K. (1990) Combinable component consensus. *Cladistics* **6:**369–372.

Bron, C., and J. Kerbosch. (1973) Algorithm 457: finding all cliques of an undirected graph. *Comm. ACM* **16:**575–577.

Brown, E. K., and W. H. E. Day. (1984) A computationally efficient approximation to the nearest neighbor interchange metric. *J. Classif.* **1:**93–124.

Camin, J. H., and R. R. Sokal. (1965) A method for deducing branching sequences in phylogeny. *Evolution* **19:**311–326.

Carpenter, J. M. (1988) Choosing among multiple equally parsimonious cladograms. *Cladistics* **4:**291–296.

Colless, D. H. (1980) Congruence between morphometric and allozyme data for *Menidia* species: a reappraisal. *Syst. Zool.* **29:**288–299.

Colless, D. H. (1981) Predictivity and stability in classifications: some comments on recent studies. *Syst. Zool.* **30:**325–331.

Cracraft, J. (1983) Commentary. Pp. 456–467 in *Perspectives in Ornithology* (A. H. Brush and G. A. Clark, Jr., eds.). Cambridge University Press, Cambridge.

Crozier, R. H., P. Pamilo, R. W. Taylor, and Y. C. Crozier. (1986) Evolutionary patterns in some putative Australian species in the ant genus *Rhytidoponera. Aust. J. Zool.* **34:**535–560.

Day, W. H. E. (1983) The role of complexity in comparing classifications. *Math. Biosci.* **66:**97–114.

Day, W. H. E., and F. R. McMorris. (1985) A formalization of consensus index methods. *Bull. Math. Biol.* **47:**215–229.

Donoghue, M. J. (1989) Phylogenies and the analysis of evolutionary sequences, with examples from seed plants. *Evolution* **43:**1137–1156.

Estabrook, G. F., C. S. J. Johnson, and F. R. McMorris. (1976) A mathematical foundation for the analysis of cladistic character compatibility. *Math. Biosci.* **29:**181–187.

Estabrook, G. F., F. R. McMorris, and C. A. Meacham. (1985) Comparison of undirected phylogenetic trees based on subtrees of four evolutionary units. *Syst. Zool.* **34:**193–200.

Faith, D. P. (1988) Consensus applications in the biological sciences. Pp. 325–332 in *Classification and Related Methods of Data Analysis* (H. H. Bock, ed.). Elsevier Science Publishers B. V., Amsterdam.

Farris, J. S. (1970) Methods for computing Wagner trees. *Syst. Zool.* **19:**83–92.

Farris, J. S. (1973) On comparing the shapes of taxonomic trees. *Syst. Zool.* **22:**50–54.

Farris, J. S. (1988) *Hennig86*, Version 1.5. Distributed by the author, Port Jefferson Station, N. Y.

Farris, J. S. (1989) The retention index and the rescaled consistency index. *Cladistics* **5**:417–419.
Farris, J. S., A. G. Kluge, and M. J. Eckardt. (1970) A numerical approach to phylogenetic systematics. *Syst. Zool.* **19**:172–191.
Felsenstein, J. (1978) The number of evolutionary trees. *Syst. Zool.* **27**:27–33.
Felsenstein, J. (1988) Phylogenies from molecular sequences: inference and reliability. *Ann. Rev. Genet.* **22**:521–565.
Felsenstein, J. (1990) *PHYLIP Manual*, Version 3.3. University Herbarium, University of California, Berkeley.
Finden, C. R., and A. D. Gordon. (1985) Obtaining common pruned trees. *J. Classif.* **2**:255–276.
Funk, V. A., and D. R. Brooks. (1990) *Phylogenetic Systematics as the Basis of Comparative Biology*. Smithsonian Institution Press, Washington, D.C.
Gordon, A. D. (1980) On the assessment and comparison of classifications. Pp. 149–160 in *Analyse de Données et Informatique* (R. Tomassone, ed.). INRIA, Le Chesnay.
Hendy, M. D., C. H. C. Little, and D. Penny. (1984) Comparing trees with pendant vertices labelled. *SIAM J. Appl. Math.* **44**:1054–1065.
Hendy, M. D., M. A. Steel, D. Penny, and I. M. Henderson. (1988) Families of trees and consensus. Pp. 355–362 in *Classification and Related Methods of Data Analysis* (H. H. Bock, ed.). Elsevier Science Publishers B. V., Amsterdam.
Hennig, W. (1966) *Phylogenetic Systematics*. University of Illinois Press, Urbana.
Hillis, D. M. (1987) Molecular versus morphological approaches to systematics. *Ann. Rev. Ecol. Syst.* **18**:23–42.
Jarvis, J. P., J. K. Luedeman, and D. R. Shier. (1983) Comments on computing the similarity of binary trees. *J. Theor. Biol.* **100**:427–433.
Kluge, A. G. (1989) A concern for evidence and a phylogenetic hypothesis of relationships among *Epicrates* (Boidae, Serpentes). *Syst. Zool.* **38**:7–25.
Kluge, A. G., and J. S. Farris. (1969) Quantitative phyletics and the evolution of anurans. *Syst. Zool.* **18**:1–32.
Le Quesne, W. J. (1969) A method of selection of characters in numerical taxonomy. *Syst. Zool.* **18**:201–205.
Maddison, D. R. (1991) The discovery and importance of multiple islands of most-parsimonious trees. *Syst. Zool.* **40**:in press.
Maddison, W. P. (1989) Reconstructing character evolution on polytomous cladograms. *Cladistics* **5**:365–377.
Margush, T., and F. R. McMorris. (1981) Consensus n-trees. *Bull. Math. Biol.* **43**:239–244.
McMorris, F. R., D. B. Meronk, and D. A. Neumann. (1983) A view of some consensus methods for trees. Pp. 122–126 in *Numerical Taxonomy* (J. Felsenstein, ed.). Springer-Verlag, Berlin.
McMorris, F. R., and D. Neumann. (1983) Consensus functions defined on trees. *Math. Soc. Sci.* **4**:133–136.
Mickevich, M. F. (1978) Taxonomic congruence. *Syst. Zool.* **27**:143–158.
Mickevich, M. F., and J. S. Farris. (1981) The implications of congruence in *Menidia*. *Syst. Zool.* **30**:351–370.
Mickevich, M. F., and N. I. Platnick. (1989) On the information content of classifications. *Cladistics* **5**:33–47.
Miyamoto, M. M. (1985) Consensus cladograms and general classifications. *Cladistics* **1**:186–189.

Nelson, G. J. (1979) Cladistic analysis and synthesis: principles and definitions, with a historical note on Adanson's Familles des Plantes (1763–1764). *Syst. Zool.* **28**:1–21.

Neumann, D. A. (1983) Faithful consensus methods for n-trees. *Math. Biosci.* **63**:271–287.

Page, R. D. M. (1989a) Comments on component-compatibility in historical biogeography. *Cladistics* **5**:167–182.

Page, R. D. M. (1989b) *COMPONENT User's Manual (Release 1.5)*. University of Auckland, Auckland.

Penny, D., L. R. Foulds, and M. D. Hendy. (1982) Testing the theory of evolution by comparing phylogenetic trees constructed from five different protein sequences. *Nature* **297**:197–200.

Penny, D. and M. D. Hendy. (1985) The use of tree comparison metrics. *Syst. Zool.* **34**:75–82.

Penny, D. and M. D. Hendy. (1986) Estimating the reliability of evolutionary trees. *Mol. Biol. Evol.* **3**:403–417.

Robinson, D. F. (1971) Comparison of labeled trees with valency three. *J. Comb. Theory* **11**:105–119.

Robinson, D. F., and L. R. Foulds. (1979) Comparison of weighted labelled trees. Pp. 119–126 in *Lecture Notes in Mathematics*, Vol. 748. Springer-Verlag, Berlin.

Robinson, D. F., and L. R. Foulds. (1981) Comparison of phylogenetic trees. *Math. Biosci.* **53**:131–147.

Rohlf, F. J. (1982) Consensus indices for comparing classifications. *Math. Biosci.* **59**:131–144.

Rohlf, F. J., D. H. Colless, and G. Hart. (1983) Taxonomic congruence re-examined. *Syst. Zool.* **32**:144–158.

Sanderson, M. J., and M. J. Donoghue. (1989) Patterns of variation and levels of homoplasy. *Evolution* **43**:1781–1795.

Schuh, R. T., and J. S. Farris. (1981) Methods for investigating taxonomic congruence and their application to the Leptopodomorpha. *Syst. Zool.* **30**:331–351.

Schuh, R. T., and J. T. Polhemus. (1981) Analysis of taxonomic congruence among morphological, ecological, and biogeographic data sets for the Leptopodomorpha (Hemiptera). *Syst. Zool.* **29**:1–26.

Shao, K. (1983) *Consensus Methods in Numerical Taxonomy*. Ph.D. Dissertation, State University of New York, Stony Brook.

Shao, K., and F. J. Rohlf. (1983) Sampling distribution of consensus indices when all bifurcating trees are equally likely. Pp. 132–136 in *Numerical Taxonomy* (J. Felsenstein, ed.). Springer-Verlag, Berlin.

Shao, K.-T., and R. R. Sokal. (1986) Significance tests of consensus indices. *Syst. Zool.* **35**:582–590.

Shao, K.-T., and R. R. Sokal. (1990) Tree balance. *Syst. Zool.* **39**:266–276.

Simberloff, D. (1987) Calculating probabilities that cladograms match: a method of biogeographical inference. *Syst. Zool.* **36**:175–195.

Sokal, R. R., and F. J. Rohlf. (1981) Taxonomic congruence in the Leptopodomorpha reexamined. *Syst. Zool.* **30**:309–325.

Stinebrickner, R. (1984) s-Consensus trees and indices. *Bull. Math. Biol.* **46**:923–935.

Stinebrickner, R. (1986) s-Consensus index method: an additional axiom. *J. Classif.* **3**:319–327.

Swofford, D. L. (1991) *PAUP: Phylogenetic Analysis Using Parsimony*, Version 3.0. Illinois Natural History Survey, Champaign.

Tolson, P. J. (1987) Phylogenetics of the boid snake genus *Epicrates* and Caribbean vicariance theory. *Occ. Papers Mus. Zool., Univ. Michigan* **715**:1–68.

Waterman, M. S., and T. F. Smith. (1978) On the similarity of dendrograms. *J. Theor. Biol.* **73**:789–800.

West, J. G., and D. P. Faith. (1990) Data, methods and assumptions in phylogenetic inference. *Aust. Syst. Bot.* **3**:9–20.

15

Congruence Among Data Sets: A Bayesian Approach

WARD C. WHEELER

When several independently derived data sets agree, producing the same cladogram, we have high confidence in the result. Such complete agreement, however, is rare and becoming rarer with the multiplication of molecular data sets.

Extreme disagreement has arisen between morphological and molecular studies. Zimmer et al. (1989) have produced results on monocot origins that contradict the morphological results of Donoghue and Doyle (1989). Field et al. (1988) have suggested relationships within and among invertebrate groups that counter the great majority of work on invertebrate systematics. Mammal studies have produced further disagreement (Goodman et al., 1985; Wyss et al., 1987), and studies of bird phylogeny (Cracraft, 1974; Sibley and Ahlquist, 1981) have had their differences. While these conflicts are not limited to comparisons between molecules and morphology, herein lie some of the most prominent cases. I do not mean to overstate the extent of the disagreement or of the acrimony accompanying it. In fact, congruence among data sets may well be more common than incongruence. Nonetheless, in these works, there are important differences between hypotheses of relationships which require explanation.

When molecular data disagree with hypotheses based on morphology, there are two common responses. Either one of the studies is dismissed or some consensus is attempted. When consensus prevails, a consensus "tree" is most often produced. There are, however, shortcomings to this approach.

The most frequently used consensus procedures are Adams (1972) and strict (Rohlf, 1982) consensus. Neither of these methods is entirely satisfying. The Adams procedure may generate groupings of taxa that are found nowhere in the initial cladograms, while the strict method is so ruthlessly conservative that all resolution may be lost by a shift in the position of a single taxon. In the strict consensus, only groups present in all input trees are included. Lately, a modified procedure has become popular (Margush and McMorris, 1981). The modified method includes groups that appear in some fraction (greater than one half) of the cladograms. In this way, "majority rule" consensus or percentage consensus trees are produced.

There is, however, a problem intrinsic to consensus analysis. These procedures tend to combine the weaknesses of the data sets; disagreements based on few data are accorded equal weight with agreements based on a great deal of information, since only topologies are compared (Miyamoto, 1985). A desirable method would combine the strengths of the data instead, making allowances for the degree to which separate data sets support different nodes. Here, I propose a method that attempts to do this using Bayesian decision theory. The Bayesian approach draws on a philosophy and an analytical procedure that directly apply to the problem of congruence in phylogenetic analysis.

THE MONTY HALL PROBLEM

Before discussing the particulars of Bayesian decision theory, I would like to start with an everyday example of the purpose and power of this type of analysis, known as the "Monty Hall Problem." Imagine yourself on the television game show "Let's Make a Deal." Monty Hall, the host, presents you with three doors. Concealed behind one of these is a fabulous prize, while the others hide objects considerably less attractive. You are invited to choose a door, guessing which one hides the best prize. Suppose you select door number 1. At this point, Monty Hall opens door number 3, revealing an undesirable bauble (by the dictates of the game, Hall never reveals the best prize). You are now faced with a dilemma: do you stay with your original pick or do you go for what's behind door number 2? The proper Bayesian would switch, and double her chances of winning.

This may, at first, seem counterintuitive. After all, the prize is equally likely to be behind any of the doors. How are your odds improved by switching? The trick comes from the use of additional information. Initially, your pick has a one in three chance of being correct. Hence, the probability that the prize is behind one of the other doors is the complement of this—two-thirds. When Monty Hall reveals that there is nothing worthwhile behind one of the two doors you did not pick, the probability of that door being the one with the desirable prize collapses to 0. The total probability, however, that the two unchosen doors contain the best prize is unchanged; and it remains two-thirds. At this point, there is a one-third chance you have picked the correct door and a two-thirds chance that the prize is behind the other door. Of course you should switch.

Put another way, if your first choice, door number 1, is correct and you stand pat, you win. But if the desirable prize lies behind door number 2 or number 3 you must switch to win. Hence, by switching you win two out of three times. It is true that your chances of finding the prize behind door number 1 do not change when Monty Hall discloses the prize behind door number 3. But your chances of finding the one prize behind door number 2 increase dramatically.

The Bayesian decision to switch is mandated by two conditions: (1) the prior information that no door is more likely than any other to contain the desirable prize; and (2) the information gathered by observing the contents of one of the doors not chosen. These are the basic components of a Bayesian analysis, the information prior to the observation and the observations themselves. This type of procedure is the only one to include both kinds of information in the decision process.

The point I will make here is that in the use of morphological and molecular data to determine the credibility of a cladogram, morphology can provide the prior evidence to interpret molecular data.

DEFINITIONS OF PROBABILITY

Since this type of analysis deals with the probabilities of phylogenetic schemes, it is necessary to first define what is meant by the probability of a cladogram. There are three ways of defining probability: logical, empirical, and subjective (Hartigan, 1983). A logical probability is a rational degree of belief in a hypothesis in light of certain evidence, whereas an empirical probability is a statement of fact about the world. The third type of definition, subjective, is similar to the logical in that it represents a degree of belief, but it is an *individual* degree of belief. Two people may assign different probabilities based on the same evidence.

An example of these different concepts of probability can be found in betting ratios. Hartigan (1983) makes the point that a logical bettor (Bayesian) wagers what they ought to, while the empiricist will bet what is profitable in the long run, and the subjectivist will bet what they are willing. In the procedure described here, I will use the logical definition of probability. The probability of a particular cladogram, then, is our degree of belief in that cladogram, in light of evidence from molecular and morphological characters.

BAYESIAN INFERENCE AND DECISION THEORY

Just as there are three types of probability definitions, there are, broadly speaking, three types of inference: "classical," likelihood, and Bayesian. I am using classical inference to refer to most of the standard methods of statistical inference, those involving the calculation of p values and confidence intervals to approximate final probabilities. Classical methods have at least two shortcomings: first, they assume that the uncollected data can affect the distribution of the estimates; and second, they base their conclusions in part on the manner in which the data were collected. Thus, if the collection procedure is not known, or is modified during the gathering of observations, *p* values have no meaning, and classical methods are invalidated. Likelihood methods, in contrast to classical methods, rely solely on the observed data; their conclusions are not influenced by the way in which the data are gathered. These methods, however, do not make use of all relevant information. Specifically, they ignore information about prior distributions. This omission is quite deliberate. Due to the philosophical difficulties in determining priors—they may be entirely subjective—likelihood approaches pointedly avoid their use, or at least assume that all priors are equal and therefore irrelevant. Bayesian procedures, not bothered by these restrictions, extend likelihood through the inclusion of prior information.

All Bayesian procedures rely on an elementary relation of conditional probability (Bayes, 1763):

$$\mathrm{pr}(\Theta = \theta_i | \mathbf{D}) = \frac{\mathrm{pr}(\mathbf{D} | \theta_i)\mathrm{pr}(\theta_i)}{\Sigma_k[\mathrm{pr}(\mathbf{D} | \theta_k)\mathrm{pr}(\theta_k)]} \tag{1}$$

The "final" or posterior probability that the parameter Θ has the value θ_i given the data $\mathbf{D}$ is constructed by multiplying the probability of the data given θ_i, $\mathrm{pr}(\mathbf{D}|\theta_i)$, with its prior probability, $\mathrm{pr}(\theta_i)$, and normalized by the sum of these terms for all values of θ (θ_k).

If the priors are known or can be reasonably calculated, all statisticians would agree that Bayesian methods should be used to determine the best estimate of Θ. Some priors seem inherently reasonable. An example would be the proportion of defective items on an assembly line. In this case, the performance of the line in the past gives a prior probability of the future production of defective items. Yet, when priors are less clearly linked to observation, there is disagreement about their use. In fact, those who hold to likelihood argue that in all but the most trivial cases, prior probabilities cannot be calculated; thus only likelihoods (or their ratios) can be used.

THE SENSITIVITY OF THE RESULT TO THE PRIOR

One way to measure the effect of prior probability is through the use of simplex space. Take, for example, a parameter Θ, which may have three values, θ_1, θ_2, or θ_3. The sensitivity of the final probability to the prior probabilities can be graphed (Fig. 15-1). The space is divided into three regions. These contain the prior probabilities that determine the choice of θ_1, θ_2, or θ_3 as the estimate of Θ. The values of the priors are plotted ($\mathrm{pr}(\theta_1) + \mathrm{pr}(\theta_2) + \mathrm{pr}(\theta_3) = 1$) and their propinquity to the decision lines noted. If this point is very close to one or several of these lines, small variations in the priors can affect the estimate of Θ, undermining our faith in the result. If, however, the priors map to areas well within a decision space,

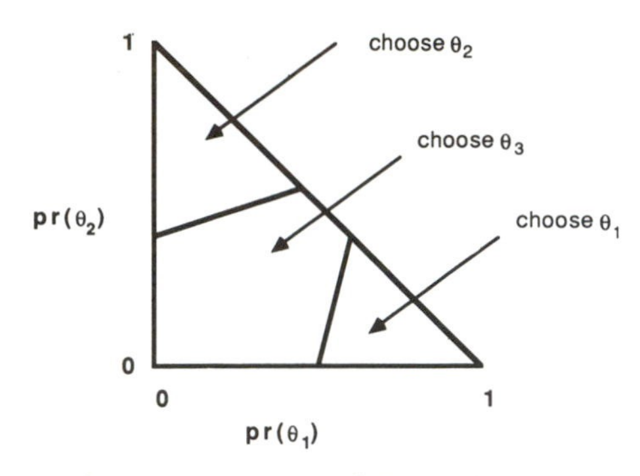

Figure 15-1 Simplex space of a simple inference problem showing the areas in which estimates 1, 2, and 3 are chosen.

we have greater confidence in the conclusions. A more precise measure of the robustness of results is offered by decision theory.

DECISION THEORY

A Bayesian decision is one which minimizes risk. In order to make this type of decision, an appropriate loss function must be specified for all possible values of θ. This loss function, coupled with prior probability and data probability values, allows us to determine the cost or risk of any decision. By definition, a decision is to accept some value θ_i as the estimate of a parameter Θ, or to embark upon a course of action θ_i out of all possible actions.

Unfortunately, the losses must be specified for all possible correct and incorrect decisions. This task may be overwhelming or impossible, given the complexity of the options and their costs. If, however, we are willing to accept that all incorrect hypotheses are equally undesirable and all correct hypotheses are equally desirable in that they are correct, a simple loss function can be used.

In this type of function, the cost of a correct decision is 0 while the cost of an incorrect choice is one (Fig. 15-2). The risk of a decision, then, is the sum of the cost of all the decisions:

$$R_i = \sum_j C_{ij}\mathrm{pr}(\Theta = \theta_i|\mathbf{D}). \tag{2}$$

The risk of the decision i (R_i) is equal to the sum over all j decisions of the product of the cost of that decision (C_{ij}—if the parameter has the value i) and its final probability. With the simple cost matrix suggested above (Fig. 15-2), the decision risk is merely the complement of its final probability. Since

$$C_{ij} = 1 \text{ if } i \neq j$$

$$C_{ij} = 0 \text{ if } i = j,$$

the risk reduces to:

$$R_i = 1 - \mathrm{pr}(\Theta = \theta_i|\mathbf{D}) \tag{3a}$$

		Decision						
		1	2	3	-	n-2	n-1	n
True Value	1	0	1	1	-	1	1	1
	2	1	0	1	-	1	1	1
	3	1	1	0	-	1	1	1
	I	.	.	.		.	.	.
	n-2	1	1	1	-	0	1	1
	n-1	1	1	1	-	1	0	1
	n	1	1	1	-	1	1	0

Figure 15-2 A simple cost matrix used in phylogenetic decisions with "*n*" decisions possible.

for a single cladogram or:

$$R_i = 1 - \sum_k \text{pr}(\Theta = \theta_k|\mathbf{D}) \quad (3b)$$

for several cladograms $\{k\}$ in i. By these manipulations, the risk of a decision with a final probability of 0.9 would be 0.1. The minimum risk occurs when the probability is maximized.

I propose that this notion of risk be used to evaluate the support for cladograms. Possibly no single cladogram will stand out as having acceptably low risk; in these cases, several cladograms should be considered until their collective risk is sufficiently low. By analogy with p values, we might choose the "risk" of type-I error to be no greater than 5%. Here, we would accept the most probable cladograms until 95% of the risk had been removed.

THE PROCEDURE

Determination of Prior Probabilities $\{\text{pr}(\theta_i)\}$

The first step in this type of analysis is to determine the prior probabilities for all of the possible values of θ. Usually, prior probabilities are calculated using known distributions of continuous or discrete variables. In this case, however, the prior probability of a cladogram in a molecular analysis is its ability to explain morphological data, which is an entirely discrete value.

There are many measures of the relative abilities of cladograms to explain data, such as the length (number of evolutionary steps) of the scheme or its consistency index (Kluge and Farris, 1969). Here, a length difference of one step (or postulated event) will be considered equal to a difference in probability of a factor of e, the base of natural logarithms. (Since these probabilities will be normalized, they need not sum to unity.) This factor of probability reflects the relative degree of belief in the different hypotheses before the molecular observations are made.

This assessment of probability may seem arbitrary at first. However, it can be justified by the link between natural logarithms and phylogenetic character weights (Farris, 1977). Of course, an individual investigator may use different methods to determine prior probabilities based on their confidence in that topology. In an unweighted parsimony analysis, all changes increment the cladogram cost by one step; similarly here, all changes are considered to be equally probable until there is evidence to the contrary.

Thus the prior probability of cladogram i with a length L_i from some previous analysis is:

$$\mathrm{pr}(\theta_i) = e^{(-L_i)}, \tag{4}$$

or to sum to unity:

$$\mathrm{pr}(\theta_i) = e^{(-L_i)}/\textstyle\sum_k e^{(-L_k)}. \tag{5}$$

Determination of Data Probability $\{\mathrm{pr}(\Theta = \theta_i|D)\}$

The second phase of the procedure involves determining the probability of the data given a certain θ_i, or topology. Any method that yields this value can be used. There are three methods, however, which seem especially appropriate: unweighted or maximum parsimony, weighted parsimony, and maximum likelihood.

The first of these may seem, initially, to be unsuited to statistical interpretation. As with the establishment of priors, a factor of e change in probability can be assigned to each event. In this way, all changes are treated equally, and a quantitative degree of belief is determined for each topology. As with Keynes' (1921) argument about the frequency of colored balls in an urn, in the absence of knowledge to the contrary, we must assume all events are equally likely. Total symmetry is a very simple model of character change, but it offers a starting point for examining the data. One effect of using the same probability factor in determining priors and data probabilities is that the result is the same as if all the data had been collected at one time and pooled. The absolute value of the natural logarithm of the final probabilities will be the lengths of cladograms constructed from these pooled data. The most probable (least risk) cladogram will be the most-parsimonious.

From this entirely symmetrical model of character transformation, the next step is to vary the weights assigned to different types of transformation. Just as these models of character transformation are used in weighted parsimony analyses, they can be used to assign data probabilities. Along these lines, Farris (1977) and DeBry and Slade (1985) have offered a Dollo-type method based on probabilistic thinking. In the analysis of nucleic acid sequences, many have suggested that transitions should be weighted differently from transversions (Brown et al., 1982). Specific cost ratios have been proposed, with obvious probabilistic interpretations. Elsewhere, I have put forward a combinatorial weight scheme for nucleic acids that yields weights for different types of character transformation (Wheeler, 1990). In each of these methods, only the different types of change receive varying probabilities. No assessments of the probability of change itself occurring are attempted (in a character transformation matrix all $a_{ii} = 0$). Instead, these procedures rely on explicit minimization of "cost" values.

$$\begin{array}{c} \\ A \\ C \\ G \\ T \end{array} \begin{array}{c} \begin{array}{cccc} A & C & G & T \end{array} \\ \begin{bmatrix} a_{11} & a_{21} & a_{31} & a_{41} \\ a_{12} & a_{22} & a_{32} & a_{42} \\ a_{13} & a_{23} & a_{33} & a_{43} \\ a_{14} & a_{24} & a_{34} & a_{44} \end{bmatrix} \end{array}$$

Figure 15-3 Markov type character transformation model showing the probability of change from one character state to another.

Likelihood methods differ, in that they seek to incorporate the rate of evolution. Not only the possible avenues of change, but the probability of change is included in the analysis. Smouse and Li (1987) use this approach in their examination of restriction fragment character change (and even discuss prior probabilities from other studies). Markov character transformation models (Fig. 15-3) chart rates of evolution as diagonal values. Since parsimony models lack these values, they offer no information on the probability of change.

With the prior probabilities and the data probabilities in hand, the final probability of each cladogram is calculated by equation (1).

Determination of Utility and Risk

The loss function suggested above and the final probabilities are used to determine the risks of various single and compound decisions [equation (3)]. A single cladogram, or several, may be accepted until risk is sufficiently minimized.

An Example

In order to demonstrate this congruence procedure, the case of the monophyly of the Eumetabola will be examined in the light of morphological (Hennig, 1969, 1981; Kristensen, 1975, 1981; Boudreaux, 1979) and molecular (Wheeler, 1989) evidence. There are 15 dichotomous unrooted arrangements of the five insect taxa (Paleoptera, Orthoptera, Hemiptera, Diptera, and Coleoptera, as shown in Fig. 15-4). In the morphological data set, there were five presence/absence characters: wing flexion (neoptery), the jugal bar, larval stemmata, holometaboly, and larval ocelli. The number of evolutionary events required by each of the possible arrangements is shown in Table 15-1. The sequence data of Wheeler (1989) were applied to these same cladograms; their lengths are also shown in Table 15-1, along with the weighted parsimony values. The procedure used to determine the weighted cladogram lengths is described in detail in Wheeler (1990). The likelihoods, or more accurately, their absolute values (Felsenstein, 1978, 1979), are also shown in this table.

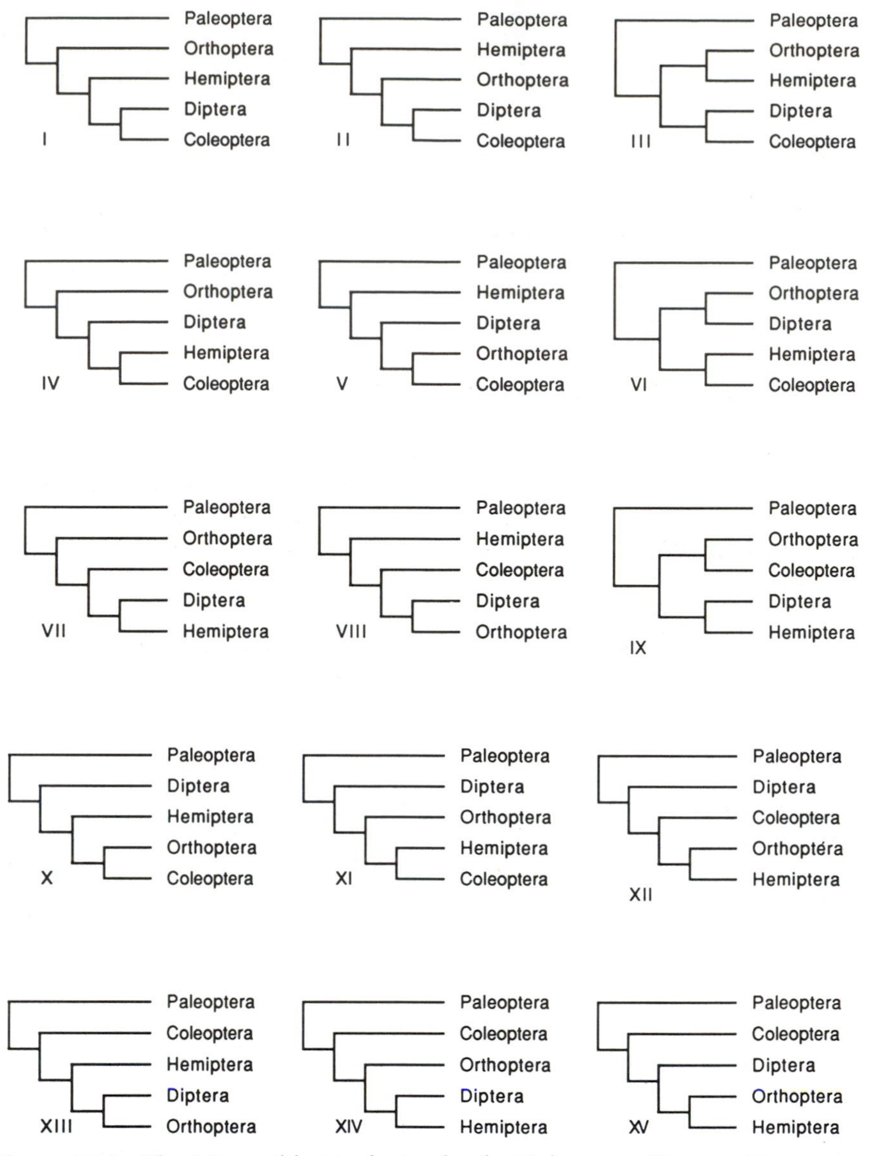

Figure 15-4 The 15 possible topologies for the Coleoptera, Diptera, Hemiptera, Orthoptera, and Paleoptea, with Paleoptera as the outgroup.

The final probabilities determined from the three types of cladogram costs and the morphologically based priors are shown in Table 15-2. Table 15-3 gives the topology decisions and associated risks.

The three types of data probabilities, maximum parsimony, weighted parsimony, and likelihood, agree for the most part. Topology III is the most favored (or at least one of the most favored hypotheses) in each of the analyses. However, the likelihood method includes topology XV and

Table 15-1 Cladogram Length

Cladogram	Morphology	Sequence Data: Unweighted Parsimony	Sequence Data: Weighted Parsimony	Sequence Data: Likelihood
I	5	133	317	494
II	6	137	327	494
III	6	132	314	489
IV	7	135	321	494
V	8	140	332	494
VI	8	136	322	494
VII	7	137	327	494
VIII	8	139	327	494
IX	8	140	334	494
X	8	137	328	493
XI	8	134	321	493
XII	8	132	316	489
XIII	8	136	322	493
XIV	8	138	329	493
XV	8	133	318	490

Table 15-2 Final Probabilities

Cladogram	Unweighted Parsimony	Weighted Parsimony	Likelihood
I	4.5×10^{-1}	1.2×10^{-1}	1.5×10^{-2}
II	3.0×10^{-3}	2.0×10^{-6}	5.5×10^{-3}
III	4.5×10^{-1}	8.7×10^{-1}	8.1×10^{-1}
IV	8.2×10^{-3}	2.9×10^{-4}	2.0×10^{-3}
V	2.0×10^{-5}	1.8×10^{-9}	7.4×10^{-4}
VI	1.1×10^{-3}	3.9×10^{-5}	7.4×10^{-4}
VII	1.1×10^{-3}	7.2×10^{-7}	2.0×10^{-3}
VIII	5.5×10^{-5}	2.6×10^{-7}	7.4×10^{-4}
IX	2.0×10^{-5}	2.4×10^{-10}	7.4×10^{-4}
X	4.1×10^{-4}	9.7×10^{-8}	2.0×10^{-3}
XI	8.2×10^{-3}	1.1×10^{-4}	2.0×10^{-3}
XII	6.1×10^{-2}	1.6×10^{-2}	1.1×10^{-1}
XIII	1.1×10^{-3}	3.9×10^{-5}	2.0×10^{-3}
XIV	1.5×10^{-4}	3.6×10^{-8}	2.0×10^{-3}
XV	2.2×10^{-2}	2.1×10^{-3}	4.1×10^{-2}

Table 15-3 Topology Decisions

Risk Level	Cladograms: Unweighted Parsimony	Cladograms: Weighted Parsimony	Cladograms: Likelihood
20%	I, III	III	III
10%	I, III	I, III	III, XII
5%	I, III, XII	I, III	III, XII, XV

excludes topology I in the 5% risk decision, while the two parsimony-based methods include I at the expense of XV.

CONCLUSIONS

At present there are two commonly used congruence procedures: global parsimony and consensus (Miyamoto, 1985; Kluge, 1989). In a global parsimony analysis, all of the characters are lumped together, and the most-parsimonious solution for the combined data is chosen as the most-favored hypothesis. Consensus procedures, on the other hand, treat the data sets distinctly; individual characters and their combinations do not play a direct role in the topology decision made from comparing the data sets. In this respect, consensus procedures treat characters as if they were members of noncomparable categories; hence, only topologies are compared. The method presented here attempts to extract the best qualities of both approaches through Bayesian analysis.

If the simple *e* factor transformation is used for both the prior distribution and the determination of the data probability, the result will be the most-parsimonious solution for the combined data. Since all characters are weighted equally, it does not matter from which analysis (morphological or molecular) they were derived. If, however, the investigator believes that character classes exist, or feels that some distinction can be made between the types of data used, this method allows him to separately determine the probabilities that influence the final decision.

The procedure draws its strength from this flexibility. These probabilities offer the investigator a quantitative assessment of the degree of belief. If several analyses favor a certain hypothesis, but each only marginally, the final probability of that hypothesis should reflect the increased confidence which comes from this corroboration.

The risk inherent in the decision to accept or reject topologies gives a measure of the level of agreement between data sets. If there is little discrimination, or the various observations differ greatly, no hypothesis will have sufficiently low risk to be accepted by itself. In these cases, several or many topologies will have to be considered to reduce the risk to a satisfactory level.

Congruence is the final test of any hypothesis. The type of procedure presented here gives a logical means of assessing the agreement and disagreement between different sets of data that bear on the same phylogenetic question.

ACKNOWLEDGMENTS

I would like to acknowledge Kirk Fitzhugh, Paul Vrana, and especially Elise Broach for reading drafts of this manuscript. I would also like to thank two anonymous reviewers. I thank Christopher Urheim for making clear to me the subtleties of the "Monty Hall Problem." This research was supported by the Alfred P. Sloan Foundation.

REFERENCES

Adams, E. N. III. (1972) Consensus techniques and the comparison of phylogenetic trees. *Syst. Zool.* **21**: 390–397.

Bayes, T. (1763) An essay towards solving a problem in the doctrine of chances. *Philos. Trans. R. Soc.* **53**: 370–418.

Boudreaux, H. B. (1979) *Arthropod Phylogeny with Special Reference to Insects.* John Wiley & Sons, New York.

Brown, W. M., E. M. Praeger, A. Wang, and A. C. Wilson . (1982) Mitochondrial DNA sequences of primates: tempo and mode of evolution. *J. Mol. Evol.* **18**: 225–233.

Cracraft, J. (1974) Phylogeny and evolution of the ratite birds. *Ibis* **116**: 494–521.

DeBry, R. W., and N. A. Slade. (1985) Cladistic analysis of restriction endonuclease cleavage maps with a maximum likelihood framework. *Syst. Zool.* **34**: 21–34.

Donoghue, M. J., and J. A. Doyle. (1989) Phylogenetic studies of seed plants and angiosperms based on morphological characters. Pp. 181–194 in *The Hierarchy of Life. Molecules and Morphology in Phylogenetic Analysis* (B. Fernholm, K. Bremer, and H. Jörnvall, eds.). Elsevier Press Science Publishers B.V., Amsterdam.

Farris, J. S. (1977) Phylogenetic analysis under Dollo's law. *Syst. Zool.* **26**: 77–88.

Felsenstein, J. (1978) Cases in which parsimony or compatibility methods will be positively misleading. *Syst. Zool.* **27**: 401–410.

Felsenstein, J. (1979) Alternate methods of phylogenetic inference and their interrelationship. *Syst. Zool.* **28**: 49–62.

Field, K. G., G. J. Olsen, D. J. Lane, S. J. Giovannoni, M. T. Ghiselin, E. C. Raff, N. R. Pace, and R. A. Raff. (1988) Molecular phylogeny of the animal kingdom. *Science* **239**: 748–753.

Goodman, M., J. Czelusniak, and J. E. Beeber. (1985) Phylogeny of primates and other eutherian orders: a cladistic analysis using amino acid and nucleotide sequence data. *Cladistics* **1**: 171–185.

Hartigan, J. A. (1983) *Bayes Theory*. Springer-Verlag, New York.

Hennig, W. (1969) Die stammesgeschichte der Insekten. *Seckenberg-Büchern*. **49**: 1–436.

Hennig, W. (1981) *Insect Phylogeny*. John Wiley & Sons, New York.

Kluge, A. G. (1989) A concern for evidence and a phylogenetic hypothesis of relationships among *Epicrates* (Boidae, Serpentes). *Syst. Zool.* **38**: 7–25.

Kluge, A. G., and J. S. Farris. (1969) Quantitative phyletics and the evolution of anurans. *Syst. Zool.* **18**: 1–32.

Kristensen, N. P. (1975) The phylogeny of hexapod "orders": a critical review of recent accounts. *Z. Zool. Syst. Evol. Forsch.* **13**: 1–44.

Kristensen, N. P. (1981) Phylogeny of insect orders. *Ann. Rev. Entomol.* **26**: 135–157.

Margush, T. and F. R. McMorris. (1981) Consensus *n*-trees. *Bull. Math. Biol.* **43**: 239–244.

Miyamoto, M. M. (1985) Consensus cladograms and general classifications. *Cladistics* **1**: 186–189.

Rohlf, F. J. (1982) Consensus indices for comparing classifications. *Math. Biosci.* **59**: 131–144.

Sibley, C., and J. E. Ahlquist. (1981) The phylogeny and relationships of the ratite birds as indicated by DNA-DNA hybridization. Pp. 301–335 in *Evolution Today* (G. G. E. Scudder and J. L. Reveal, eds.) Carnegie-Mellon University, Pittsburgh.

Smouse, P. E., and W.-H. Li. (1987) Likelihood analysis of mitochondrial restriction-cleavage patterns for the human-chimpanzee-gorilla trichotomy. *Evolution* **41**: 1162–1176.

Wheeler, W. C. (1989) Evolution and systematics of insect rDNA. Pp. 307–321 in *The Hierarchy of Life. Molecules and Morphology in Phylogenetic Analysis* (B. Fernholm, K. Bremer, and H. Jörnvall, eds.). Elsevier Science Publishers B. V., Amsterdam.

Wheeler, W. C. (1990) Combinatorial weights in phylogenetic analysis: a statistical parsimony procedure. *Cladistics* **6**: 269–275.

Wyss, A. R., M. J. Novacek, and M. C. McKenna. (1987) Amino acid sequence versus morphological data and the interordinal relationships of mammals. *Mol. Biol. Evol.* **4**: 99–116.

Zimmer, E. A., R. K. Hamby, M. L. Arnold, D. A. Leblanc, and E. L. Theriot. (1989) Ribosomal phylogeny and flowering plant evolution. Pp. 205–214 in *The Hierarchy of Life. Molecules and Morphology in Phylogenetic Analysis* (B. Fernholm, K. Bremer, and H. Jörnvall, eds.) Elsevier Science Publishers B. V., Amsterdam.

Index